Formula 1 Technology

Other SAE books of interest:

Race Car Engineering & Mechanics
by Paul Van Valkenburgh
(Order No. R-308)

Race Car Vehicle Dynamics
by William F. Milliken and Douglas L. Milliken
(Order No. R-146)

The Last Great Miller
The Four Wheel Drive Indy Car
by Griffith Borgeson
(Order No. R-244)

The Successful Race Car Driver
A Career Development Handbook
by Robert Metcalf
(Order No. R-254)

For more information or to order this book, contact SAE at 400 Commonwealth Drive, Warrendale, PA 15096-0001; phone (724) 776-4970; fax (724) 776-0790; e-mail: publications@sae.org.

Formula 1 Technology

Peter Wright

Illustrations by Tony Matthews

SAE INTERNATIONAL®

Society of Automotive Engineers, Inc.
Warrendale, Pa.

Illustrations and photographs by Tony Matthews, unless stated otherwise in the captions.

Society of Automotive Engineers, Inc.
400 Commonwealth Drive
Warrendale, PA 15096-0001 U.S.A.
Phone: (724) 776-4841
Fax: (724) 776-5760
E-mail: publications@sae.org
http://www.sae.org

ISBN 0-7680-0234-6

Library of Congress Card Number: 00-104452.

SAE Order No. R-230

Table of Contents

Acknowledgments

Contributions to this book, made by individuals, come from three sources: First, those people whom I have known and worked with over the years, who have inspired and taught me everything I know about cars and motor racing, and who have enabled me to understand that which I think I understand. Second, those people who have provided technical or practical input to the book, explaining areas in which my understanding is lacking, or providing access to the normally secretive parts of Formula 1. Finally, but not least, those people who have been part of the actual writing, revising, editing, and publishing process.

In the first category, I would include the following, in alphabetical order:

Elio de Angelis, Richard Aubrecht, Ralf Bellamy, Eric Broadley, Colin Chapman, Dan Clarke, Peter Collins, John Davis, Gérard Ducarouge, Bernie Ecclestone, Ian Foley, Gérard Fresson, Ken Garnjost, Steve Green, John Harvey, Johnny Herbert, Frank Irving, Peter Jackson, Nigel Mansell, Roland Magnuson, John Miles, Bill Milliken, Doug Milliken, Peter Milliken, Max Mosley, Chris Murphy, Martin Ogilvie, Alec Osborne, Tony Rudd, Terry Satchell, Ayrton Senna, Willy Southcott, Nigel Stepney, Alex Stokes, Rick Townend, Sid Watkins, David Williams, Conway Young, and Alex Zanardi.

In the second category, also in alphabetical order, are:

Tony Best, Peter Digby, Pierre Dupasquier, Mike Endean, Patrick Head, Mario Illien, Bill Milliken, Doug Milliken, Tony Purnell, Pat Symonds, John Taube, and Jean Todt.

Finally, among those in the third category is Tony Matthews, who sees race cars in a way that few others see them and whose illustrations in this book speak for themselves. Tony's patience with my changing requests made the whole process of illustrating this book a pleasure. Also in this category is Hannah Woolhouse whose comments, criticisms, and corrections have made this an easier book to read.

Foreword

Anyone interested in Formula One has endless sources of information, including specialist publications, books, television programs, and the daily press. However, none of these gives much idea about how modern Formula One cars really work. These sources touch on various technical details, but the overall picture remains a mystery. As a result, some of the most interesting aspects of Formula One are lost to all but the initiates who actually work in the paddock.

This book is likely to change all that. It is the first systematic explanation of the technical side of modern Grand Prix racing. As such, it will be essential reading for any serious follower of Formula One and of immense help to aspiring drivers and engineers in other forms of racing.

The FIA receives letters and e-mails from all over the world with suggestions, comments, and criticisms. Unfortunately, most of these miss the point because of lack of technical information and understanding. This book is capable of providing what has been missing. I think it will have a significant impact.

Max Mosley, President
Fédération Internationale de l'Automobile
July 2000

Introduction

History shows that engineering (using an idea without necessarily understanding it) usually gets the drop on science (understanding an idea without necessarily having any idea what to do with it).
—Philip Whiteman

I am extremely fortunate to have been involved in motor racing during the last 35 years of the first half-century of Formula 1. During that time, I have worked directly for Formula 1 teams for 22 years, and the remainder of the time on relevant technologies: aerodynamics, composites, and closed-loop control systems. I have witnessed firsthand the most technically stimulating and challenging period, and I have seen the industry grow enormously in both size and sophistication. It has been a period during which the technology and engineering has evolved from mechanical systems—most of which you could hold in the hand, look at closely, and see how they should work, even if you could not always understand why they did not—to systems dominated by silicon wafers so small you hardly see them and which often cannot be totally understood by a single human brain.

When I joined BRM straight from university in 1967, it was a World Championship winning team which, unusually for a British one, designed and built its own engines and transmissions. Tony Rudd, BRM's chief engineer and team manager, led an engineering department that consisted of one engine designer, one chassis designer, and one gearbox designer, plus two draftsmen. I became the R&D department once it was established that anything I designed was too expensive to make. All calculations were made on slide rules or using logarithm tables, which inhibited carrying out extensive analysis, especially stressing. Suspension geometry was plotted on the drawing board using fishing line; the instantaneous center was always off the board and sometimes stretched to the other side of the drawing office. In the workshops, highly skilled panel beaters wheeled magnesium skins to form the bodywork, and machinists achieved astonishing tolerances on old milling machines and jig-borers. Engine test beds were located on an old airfield several miles away, where a pair of ex-Tiger Moth, de Havilland Gipsy Major engines was used to drive propellers to vent the test beds. This was the state of the Formula 1 art.

Today, Formula 1 teams such as McLaren or Williams are housed in gleaming, modern buildings and are staffed with 200 to 300 persons (Fig. 0.1). The 30-person design department is equipped with CAD stations, feeding design information directly to CNC machining centers, sometimes served only by robots. The 25 engineers in R&D are linked to design and racing by an integrated network of databases and analysis software connected to the race cars, rigs, and their on-site 50% wind tunnel. The composite department, with its autoclaves and five-axis pattern cutting machines, dominates manufacturing. Engine manufacturers, such as Ford-Cosworth and Mercedes-Illmor, are approximately the same size, but they have more machining capability and rows of engine test beds instead of wind tunnels and seven-post rigs. The 500-person Scuderia Ferrari combines both team and engine manufacturer. When the team, its engine supplier, and major sponsor go to a Grand Prix, the 30+ tons of equipment is transported in six or more articulated trucks. Three or four double-decker hospitality vehicles look after guests and the 60 team personnel running the cars. BRM took only one converted bus to races 30 years previously. Instead of the roomy pit complexes at modern circuits, the paddock where the cars were worked on, was exactly as the name implies: a grass paddock.

Whereas one man such as Tony Rudd could understand the whole car—chassis, engine, and gearbox—and could relate to his drivers because he too had raced, the size of the engineering task is now so large and complex that specialists are required for every aspect. Room remains for an overall engineering view, and those individuals who have the instinct of McLaren's Adrian Newey for what makes a race car fast, still succeed the most often. The technology of Formula 1 has, however, almost completed the transformation from the black art of the 1950s and 1960s to the science of the 2000s. Black art allows success to come to those who have some science, are able to extrapolate intuitively, and have the luck to be able to just "do it." Computer science enables every solution to be evaluated, with ultimate convergence on the solution. The way a race car works is nearly understood, but pockets of complexity remain and the computer models have not yet unraveled their inner workings. Turbulent rotating airflows, tires, and the motion, thermodynamics, and chemistry of the gases inside a combustion chamber that fires 150 times per second are among them.

Until the 1950s, Grand Prix cars were configured similarly to road cars. The engine placement, size, and type were similar, although more powerful; the chassis layout, suspension, and wheel and tire sizes were sometimes even derived from road cars; bodywork was more streamlined, mudguards were dispensed with, and the passenger seat was discarded. A sporting driver of road cars would recognize the controls and be quite able to drive the cars. In the 1960s, 1970s, and 1980s, a series of inventions and technological advances transformed Formula 1 cars such that the average motorist would have trouble identifying many of the components and probably would be unable to drive the cars, even if he could withstand the physical forces it imposed. The performance of the cars has increased greatly and would have done so even more but for regulatory restrictions on both the type and quantity of the technologies permitted. This book examines what, how, and why this has happened during this period.

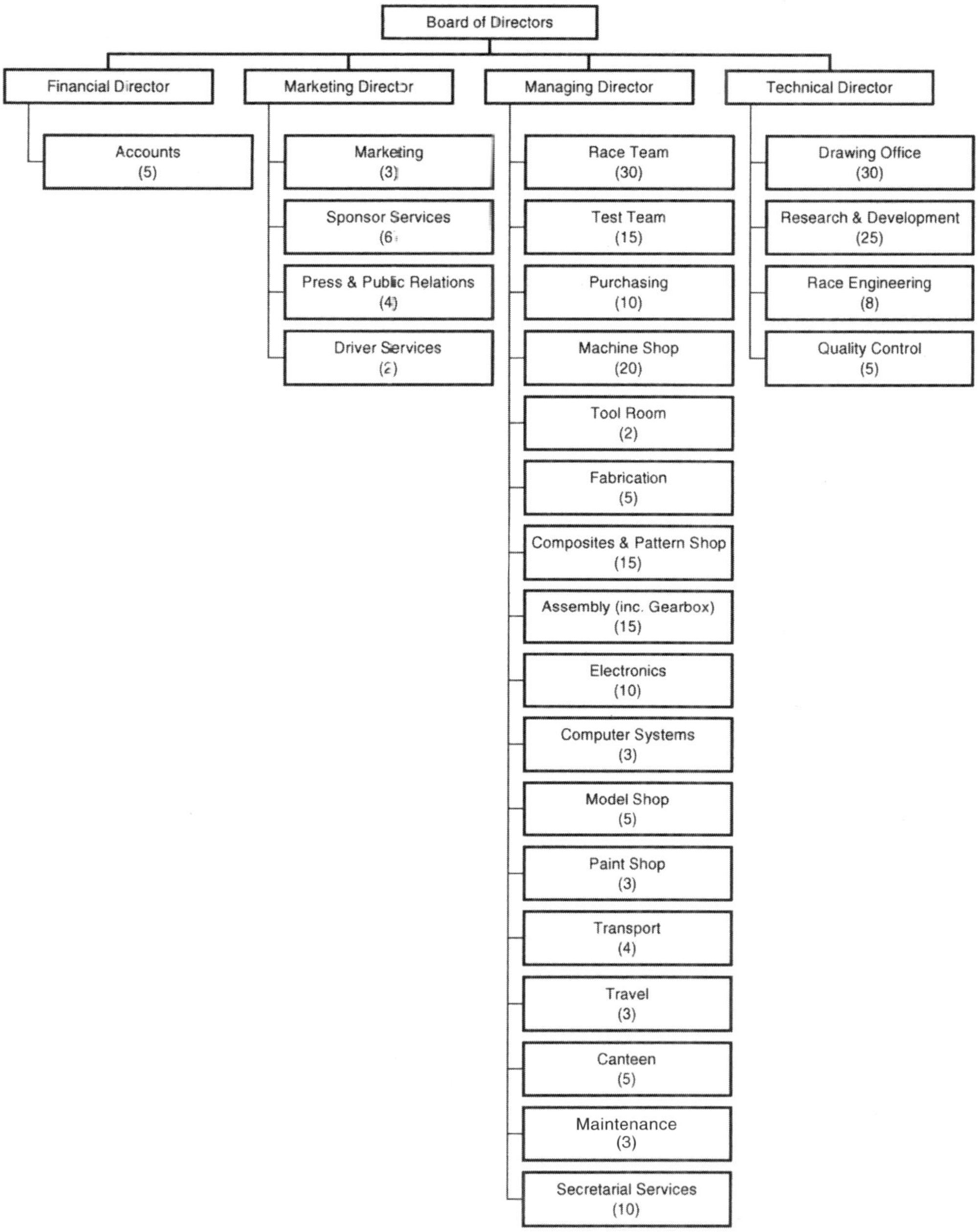

Fig. 0.1 Organizational structure of a typical British Formula 1 team.

In his book *Innovation* (Ref. 0.1), William Kingston defines the stages of innovation. The originator who postulates the idea is the dreamer. The person who describes it in visual terms is the artist. Finally, the person who takes the visual form and produces a working demonstration of the idea is the inventor. Perhaps Kingston also could have given credit to the person who authorized the expenditure that enabled the idea to be exploited successfully and become known. That person is often the one who has the vision and means not only to employ people with the ability to innovate, but also to enable the inventions to reach fruition. Formula 1 is full of claims to the ideas that have worked, but little has been patented and thus those claims have never had to be tested in law. In most cases, if A had not invented something, B would have done so shortly afterward. The invention occurs when the need is apparent and is thus almost inevitable. I have tried to avoid the issue of which individual deserves the credit for the key ideas and technologies, but rather to concentrate on which organization pioneered those ideas in racing and how and why the inventions came about when they did.

In his book *Guns, Germs and Steel* (Ref. 0.2), Jared Diamond, professor of physiology at the University of California Medical School, has written about the conditions conducive to the development and diffusion of technology within societies. The conditions he describes are as relevant to the Formula 1 environment as to the primitive societies that invented the wheel and the alphabet. He states that most inventions are borrowed and adapted from another society in close proximity, because this is nearly always easier than reinventing independently and it results in a more rapid spread of the idea. To adopt an invention, a society must be willing to see, learn from, and be receptive to the idea; otherwise the society will tend to become disadvantaged by the lack of it and will eventually be overwhelmed. Once something has been invented and examples exist, it becomes much easier to reinvent it independently. For example, the process for making fine porcelain in Europe was developed more or less independently by Meissen, Sèvres, Wedgewood, and Spode, nearly a thousand years after the Chinese had perfected it. The Europeans were assisted by having examples of Chinese porcelain to inspire them, although they did not know the formula for the clay or the details of the firing process. Diamond claims that the development of technology is an autocatalytic process; that is, it speeds up at a rate that increases with time because the process catalyzes itself. This is certainly what has happened in Formula 1. When active suspension required the installation of servo-hydraulics and a chassis computer on the cars, numerous other chassis control systems quickly followed, adapting the active technology and hardware. Diamond's analysis seems to be particularly relevant when considering why the British motorsport industry has been so successful.

Compared to the almost *formula libre* of the early years of Formula 1, and indeed before that also, it is extremely regulated formula. Before the development of wings and ground effect, there was only a perceived need to regulate engine size and weight. Performance did not escalate out of hand, and safety was not an issue. Motor industry, or indeed national participation, was primarily on the basis of promoting and displaying technical supremacy. The cars resembled the low-volume, exotic road-going products of the manufacturers, and the public following was so small in today's terms, in the absence of television, that the racing results had little marketing effect on mass-produced cars. Aerodynamic downforce, wide and slick tires, turbocharging, aerospace materials, and computer-control systems added so much performance potential that regulation to control speeds was inevitable. Other influences, including safety awareness, costs, the sporting spectacle, and the need to modify circuits to suit television, all have contributed to the current Fédération Internationale de l'Automobile (FIA) Technical Regulations and the way the permitted technologies have evolved. Most vehicles—whether they are road cars, trucks, civilian and military aircraft, or ships and spaceships—compete with each other in absolute terms, subject only to safety and environmental regulation. Their economics are an inherent competitive factor. Race cars compete in relative terms, almost unencumbered by cost constraints but with the performance technologies greatly limited by regulations. It is a somewhat unnatural environment, necessary to maintain motorsport as a contest among humans and one that has had a significant effect on the way it has evolved technically.

Only five main parameters determine the performance potential of a race car: power, weight, tire grip (coefficient of friction), drag, and lift (or downforce). Until Chaparral, greatly assisted by Chevrolet R&D, demonstrated the significance of lift by generating downforce aerodynamically, it had been largely ignored and designers had concentrated on only the first four parameters. Tire grip did not increase significantly until the 1960s, when tread widths started to increase; wear rate of the narrow treads was a more critical factor until then. Formula 1 technology is about how to maximize these five parameters as much of the time as possible, and to enable the driver to use as much of them as possible all of the time. Hundreds of technologies are involved, not merely in the makeup of the car and its systems but also in the means by which the components are manufactured and in the systems for testing and racing the cars. This book is about some of the most important of these technologies.

Making up these hundreds of technologies are thousands of details that go into the design, the manufacturing, and the development processes involved. This book is not about these details because they embody the main intellectual property of the companies employing them. Formula 1 has become so competitive and the stakes have become so high that the companies involved have procedures that attempt to protect their secrets. Even if they allowed the cars to be photographed regularly with the body panels removed, little would be revealed because the important secrets are embedded in the fine detail of piston crown shapes, anti-friction surface coatings, damper valving, hydraulic circuits, and software. None of these are ever disclosed. The cars are retained at the end of a season in team museums or, if sold to collectors, they are stripped of their secret components. The opportunity to inspect and describe a Formula 1 car, as could be done 20 years ago, is gone. Many companies involved in Formula 1 instruct technical personnel never to discuss anything technical with either the press or people from competing companies. It is all too easy to reveal information useful to a competitor simply by indicating an interest in a particular technology.

The basic principles and fundamental technologies described in the following chapters are not secrets. They are known and understood by most designers and engineers in Formula 1 and any of the major racing series. They may not be known and understood by technically oriented people outside racing, even within the automobile industry. This book is written for those who are interested in and follow Formula 1, and who wonder what the designers are trying to achieve and why they follow a particular trend. I have tried to avoid using formulae and equations to describe the concepts upon which the technologies depend, but that is not always possible. Race cars must follow Newtonian laws, and I believe that stating simple equations is sometimes necessary to illustrate how fundamental are some of the concepts that reside behind all the technical sophistication.

The development of three of the technologies—ground-effect aerodynamics, the use of carbon-fiber reinforced composites in primary structures, and electro-hydraulic control systems (active suspension)—is described in more detail because I had firsthand knowledge of the initial stages of their development. All three started as classic black-art programs and matured into sciences with good theoretical bases now in place for them. Each involved an initial concept demonstrator phase, lasting approximately six to twelve months and having very little financial outlay. They subsequently took another six to twelve months to reach the racetrack. Ground effect required an additional six months before the benefits really showed and, for some inexplicable reason, another year or two before the competition fully caught up. Carbon-fiber reinforced plastic (CFRP), and the processes that enabled it to be used to construct the monocoque, surfaced simultaneously in two teams, and the benefits were immediately apparent to all. Active suspension lapsed due to the death of Colin Chapman and a change in the aerodynamic regulations. It resurfaced four years later, winning Grand Prix's for two teams, but required another five years before dominating Formula 1. Two years later, it was banned. The life cycle for a Formula 1 car is approximately 18 months, but any significant new technology (such as the aforementioned ones or radial tires and turbocharging) tends to require at least three years to mature into race-winning systems. The time scales in Formula 1 are short compared to those of similar industries. That is a function of the participants' willingness to take risks, as well as the gains to be won by introducing something new but unproven which may give an edge in performance over competitors.

As with so much nowadays, the microprocessor and the complex and repetitive tasks it is set to perform have had the biggest effect on technology in Formula 1. Starting in the late 1970s, they first appeared on Formula 1 cars to control engines and manage data acquisition systems. Firmly established on the cars in the 1980s, they also began to play a major part in the design and development processes, as CAD, CAE, and CAM software and specialized applications developed for the aerospace industry began to be exploited by the remainder of industry. By the 1990s, the computer-controlled functions permitted on the cars were severely limited by regulations; nonetheless, the computer had become an integral part of the design, development, manufacturing, and racing processes. The escalation in performance during the last decade—in spite of the FIA applying continuous performance-limiting regulations and no major new, fundamental, technical advances—can be attributed to the ability of the computer to provide understanding through simulation and modeling, and to optimize highly complex systems.

One of the most perplexing aspects of the evolution of motor racing technology is why it took so long for the benefits of aerodynamic downforce to be appreciated. Figure 0.2 shows an airfoil section being tested on a car around the turn of the last century. Although the wing is mounted in such a way as to produce lift, and the rear-facing passenger was noting the magnitude of that lift, another 60 years passed before it was realized that a similarly generated downforce could radically improve the overall performance of a race car. Why did the technology not cross disciplines? The scientific approach used by Chevrolet R&D to measure and analyze race car performance, and the ability of the computer to simulate the effect of new concepts, finally made the now obvious connection. In doing so, Chevrolet heralded the start of the scientific age of motor racing (Refs. 0.3 and 0.4).

Fig. 0.2 Testing an airfoil section on a car, circa 1900. (Courtesy of the Cowin Collection)

In spite of the impact of the computer, anyone who has driven a rear-wheel drive car at or near its limit will have experienced most of the topics discussed in this book, except perhaps for the startling increase in high-speed grip generated by ground-effect downforce. The physics are the same, but the magnitude of some of the accelerations experienced is approximately four to five times greater. Often, the first experience of a car driven at the limit stimulates an interest in race cars, a desire to understand what makes up that experience, and how a race car driver can deal with a much greater embodiment of it. Some go on to race, while others prefer to delve into the science and engineering of the cars. In writing this book, I have tried to make Formula 1 a little bit more accessible to those people.

In the introduction to his autobiography, *It Was Fun!* (Ref. 0.5), Tony Rudd wrote,

> I must, however, repeat the advice I gave to Peter Wright…when some 25 years ago he applied for the job of being my personal assistant—"It will be fun, but there's not much money in it."

Tony was absolutely correct until around the 1980s. I recall Gérard Ducarouge in 1987, then technical director of Team Lotus, saying after an eight-hour day of testing and well into a further eight hours of night work on the cars,

> It isn't fun anymore.

By then, a career as an engineer in Formula 1 was already pretty good, compared to other engineering jobs. Now, 34 years after Tony gave me that advice, I suspect that most of the fun has gone out of Formula 1 design and engineering, but there is a lot of money in it now.

Chapter 1

The Racing Objective

One cannot really enjoy speed to the absolute limit
if there is a destination involved.
—Stirling Moss

Mankind has been hooked on racing for a very long time. Although he has progressed from racing his fellow man—first on foot, then on horseback prior to inventing horse-drawn vehicles, and finally using the automobile—one feature of all races remains the same: The competition is to travel from one point (the start) to another (the finish) as quickly as possible. Whether the race is to accelerate from a standstill over a straight quarter-mile track, or to find the fastest route through the desert from Paris to Dakar, or to circulate a number of times around a closed course, the race is based on only two parameters: position and speed. The fastest over the chosen distance is the winner, and the record books record the race distance and the average speed.

While competing in a race, the driver's task is to control his velocity, both magnitude and direction, and its mathematical integral—position. The designers and engineers responsible for his chosen vehicle have a different but related task: they must provide the driver with the means to control and change the velocity at the greatest rate possible—acceleration. Their vehicle is the system by which artificial forces (i.e., non-human forces) are generated in order to accelerate the driver's body while he attempts to use the vehicle control systems to maintain the highest possible acceleration level, in the appropriate direction, at all times. For maximum acceleration, the force generation systems must be as powerful as possible and their mass, added to that of the driver and including the fuel necessary, as small as possible.

The technical history of motor racing, particularly Grand Prix racing, is a record of the invention and development of the force-generating systems. To evaluate how successful they have been and to assess the performance of the current state-of-the-art technologies, a method of describing their performance is needed. The performance or maneuver envelope, to state its full name, is the means by which the performance of a vehicle can be described. Originally developed to describe the performance of aircraft, the concept can be adapted to cars, in which case it takes a three-dimensional form, the surface of which defines the maneuvering limits of a vehicle in appropriate terms. In the case of a car, these are typically speed, longitudinal acceleration, and lateral acceleration. When all possible combinations of these parameters attainable by the car on a given surface and in given ambient conditions are plotted on mutually perpendicular axes, they form a volume, the surface of which envelopes all operating points of the car.

Speed, longitudinal acceleration, and lateral acceleration are normally chosen to describe the maneuver potential of a car because speed and the ability to modify its magnitude and direction by generating an acceleration vector are the essence of its motion, as we have seen previously. The combinations of these parameters that define the surface—the limiting conditions—provide essential insight into the maximum forces that are generated by the car through its systems—engine power, aerodynamic lift and drag, brakes, and tires—and the influence of its inertia.

The significance of the maneuver envelope to a race car is that it defines absolutely the potential performance of the car on a given track. How much of that potential can be realized depends on how near the surface of the envelope the driver can operate the car and for how long. Any time he operates within the surface, he is driving at less than the potential, but that may not be due to the driver alone. The ability to remain on the surface is a function not only of the driver's skill and bravery, but also of the stability and control characteristics of the car. Not only must the designers and engineers provide a car with the largest possible envelope, with appropriate tradeoffs between one part and another, but they must ensure that it is usable. Examination of running data to determine how close to certain parts of the envelope the driver is able to operate reveals a great deal about the characteristics of both the car and the driver.

The trend in almost all forms of motorsport is to reduce the size of the envelope by constraining the key performance features of the car with the Technical Regulations of the sanctioning body. The trend in track design, other than ovals, has been to deny access to parts of the envelope in the cause of safety, especially high speed combined with high lateral acceleration. Only where the attainment of ultimate performance is more important than competition is the envelope virtually unconstrained, as in drag racing and the quest for the land speed record. The challenge of pushing out "the edges" of the envelope remains, within the technical constraints; however, the emphasis is more and more on providing the driver with an envelope at the edge of which he is able to spend as much time as possible.

What does the envelope look like?

A plot of longitudinal acceleration against lateral acceleration, at any given speed, is the classic "g-g" diagram, as shown in Fig. 1.1. A full description and the theory behind the "g-g" diagram can be found in Ref. 1.1.

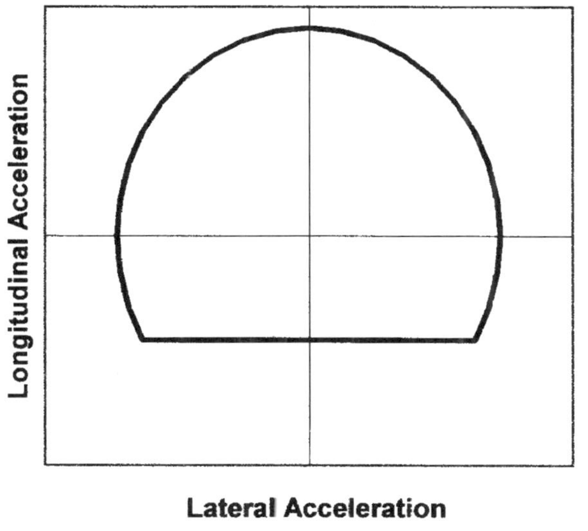

Fig. 1.1 A "g-g" diagram for a given speed. (Author's illustration)

For a car without aerodynamic download (or lift), the braking and cornering sectors do not change much with speed; only the acceleration sector reduces, due to power limitation as speed increases. For virtually all modern race cars that employ aerodynamics to generate downforce, the braking, cornering, and traction limited acceleration sectors are increased in size, and the power limited acceleration sector is reduced due to the greater drag as speed increases. This is fundamentally the aerodynamic compromise that is sought during the setting up of a car: which sector of the envelope to increase at the expense of another.

Thus, for a race car, it is not enough to merely draw a single "g-g" diagram; one is needed for every speed up to the maximum. However, before doing so, it is necessary to understand what influences the shape of the diagram, at any given speed, and to have a method of constructing it from the known physical characteristics of the chassis, suspension, engine, transmission, brakes, tires, and aerodynamics.

The "g-g" diagram shown in Fig. 1.1 is idealized and derives the elliptical part of its form from the friction ellipse that describes the combined longitudinal and lateral force capability of a tire. To calculate the total maximum lateral force capability of the four tires, it is necessary to sum them, having first calculated the vertical load, slip angle, camber, and longitudinal force on each of them, which depend on the lateral forces on each one. This is where vehicle models enter, with the software to solve them in a reasonable time. A vehicle statics model, such as the Milliken Moment Method (Ref. 1.1), takes detailed chassis, tire, aerodynamic, and suspension and steering kinematic and compliance characteristics, and calculates the maximum trimmed lateral acceleration at a given speed and longitudinal acceleration. To obtain enough points to draw a reasonably smooth curve, at one speed, requires approximately 30 cases to be calculated.

The truncation of the ellipse on the acceleration side is due to the low-speed traction limit of two-wheel drive. Four-wheel drive theoretically would give a full ellipse, provided sufficient power existed. At higher speeds, when the overall gear ratio is higher and traction is no longer the limiting factor, the truncation is caused by limited torque at the wheels.

Tires do not have perfect friction ellipses, and their maximum force characteristics vary with vertical load and camber. Also, each tire is presented to the road at a different load, camber, and slip angle, and each driven wheel receives a different torque from the differential. Thus, the real shape is not the simple form shown in Fig. 1.1. The vehicle designer must understand the limiting factors at each point on the curve, and what stability and control characteristics the car has there, in order to ensure that the optimum compromises have been made and that the driver will be able to take the car to that point and control it there. The vehicle model used for calculations should provide this information as part of its output.

Thus, at a given speed, the "g-g" diagram provides a two-dimensional form, the edges of which are the limiting operating points for the car and which enclose all possible operating points. Varying speed adds a third dimension, which is plotted on the Z axis, and the edges now form a surface—the performance envelope. If the "g-g" diagrams are to be calculated at, say, 10 kph (6.2 mph) intervals from 0–Vmax (say, 350 kph [217 mph]), to develop the surface, more than 1,000 cases would have to be computed. To illustrate the salient speed effects, it is quicker to look at the effect of speed on maximum braking, acceleration, and cornering (i.e., the points at which the "g-g" plot intersects the axes).

Braking and cornering exhibit similar characteristics. The vertical load on the four tires increases as the square of speed, due to aerodynamic downforce, but the force-generating characteristics of the tires do not increase linearly with vertical load. Also, the load transfer, both longitudinally under braking and laterally under cornering forces, increases with increasing accelerations. Thus, the maximum g-levels attainable are a function of aerodynamics, center of gravity (CG) height, front and rear track, wheelbase, springs and anti-roll bars, suspension geometry, and chassis torsional stiffness, to name only the first-order effects. Figure 1.2 gives an impression of the sensitivity of maximum braking and cornering accelerations to speed.

Figure 1.3 illustrates the same effect on maximum acceleration, exhibiting two distinct regimes. Starting from rest, the maximum acceleration increases as speed increases, due not only to aerodynamic download but also to greater load transfer to the rear wheels as the acceleration increases. Four-wheel drive would be affected only by the aerodynamic download; front-wheel drive would have a negative component due to load transfer. When the available traction exceeds the torque available at the rear wheels, acceleration reduces all the way to zero at the maximum speed, as the drag increases with the square of the speed, and torque at the wheels

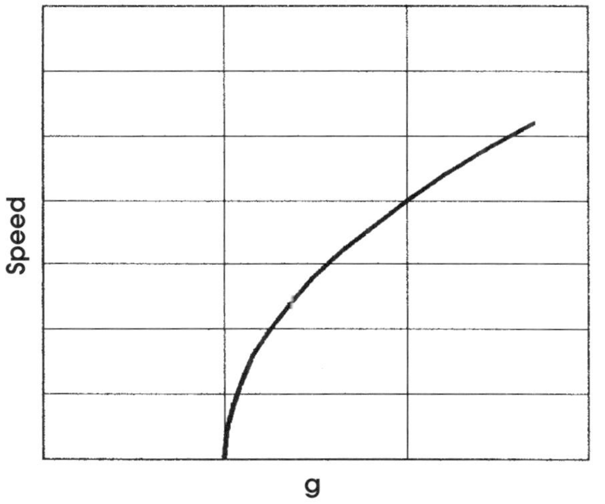

Fig. 1.2 Effect of speed on maximum braking and cornering accelerations. (Author's illustration)

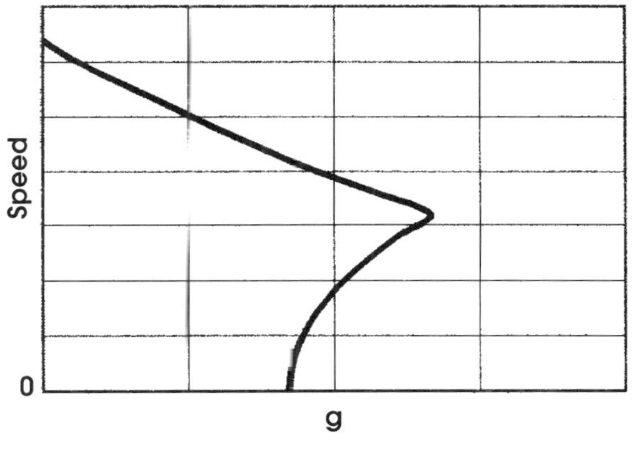

Fig. 1.3 Effect of speed on maximum acceleration. (Author's illustration)

reduces. The shape of this part of the curve is not necessarily as smooth as depicted in Fig. 1.3 because the torque available is dependent on the torque curve of the engine divided by the overall gearing, in the gear appropriate to a given speed.

The "g-g" diagram at higher speeds, where acceleration is power limited rather than traction limited, resembles the one shown in Fig. 1.4.

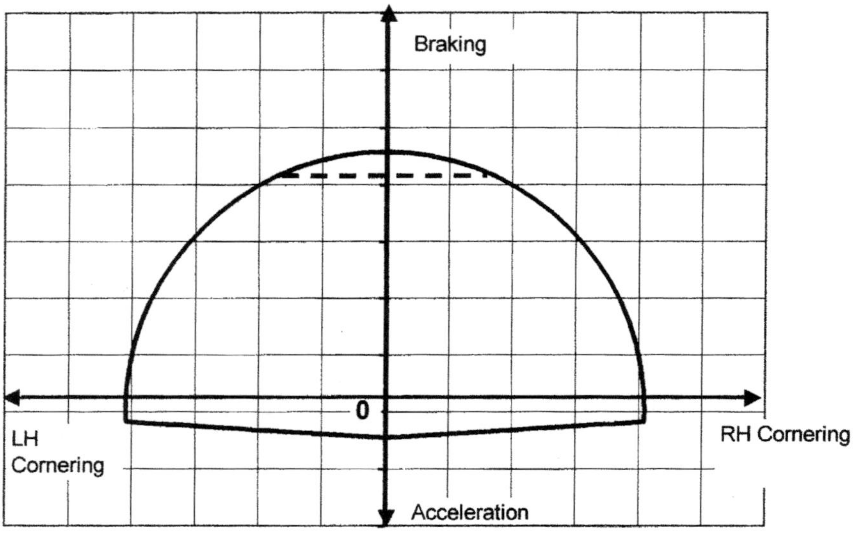

Fig. 1.4 A "g-g" diagram: high speed, power limited. The dotted line indicates the effect of brake power limitation.

The acceleration limit is no longer constant at all levels of lateral acceleration due to the power consumed in overcoming the cornering force. For example, this would be approximately 125 bhp for a Formula 1 car in a 240 kph (149 mph) corner, simply to overcome the increase in the resistance of the tires running at slip angles. This characteristic is obvious in Indy and NASCAR racing on ovals, where the top drivers will drive a lap without lifting the throttle. The speed falls as the car enters the corner and the force builds up, and increases again only when the cornering force is reduced on exiting the corner.

The full braking limit of the four tires may not be available at high speed and download, if the brakes are not powerful enough or the driver's leg muscles are inadequate. The broken line would now be the limit. When power brakes appeared on some Formula 1 cars in 1993, peak braking g-levels rose from approximately 4g to nearly 5g.

Putting the "g-g" diagrams together, with speed as the Z axis, generates a three-dimensional surface, as illustrated in Fig. 1.5.

This is what the performance envelope of a Formula 1 car might resemble. It is not a static thing; it changes as the environment in which it is operating changes. As wind, temperature, track surface, and profile change, so do the size and shape of the envelope. When it rains suddenly, the longitudinal and lateral acceleration dimensions may halve instantly, as many drivers have discovered. A car that has good grip on a cold March test day at Silverstone often loses it on a hot afternoon at Interlagos, two weeks later. The altitude and temperature

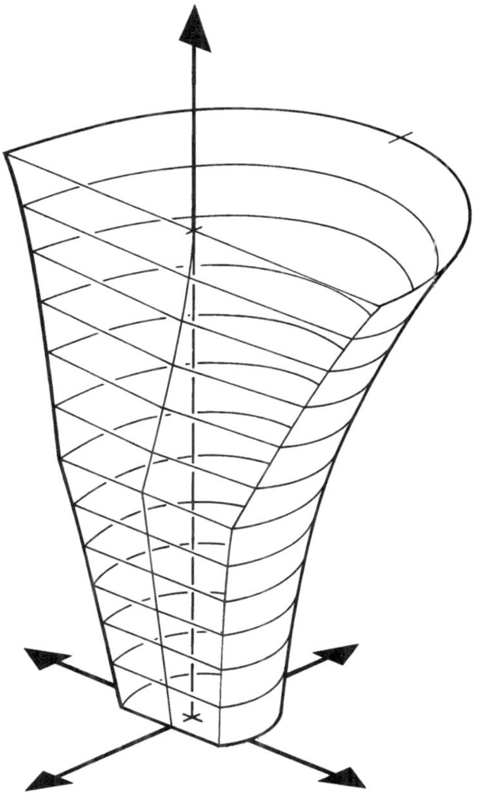

Fig. 1.5 A "g-g-V" diagram.

difference will have caused a 20% drop in air density, with a 20% loss of downforce. Power and drag also are reduced 20%; thus, acceleration, but not top speed, is also lowered. The "g-g-V" diagram is now 20% smaller all around, but just as tall.

As discussed previously, it is unlikely that a real car would have such a smooth, idealized envelope shape. However, let us assume that the designer has really got it right and it is very close, and let us relate it to the performance of the car on the track by examining how the driver attempts to exploit its full potential. Many parts of the surface are never accessed on a circuit, due to the design of the corners. However, from the point at which the driver applies the brakes for any corner, until the car is back onto the next straight, his task is to keep the car operating on the surface of the "g-g-V" diagram. Ideally, as soon as the driver hits the brakes, maximum braking-g for that speed should occur. In practice, the driver will work up to the limit, feeling for it as the brakes and tires rise to their peak operating temperature. During this time, the speed falls and so does the maximum braking-g available (Fig. 1.2); therefore, the driver must roll off the brakes to avoid locking them. This is represented by points A→B in Fig. 1.6.

At some point during the braking phase, chosen by the driver based on the line he wishes to take through the corner, he will turn in (points B→C) trying to maintain the car operating on the edge of the quadrant between braking and cornering, which is diminishing all the time as speed continues to fall. This is the area of the diagram that determines the driveability of the car and separates the men from the boys more than any other. However good the car is, it will be in its least stable condition while braking and turning inward. The "g-g" data plots of unstable cars show a distinct reluctance by the driver to venture anywhere near the limit in this sector of the "g-g-V" diagram. If the setup of the car is ideal, suddenly the driver is out at the surface all the way around from straight-line braking to steady-state cornering. Figure 1.7 shows actual "g-g" data for a current Formula 1 car.

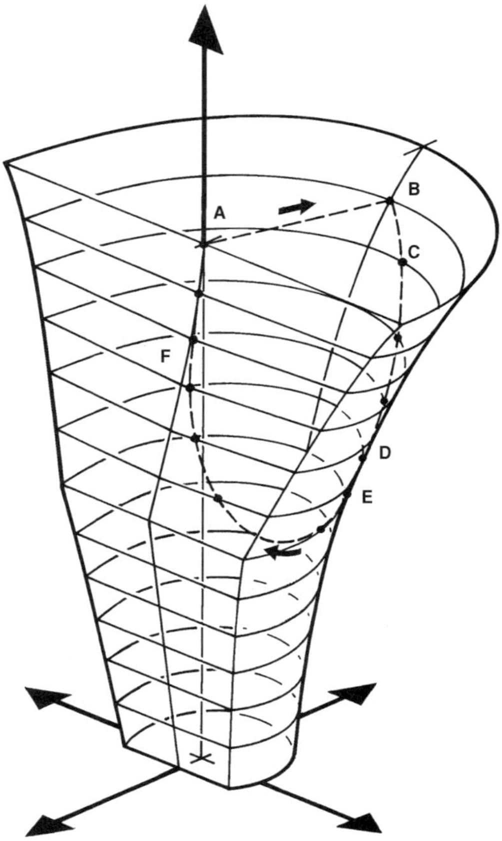

Fig. 1.6 Braking, cornering, and accelerating—an ideal trace on the "g-g-V" diagram.

During steady-state cornering (points D→E in Fig. 1.6), it is not so difficult to hold the car at the edge of the envelope. However, as power is applied and cornering is reduced (points E→F), it is a continuous hunt for the ever expanding limit, using steering and throttle. Rear-wheel drive cars also tend to be unstable in the cornering-to-acceleration sector but are more controllable because the driver can influence both axles. Back in a straight line, full throttle and fast gear changes will keep the operating point out at the surface (points F→A).

The maximum values of speed, acceleration, braking, and cornering—the intersections of the surface with the three axes—are determined primarily by the engine power, braking power, weight, aerodynamic downforce and drag, and tires. These features are regulated by the Technical Regulations because they dominate the performance of the car, with most of the time during a lap spent in these areas of the performance envelope. With tires of similar performance on all Formula 1 cars, and brakes and weight virtually the same, the spread of car performance down the grid is mostly a function of engines and aerodynamics. When two teams have the same engine and put the same amount of effort into aerodynamic development and testing, the difference between their cars is more likely to lie in the sectors of the "g-g-V" diagram where driveability (stability and control) is the limiting factor, i.e.,—combined braking and cornering, and acceleration and cornering.

The Technical Regulations are also brought to bear on these areas of the envelope in the banning of driver aids. ABS and traction control, four-wheel steering, and active differentials and differential brakes ("fiddle brakes") are not envelope expanders. However, they do help the driver to operate consistently nearer to the edge.

A close analogy can be made between the lifting force of an aircraft and its control around the pitching axis, and the lateral force on a car and its control around the yaw axis. However, one aspect of an automobile differs significantly from that of an aircraft. On most aircraft, the

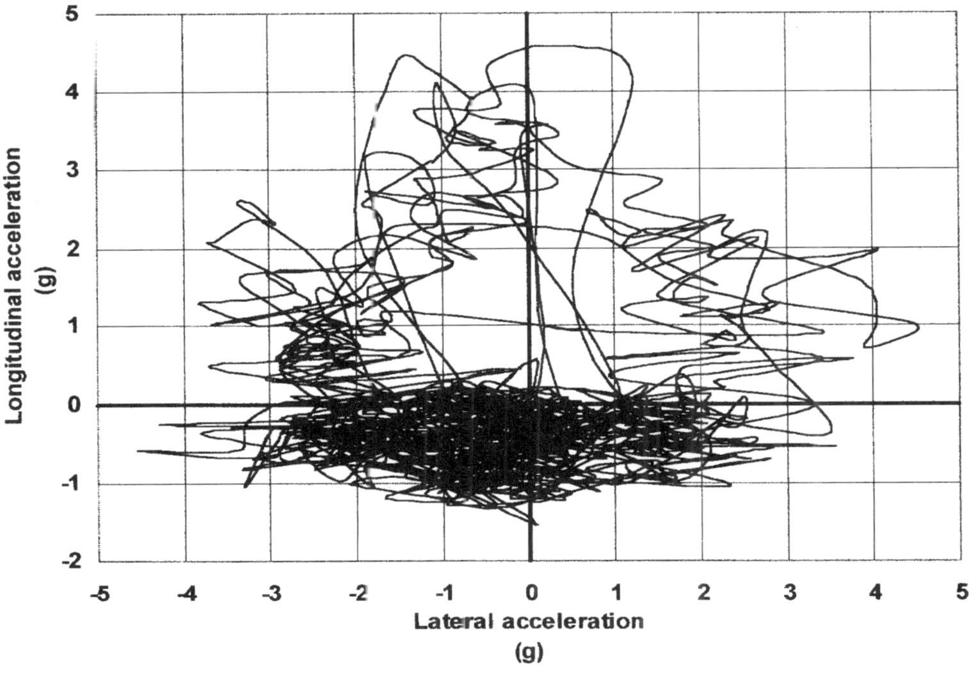

Fig. 1.7 Typical "g-g" diagram of a current Formula 1 car—all speeds. (Author's illustration)

device used to generate control forces in pitch—the tail plane including elevators—does not contribute significantly to the overall lifting force generated by the wings. As a result, when the wings reach the limit of their lift-generating capabilities (i.e., when they stall), the tail plane and elevators are still able to generate control forces, and recovery is possible.

On a car, the lateral force is generated by the tires on both front and rear axles together, while control and stability also employ these same two pairs of tires to generate the required forces. When the limit of the tires is reached (i.e., they slip or skid), control is lost. Thus, when the car is at the edge of the "g-g-V" diagram, it is "out of control." This feature of cars, and the need to operate them as close as possible to the edge of the envelope where they become uncontrollable, is what makes motor racing a human sport amid all the technology. Stirling Moss described driving a car at the limit as being akin to a bowler bowling a ball. All the spin, speed, swing, and pitch are set before the ball leaves the bowler's hand; after that, he cannot influence the trajectory. A driver, driving at the limit, sets up the car on entry to a corner. If it is truly at the limit, the driver cannot subsequently modify the trajectory. He is a passenger on the "ball."

Equipped with the "g-g-V" diagram, it is possible to assess the technologies that set out to influence the magnitude of accelerations and to quantify their maximum potential for generating speed on a given course. However, it is not until the car has run that it is possible to determine how much of that potential can be realized by the technologies that provide the driver with the control he needs. Computer simulations and the measurement of the dynamic characteristics of a car are not sophisticated enough yet to predict faithfully the transient and disturbance responses of a Formula 1 car and to provide this information.

Chapter 2

Evolve or Die

A definition of insanity is doing the same thing tomorrow as you did today and expecting a different result.
—Bob Proctor

Grand Prix racing had been taking place for 44 years before the FIA instituted a World Championship in 1950, setting the scene for what is universally recognized today as Formula 1—the world's premier motor racing formula. Throughout the almost 100 years of this top category of motor racing, the "formula" or technical rules under which it has taken place have often changed, but the essential technical challenge has remained the same. The physics of an automobile's ability to be driven as rapidly as possible around a closed circuit do not change. However, as man's ingenuity progresses, there has been a need to trim and restrain the technologies he has developed in order to keep the cars within the bounds of human endurance and safety. There is also a need for motor racing to be a sporting contest among drivers and to keep the cost of staging this contest to that which is affordable by spectators and the other parties that have commercial interests in motorsport.

The startup of the FIA World Championship heralded a significant shift in the technical development of the cars competing in it. Until that point, those members of the motor industry who believed that racing not only improved the breed but also created the image that sells the offspring, fought Grand Prix racing. The cars that were raced bore a passing resemblance to the cars in the showroom, and manufacturers showed little inclination to deviate from the classic front-engine/rear-wheel drive arrangement, with narrow tires and aerodynamics designed to minimize drag with little or no regard to lift. It took World War II and a relatively free formula to disturb that *status quo*, creating an environment that stimulated radical changes in the design of race cars and which has gathered pace since that time. The combination of the technical stimulus of war, the entry of individuals into Grand Prix racing who were not bound by any conditions or preconceived ideas of how a car should be designed, and commercial sponsorship funding is described in detail in Chapter 17. Between the late 1950s and the end of the 1970s, the fundamental parameters governing the performance of any race car were redefined and their limits probed.

Until 1958, Grand Prix cars had evolved along a fairly straight path, based on achieving the highest power in the lowest-weight, lowest-drag car possible within the "formula" of the time. The circuits encouraged acceleration and top speed as the most important performance measurements, and thus the main effort went into only four of the five primary parameters that govern the performance of a race car; that is, power, weight, drag, and tire grip. Aerodynamic downforce was ignored.

The hundreds of secondary parameters that make up the characteristics that determine how much of the potential performance a driver can extract from the car on a given circuit and on a given day (e.g., weight distribution, CG height, brakes, gear ratios, suspension kinematics, structural stiffness, and tire characteristics) were steadily developed by designers and engineers, guided partly by science and partly by instinct. Embodied in these systems were the thousands of details that determine whether the concepts actually deliver their characteristics reliably. Limited only by the expertise and imagination of the designers and engineers, the properties of the materials available, and the skills of the craftsmen and mechanics making the parts and assemblies, these details survive and allow us to see into the minds of the designers and engineers responsible for the cars. Valve-actuation systems, gear-selection mechanisms, tire construction, fuel and oil chemistry, body shapes, and similar factors all reflect the thousands of hours involved in their creation; however, they do not directly reveal the thinking behind them.

Of all the cars raced until 1958, the 1955 Mercedes W196 probably represents the pinnacle of classic, front-engined Grand Prix design (Fig. 2.1). The eight-cylinder, 2.5-liter, normally aspirated, 280 bhp engine was canted-over to reduce frontal area and to lower the CG height. Combined with a rear-mounted gearbox, the car achieved an empty weight of 650 kg (1,433 lb). With the driver onboard, the front-to-rear weight distribution varied between 47:53 with no fuel, to 40:60 with 175 kg (386 lb) of fuel (220 liters [58 gal]) in its tail-mounted fuel tank. The large variation in weight distribution, which could have been avoided by putting the fuel in pannier tanks as on the Lancia D50, was considered a reasonable compromise to achieve the most rearward possible weight to compensate for the front-mounted engine.

The W196 engine reflected everything Mercedes had learned in racing before the war, with technologies developed for its aero-engines added where appropriate. It was fitted with Bosch's mechanical, direct-injection fuel system to overcome the low-rpm mixture problems of large-bore carburetors. The engine could be revved to more than 8500 rpm, thanks to the ability of the desmodromic valve actuation to overcome the inertia forces of the two large valves, and the splitting of the crankshaft into two, with the power taken off from the middle, thus avoiding destructive torsional vibrations.

The tubular space-frame chassis was suspended on double wishbones at the front and a low-pivot swing-axle arrangement at the rear. Torsion bars and hydraulic dampers finally made leaf springs and friction dampers obsolete and, combined with the stiff frame, allowed suspension engineers actually to take control of both sprung and unsprung masses. Immense, servo-assisted

Fig. 2.1 1955 Mercedes W196.

drum brakes mounted inboard at the front and rear provided consistent stopping power without compromising unsprung mass.

The body shape was dominated by the front engine and its cooling requirements, the open-cockpit, and the tail-mounted fuel tank. Mercedes, with access to a model wind tunnel, experimented with bodywork that fully enclosed the wheels. Having conducted land speed record attempts prior to World War II, Mercedes fully understood the mechanisms for drag reduction.

Wire wheels and treaded tires no wider than those used on today's small saloon cars (6.0 × 16 front, 7.0 × 16 rear) were hard pushed to deliver the forces to the road. Expected to last the whole race, wet or dry, tread compounds were also similar to today's road tire compounds, with a coefficient of friction around 1.0.

From the engine construction techniques and materials, to the formulation of the fuel, the technologies employed were the fruits of the accelerated research of a company involved in wartime aviation, and no expense was spared in their application. Although Mercedes retired from Grand Prix racing before the end of the front-engine era, and cars such as the Maserati 250F dominated until the Cooper heralded the revolution, the W196 was the most technically advanced Grand Prix car, and Mercedes undoubtedly would have continued development if it had remained in racing. We can only speculate about the point at which Mercedes would have moved the engine to a location behind the driver, and whether it would have been Mercedes that pioneered this inevitable development.

The performance envelope of these ultimate front-engined Grand Prix cars can be estimated only because reliable data is not available from them. With approximately 280 bhp, a light racing weight of approximately 750 kg (1653 lb) and tires limited to a $\mu = 1.0$, they were traction limited to around 135 kph (84 mph) and had top speeds of more than 260 kph (161 mph) (the streamlined Mercedes W196 could achieve 290 kph [180 mph]). Maximum braking and cornering accelerations were around 0.95g. If we assume zero lift (probably not true for the streamlined version), these accelerations were available throughout the speed range, at least until the cars became power limited in corners (Fig. 2.2).

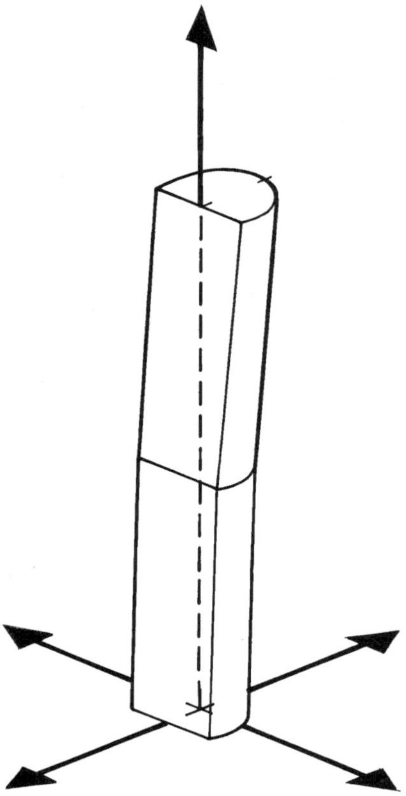

Fig. 2.2 A "g-g-V" diagram for the Mercedes W196.

The Grand Prix cars of the 1950s were products of the technical might of the German motor industry and the artistic craftsmanship of the Italians. The shock of having the results of more than 50 years of steady evolution be rendered obsolete by a few British individuals in only a few years is easy to comprehend.

Without the influence of a successful, indigenous, racing-orientated motor industry, and having honed their skills and gathered experience in power-limited formulae—the 750 Motor Club and Formula 3—John Cooper, Colin Chapman, Eric Broadley, and their competitors looked again at the basics. They reasoned that if power is limited, the same performance can be achieved by reducing weight and drag. The circuits on which they raced, laid out on redundant airfields, put a premium on braking, handling, and traction, rather than straight-line acceleration and top speed. When they graduated to Grand Prix racing, they applied these same principles to Grand Prix cars. The available engines, derived from road car or industrial units as opposed to being purpose-designed, were no match for the years of continuous development by Ferrari and Maserati; however, they were light in weight. Applying aircraft construction and design techniques, they built stiff, lightweight tubular space-frame chassis to which they attached soft, well-damped suspension. Rear-mounted engines achieved nearly ideal weight distribution for optimum traction and balance, and the whole was clothed in a body, based on World War II low-speed aeronautical aerodynamics. They entered the downward spiral of weight—the lower the weight of the car, the lighter the components can be—light disc brakes, light suspension, light gearboxes, etc. Most important, with lower fuel consumption and small tanks mounted at the center of gravity, the cars were easy on their tires and retained their handling balance as the fuel load was consumed. They literally ran rings around the heavy, powerful front-engined cars.

By the mid- to late 1960s, tubular space frames had given way to stressed-skin aluminum monocoques, based again on aeronautical techniques. Tire technology had progressed to low-aspect-ratio, wide and slick tires. Tread compound chemistry was pushing μ above 1.5. Most revolutionary of all, Chaparral, supported technically by Chevrolet R&D, had demonstrated once and for all that speed around the corners is equally if not more important than speed on the straight, and that speed on the straight is as much a function of the speed through, and the traction out of corners plus the ability to brake late for the next corner, as a function of power and drag. The gain in increased grip from the tires, as a product of aerodynamic induced downforce (negative lift), more than offset the accompanying increase in drag. Formula 1, wondering momentarily why they had not realized the obvious before this time, grabbed the new technology and began to explore how to generate the maximum downforce with the least drag. The five fundamental parameters in determining the performance were complete: power, weight, aerodynamic downforce, drag, and tire grip.

Wings exploded onto the scene and were soon reined in by the FIA. Body shapes assumed new forms, obviously not optimized for low drag. Although not the ultimate car of this era, the Lotus T72 (Fig. 2.3) was the most successful and epitomized it. Keith Duckworth had designed the Ford-Cosworth DFV engine for the predecessor of the T72, the Lotus T49, which was probably the first engine designed with the whole car in mind. Stressed as part of the structure, neither its size, shape, nor weight compromised the layout and dynamics of the car. Never the most powerful engine, it matched the creed of the British designers and was made so widely available and affordable that it soon dominated the Formula 1 grids. Car constructors could concentrate all their efforts and resources into chassis and aerodynamic development, relying on sufficient "bolt-on" power from the DFV, instead of needing the massive engineering infrastructure necessary for the build and development of a successful Formula 1 engine.

With the T72, Colin Chapman not only started the trend toward using the body itself to generate downforce and the mounting of the engine cooling radiators on the sides of the car, but also took the next steps in suspension design to attempt to obtain the most from the new, wide tires. Nearly three times as wide as the tires on the Mercedes W196, the 20-inch wide slick treads would generate 2g when loaded by the aerodynamic downforce. The suspension forces, jacking, dive, and squat were now great enough to significantly raise, pitch, or roll the chassis, further complicating the dynamics of the car. Chapman built anti-dive and anti-squat geometry into the T72 suspension and introduced pull-rod suspension in the quest for rising-rate characteristics combined with aerodynamically clean suspension linkages. He also fitted the brakes inboard at both ends of the car to reduce unsprung weight and to isolate brake heat from the tires. He was determined to be able to use softer compound tires than those of his competitors. Not all of these features were successful, but the emphasis was correct and the roots of today's suspension design can be traced to the Lotus T72.

By the end of this period, the cars had 500+ bhp and a weight similar to the last of the front-engined cars—approximately 450 kg (992 lb) empty. The wide tires and downforce-generating aerodynamics had so increased the drag that top speeds also were similar. However, the payoff

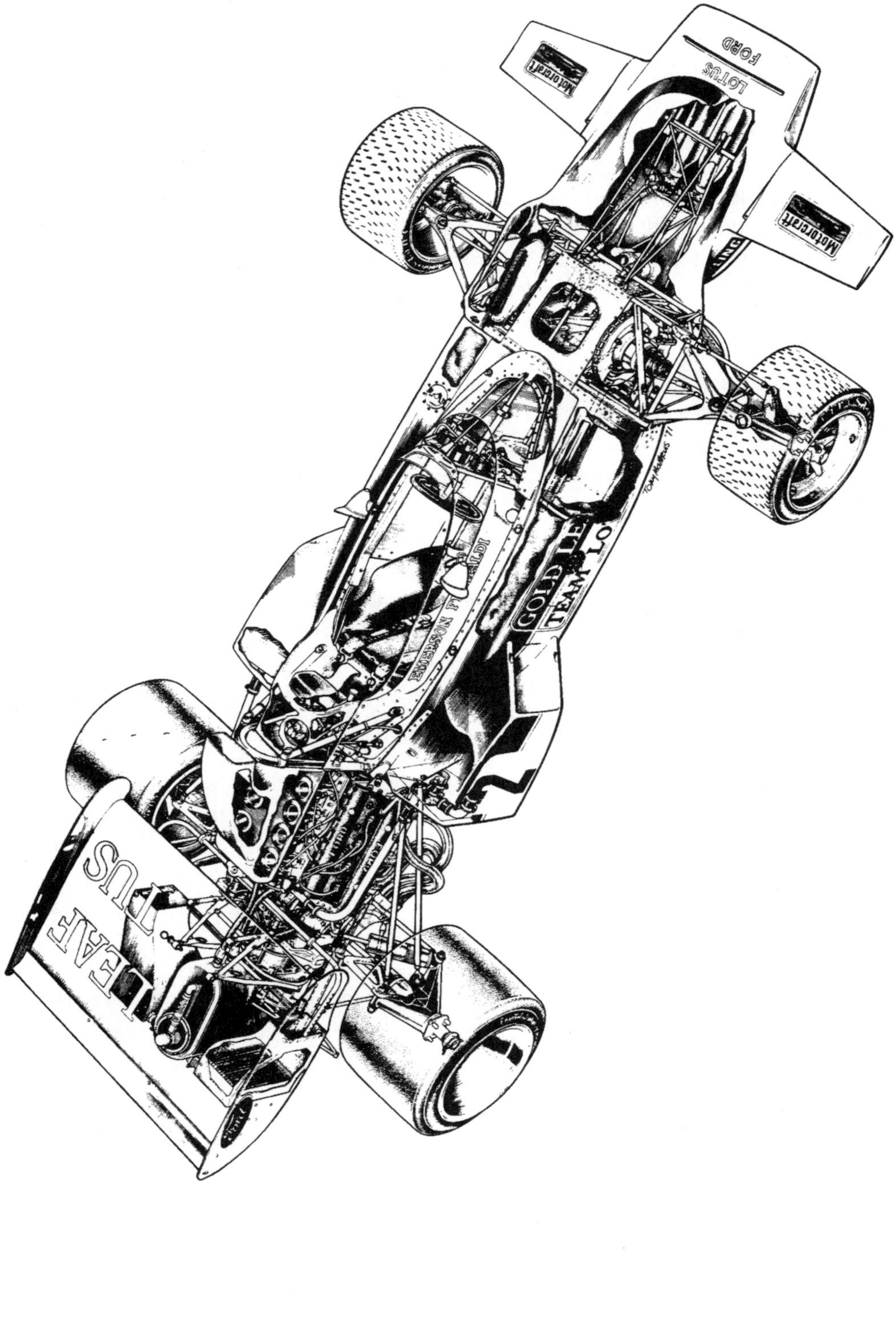

Fig. 2.3 1971 Lotus T72-Ford.

was the doubling of maximum acceleration, braking, and cornering g-levels (Fig. 2.4). The "g-g-V" diagram doubled in girth, with no increase in height. The increased power had been directed toward increasing acceleration, cornering, and braking levels, rather than top speed.

The next seven years (1977–1984) saw two new technologies—ground effect and turbocharging—both of which influenced the magnitude of the performance parameters. In the same period, five other key secondary technologies were developed, without which the two prime ones could not have been fully exploited. These five were carbon-fiber reinforced plastics, computer-control systems, carbon-carbon brakes, radial tires, and fuel chemistry. Such was the surge in performance that these technologies brought to Formula 1, that it now entered the era when it became necessary for the FIA not only to put in place a program of safety measures, but also to limit the magnitude of the performance gains available.

The magnitude of the downforce that could be generated by the regulated wings was ultimately determined by the wing area that could be achieved and the poor efficiency (L/D) of the low-aspect-ratio devices. When Chapman had set out to use the whole plan area of the T72 to generate downforce, he concentrated on creating a positive pressure on the top surface of the wedge-shaped body. In contrast, ground effect exploited the potential for generating a low pressure on the underside of the car and was so effective that the g-levels possible in a Formula 1 car doubled again within only a few years, and the FIA was forced to step in and limit the effect. It banned skirts and stipulated that the undersurface of the cars must be flat. It took a few years to fully achieve this policy because the ingenuity of the designers evaded the regulations until all loopholes were closed. The history of the development of ground-effect aerodynamics is recounted in Appendix A.

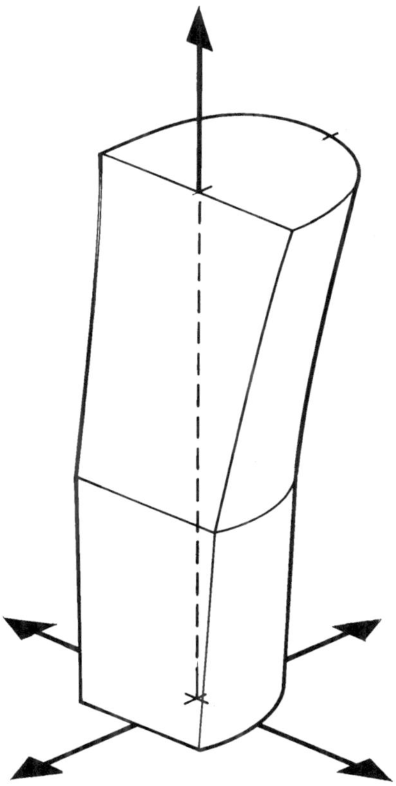

Fig. 2.4 A "g-g-V" diagram for the Lotus T72.

When the 1.5-liter formula was changed to a 3-liter normally aspirated (NA) formula in 1966, the alternative possibility of a 1.5-liter supercharged engine was written into the regulations. The 2:1 equivalency was based on engine-driven mechanical supercharging, and no one chose to follow this route. However, Renault changed all that in 1977 when it decided to enter Formula 1 with a turbocharged 1.5-liter V-6 engine, based on its Le Mans experience. Turbocharging had been developed in the United States during World War II for aircraft engines and was applied to road car engines in the 1960s. Using the residual energy in the exhaust gases as they expanded through a turbine to power the intake charge compressor enables the specific output of a piston engine to be more than doubled. Within two years, Renault had developed a high boost pressure, inter-cooled engine that produced significantly more power than the normally aspirated engines of twice the capacity. Poor throttle response inhibited drivers on twisting circuits, but the rate at which the Renault passed other cars on the straight was noticed and others soon followed this route. The power race that developed could be supported only by major manufacturers, and turbocharging brought the renewed support of Formula 1 by the motor industry.

The first two World Championships won with turbo-engines (BMW in 1983 and TAG-Porsche in 1984) had Bosch electronic fuel injection systems in common. Turbo-engines are difficult to fuel across a wide range of rpm, throttle, and boost settings, and computer-controlled ignition and fueling are essential to release the full power potential and to solve the throttle-response problem. They also are difficult to cool. At the peak of their development, four times the power was generated within a given cylinder size, compared to a normally aspirated racing engine. Fuel chemistry came to the aid of the engineers, with specialized blends of fuel being developed to provide internal cooling and high energy release, thus keeping down the volume and mass of fuel to be carried. These fuels bore little resemblance to pump petrol as recognized by the users of road cars.

Computers first found their way onto race cars to manage data acquisition systems during the 1970s. They appeared on engines in 1977, and the first chassis applications were for active suspension systems (Lotus in 1983). Since that time, their application has been limited only by regulation.

The sensitivity of ground-effect aerodynamics to the attitude and height of the undersurface of the car relative to the road forced suspension and tires to become very stiff. The resulting increase in suspension frequencies revealed serious structural deficiencies in the riveted and bonded aluminum monocoque structures that formed the chassis of the cars. Carbon-fiber reinforcement had been in use for several years to stiffen body panels. As designers became more familiar with the material and the aerospace industry developed the design and manufacturing processes to apply it, it was inevitably used for the primary structure of the cars. The first carbon-fiber reinforced plastic (CFRP) monocoques appeared in 1981; McLaren's was manufactured by the American aerospace company Hercules, and Lotus's was built in its own composite facility. After concern over impact performance had been allayed (put in question by Rolls-Royce's problem with bird damage to CFRP fan blades on its RB211 engine), CFRP monocoques became universal. Since then, CFRP has largely replaced fabrication and is knocking on the door of aluminum and magnesium structural castings. High temperatures, the difficulty in feeding in concentrated loads, and the regulations are the only factors keeping its application in check. Energy-absorbing structures, built of CFRP-skinned aluminum honeycomb-cored panels, have made an enormous contribution to driver protection in the event of an accident (Appendix B).

Carbon fibers combined with a carbon matrix revolutionized brakes and clutches during this period. The high-temperature friction and structural performance of carbon-carbon has permitted these devices to shrink in size and weight while pushing up the energy dissipation characteristics to match the grip of the tires under braking, and the high torque and rpm of engines.

Renault's entry into Formula 1 brought in Michelin and radial racing tires, already under development by the two companies in sports car racing. Their superiority and consistency made cross-ply tires obsolete in Formula 1, probably forever. The lack of movement between the radial plies, compared to cross plies, resulted in lower temperatures and enabled softer compounds to survive. The belt, essential in a radial tire, enabled high-aspect-ratio tires to present the maximum tread area to the ground at all times. Qualifying tires were capable of a coefficient of friction of 2.2, although 1.7 was more typical for races.

By 1984, turbo-engines were putting out approximately 100 bhp more than 3-liter NA's in racing trim and at least 200 bhp more for qualifying and, if necessary, for overtaking. Combined with the downforce levels being achieved with ground effect, 5g was possible at high speed, in braking, and in cornering. Top speeds had not risen dramatically on most circuits where downforce was more beneficial than ultimate speed (Fig. 2.5). Again, the girth of the "g-g-V" diagram had doubled, with little change in height. This time, the doubling had occurred in only seven years. Power had again been directed to increasing horizontal accelerations, not top speed.

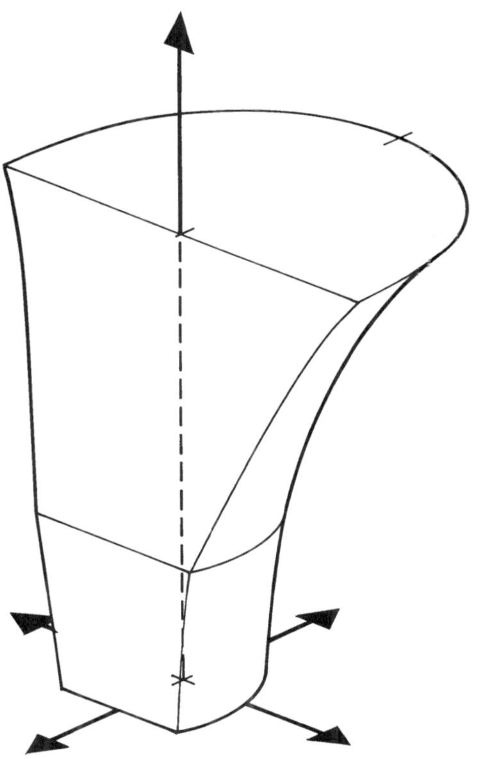

Fig. 2.5 A "g-g-V" diagram for the Lotus T95.

Figure 2.6 shows a typical car of this era—the 1983 Lotus T95, powered by the 1.5-liter, turbocharged Renault engine.

We enter the next era, from 1984 to 1993, with the FIA having limited ground-effect downforce by the banning of skirts and the imposition of flat undersides, and with boost pressure limitations imposed to facilitate the transition of Formula 1 back to normally aspirated engines of approximately 700–800 bhp. These measures effectively limited further expansion of the "g-g-V" diagram, and designers and engineers turned to technologies that enabled the driver to use more of the potential performance. This is the era of the control system. Applied in ever more imaginative and sophisticated ways, computer control took over all aspects of power-train optimization, vehicle dynamics, and driver control. While not affecting the fundamental performance parameters, these systems ensured that all parameters were optimized for a given circumstance, and they assisted the driver in the control task of achieving the maximum potential at all times. They included:

- Active suspension
- Automatic and semi-automatic gear changing
- Engine control: ignition, fuel, throttles, and intake lengths
- Drive-by-wire
- ABS
- Traction control
- Four-wheel steer
- Differential control

These technologies enabled performance to be maintained in spite of ever more restrictive regulations. Faced with the prospect of computers relieving drivers of the task of controlling and even racing the cars, with the inevitable collapse of motor racing as a sport, the FIA stepped in again and banned driver aids—anything that assisted the driver in the task of control. At the same time, the control of aerodynamics through active suspension was banned. Only engine parameter controls (returning to carburetors was not a realistic option), drive-by-wire (the engine manufacturers lobbied hard for this), semi-automatic gearboxes (the cost savings involved in the protection afforded to engines by preventing over-revving were too great to be given up), and differentials (on the false basis that they were purely mechanical analogs) survived.

Fig. 2.6 The 1983 Lotus T95-Renault.

Performance during this period was affected as much by there being no competition among tire companies as by the Technical Regulations. However, by the end of the period, braking was back up to 5g, and cornering was not far behind at 4.5g (Fig. 2.7), in spite of a tire width restriction. Figure 2.8 shows the definitive car of this era—the 1993 Williams-Renault FW-15C.

The final period, through to the present day, is a period of optimization following the fatal accident of Ayrton Senna's Williams at Imola in 1994. The aftermath of this highly publicized accident saw severe performance limitations applied to the cars:

- Power: 3 liters NA and pump fuel
- Downforce: Stepped flat undertrays and limited diffusers
- Circuits: Elimination of long, fast, high-g corners

Tires escaped further limitation until a second tire company, Bridgestone, joined Goodyear in Formula 1 and the FIA committed itself to limiting performance to 1996 levels, by mandating circumferential grooves in the tires, the number of which could be adjusted according to the rate of development of the rest of the car.

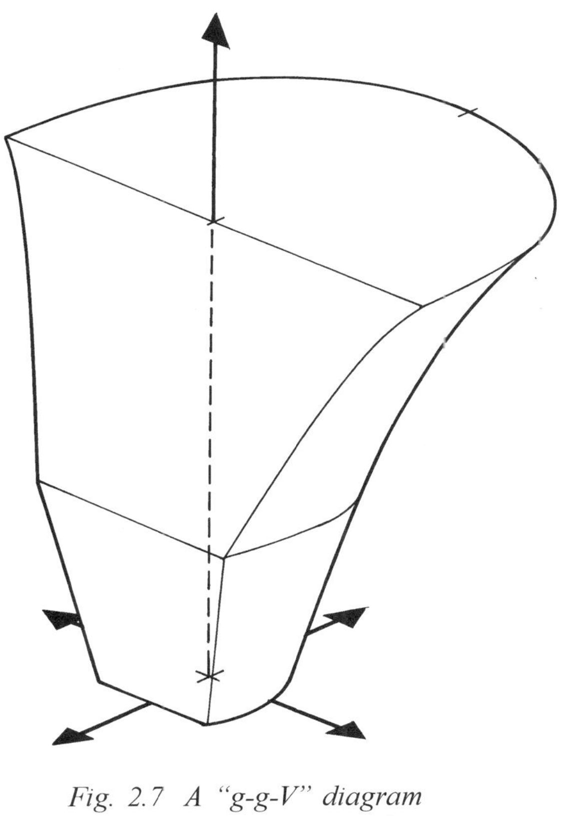

Fig. 2.7 A "g-g-V" diagram for the Williams FW-15C.

With most features of the car now constrained, teams have turned to optimization. Computer-aided design and engineering, sophisticated test rigs and equipment, large graduate engineering staff, and massive budgets have led inevitably to a convergence in the design of the cars as the sport enters the new millennium.

The cars weigh in at 600 kg (1322 lb), including the driver and up to 70 kg (154 lb) of ballast. (With the engine weighing less than 100 kg (220 lb) and the unballasted car without driver weighing approximately 450 kg (992 lb), both are less than the equivalent weights for a Formula 3 car and engine.) The cars are powered by 800+ bhp V-10 engines, revving to 18,000+ rpm, with every parameter controlled by computer. Six- or seven-speed semi-automatic gearboxes change gear in 20 milliseconds at the flick of a steering-wheel-mounted lever. The aerodynamic configurations

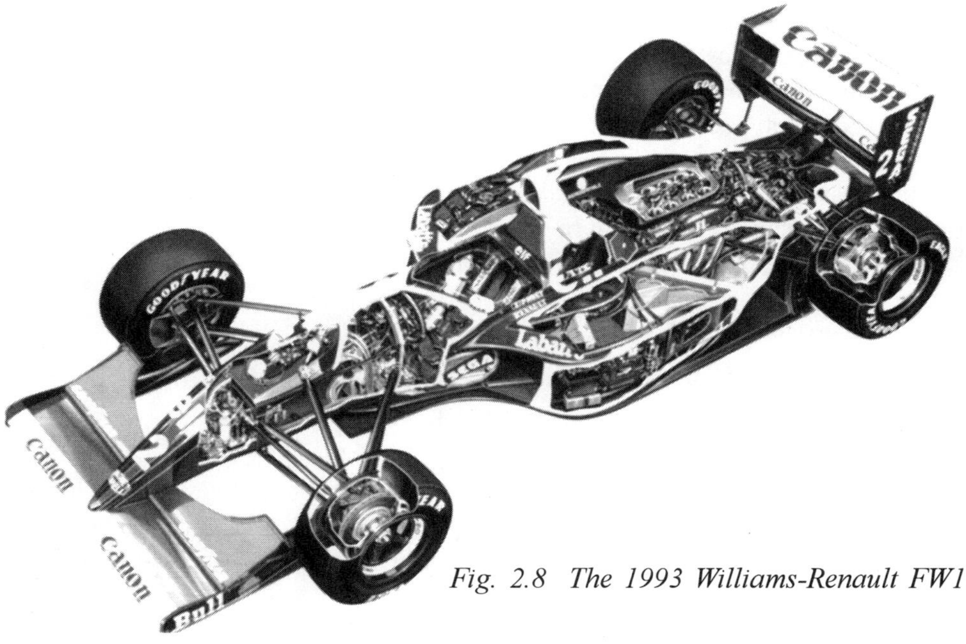

Fig. 2.8 The 1993 Williams-Renault FW15C.

show only detail variation, and with monocoques, suspension, body, gear-case, and many small stressed parts made of CFRP, the cars are distinguishable mainly by their sponsor livery. However, what makes the difference are thousands of details that are now designed and analyzed by CAE software, tested in $15 million wind tunnels or in $5 million rigs, manufactured by computer-controlled processes, and tested in laboratories and on the track while being monitored by hundreds of sensors. Small performance increments cost more and more to achieve each year, and only the teams with the best funding, facilitation, and staffing, all supported by industry partners, can compete.

Performance has nearly been held in check at approximately 4.5g (Fig. 2.9), although the restraint of aerodynamic downforce and tire width has allowed the top speeds to rise to approximately 300 kph (186 mph) on a typical track, and more than 350 kph (217 mph) on circuits with long straights. Figure 2.10 shows the 1998 Ferrari F300.

In the 45 years since the Mercedes W196, the five main performance parameters have changed, some dramatically:

- Power: × 3
- Weight: Little change
- Downforce: Infinite change
- Drag: × 1.8
- Tire grip: × 1.7

Without regulation, these changes would have been even more dramatic, and performance would have escalated to the limits of human drivers. Plots of power and downforce show how, as each has risen to a new peak, it has been knocked down by regulation changes (Fig. 2.11). Even the technology of regulation is more sophisticated now, and it is applied steadily to keep pace with development and maintain what has been decided as a safe performance.

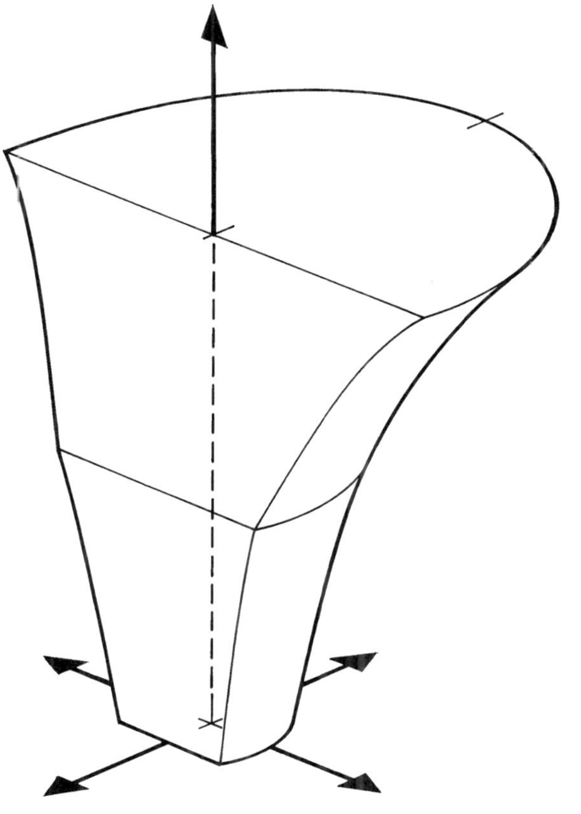

Fig. 2.9 A "g-g-V" diagram for the Ferrari F300.

Perhaps the most revealing aspect of Formula 1 development that these plots show is the difference in the slope of power development per year for the two periods 1967–1979 and 1995 to the present, during both of which 3-liter NA engines were being developed on pump fuel. In the 1970s, power increases of 6 bhp/year were usual. In the late 1990s, they became closer to 30 bhp/year. Only if it were possible to replace power increase per year with power increase per dollar (which it is not!) would it become clear that the quality or ingenuity of the engineers has not changed; rather, the quantity has changed.

The largest "g-g-V" envelope occurred around 1984 (Fig. 2.12) and has been held to approximately these dimensions since then. How much larger it would have been if the peaks in power, downforce, and tire grip had coincided is academic. In spite of the envelope no longer expanding, Formula 1 cars become faster every year. The reason for this is the driver's ability to use more of the available envelope. Back in the late 1980s, when the cars had more than 1200 bhp for qualifying, combined with unrefined flat bottoms that exhibited such high pitch sensitivity that they required 2000+ lb/in. springs, and tires generating a coefficient of friction of 2.2, the Technical Director of Ferrari, Harvey Postlethwaite, was once heard to mutter, "I wish drivers would stop complaining about the cars. If they think these monsters are ever going to be nice to drive, they are wrong!"

As engineers come to grips with vehicle dynamics models, supported by sophisticated data systems and more rig-derived information about the dynamic characteristics of all parts of the

Fig. 2.10 The 1998 Ferrari F300.

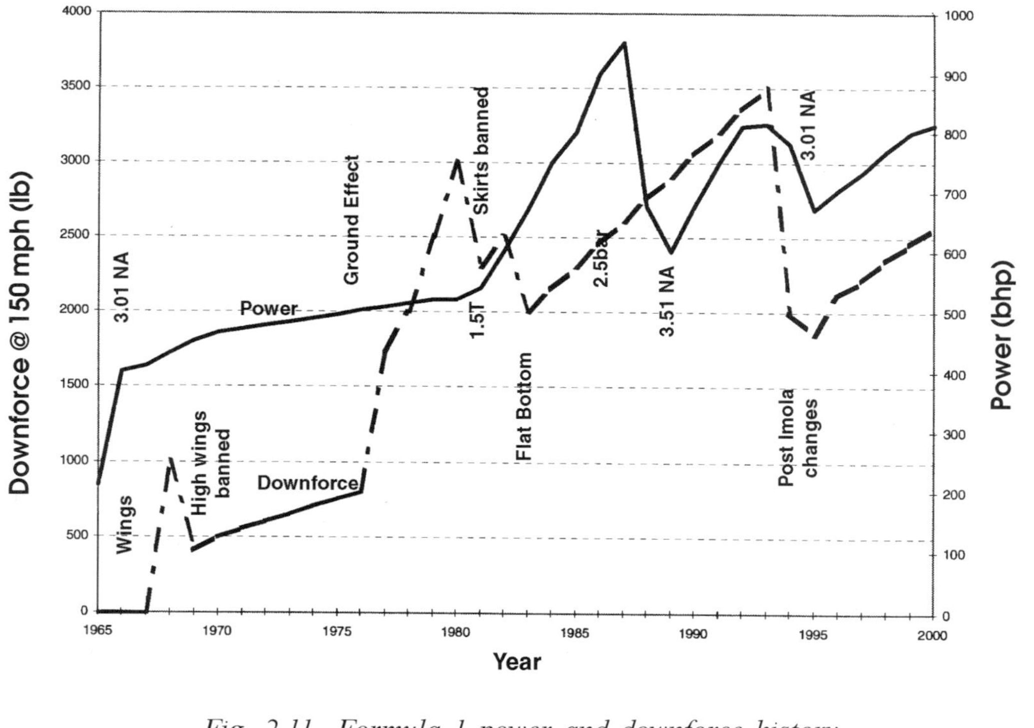

Fig. 2.11 Formula 1 power and downforce history. (Author's illustration)

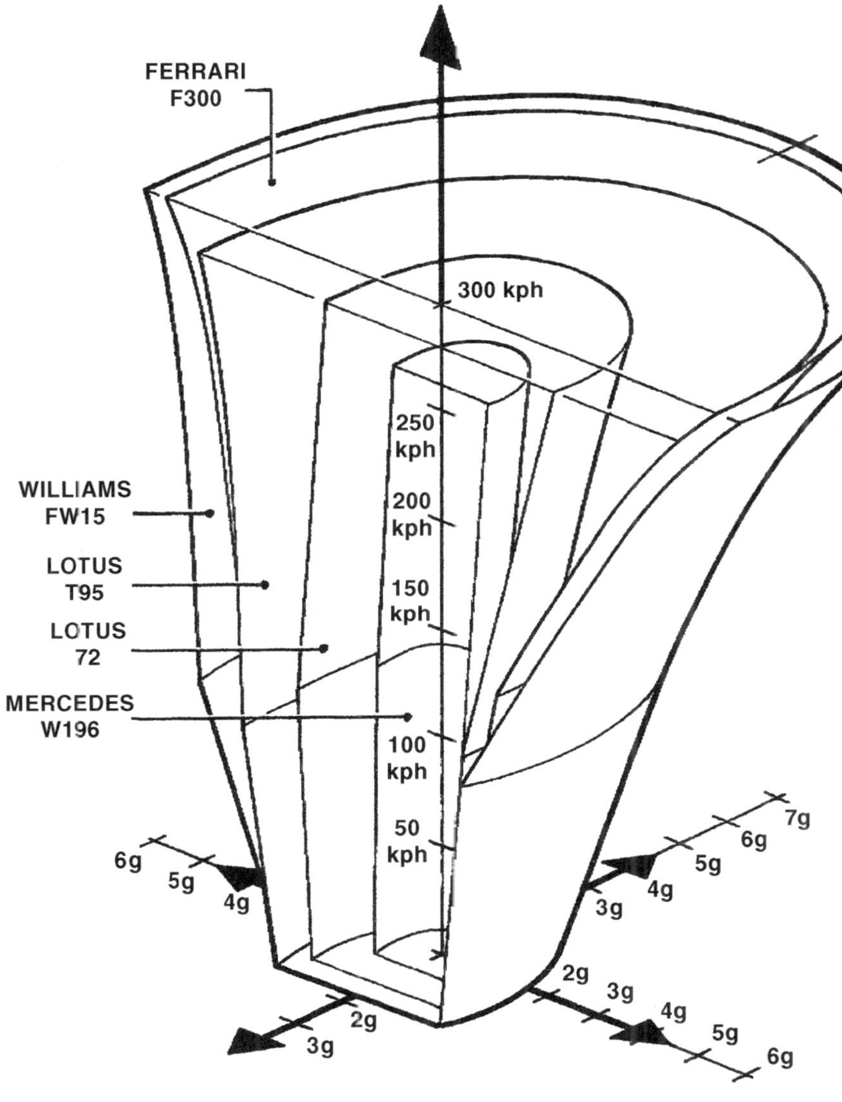

Fig. 2.12 The "g-g-V" diagrams for each era described.

cars, the cars are becoming "nice" to drive, and that is bringing a new set of problems. How can Formula 1 be a sport with worldwide hero appeal, if anyone could drive the cars at the limit? Resolving the conflict among technology, sport, and entertainment is the challenge for the new millennium.

Chapter 3

Technology Trees and Systems Integration

Technology is just a tool.
—Bill Gates

Chapters 1 and 2 defined the potential performance envelope of a Formula 1 car. The evolution within the regulations of five key performance parameters (power, weight, aerodynamic downforce, drag, and tire grip) were also reviewed over the 50+ years of Formula 1, leading to the startling performance of today's cars. Before examining in detail some of the hundreds of technologies that provide the means to realize that performance—to take it from theory to reality—we must assemble them in an order that allows their roles and significance to be evaluated, and identify which are key technologies relevant to most of the engineering disciplines involved. The technology tree is a means of doing this, by assembling technologies into levels of ever greater subdivision and detail (trunk—branch—sub-branch—twig—leaf—...—cell). Every aspect of designing, manufacturing, testing, and racing a Formula 1 car involves many technologies that contribute to the success of a given example, racing against similar but different examples. Like a tree in a forest, the sum of those contributions (e.g., each cell, leaf, and twig) determines how that individual entity develops and whether it succeeds.

Although trees grow and branch upward, convention has it that we represent them inverted, as with family trees, being more representative of the root system. The trunk of the technology tree is the Formula 1 car, racing on the track, and the three main branches are: (1) design and engineering, (2) manufacturing, and (3) racing (Fig. 3.1).

To determine the sub-branches of design and engineering, it is helpful to look again at the engineering objective for a Formula 1 car: speed around a defined path, achieved by maintaining the maximum acceleration in the appropriate direction at all times. If acceleration is broken down into its Newtonian components, the branch can be subdivided into those technologies that contribute to inertia and those that contribute to the generation and control of forces (Fig. 3.2).

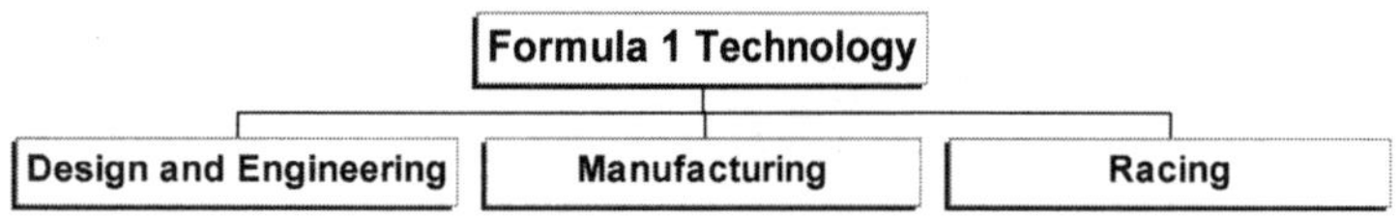

Fig. 3.1 Formula 1 technology headings.

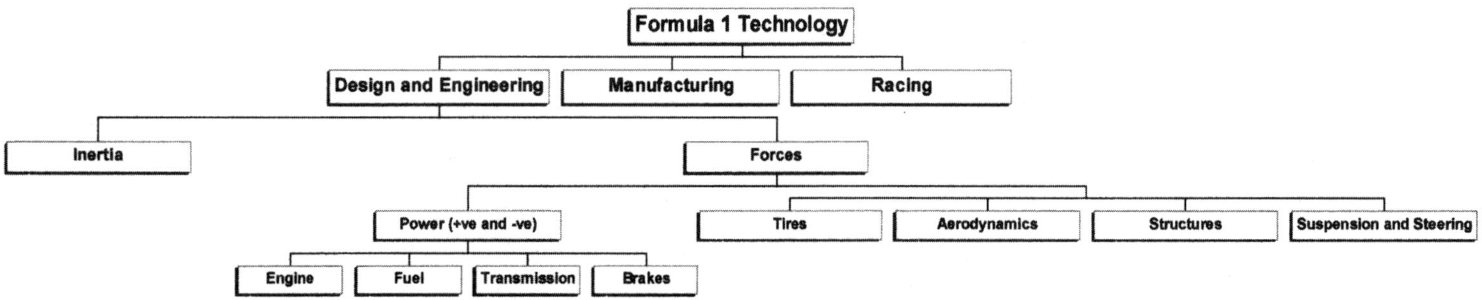

Fig. 3.2 Design and engineering technologies.

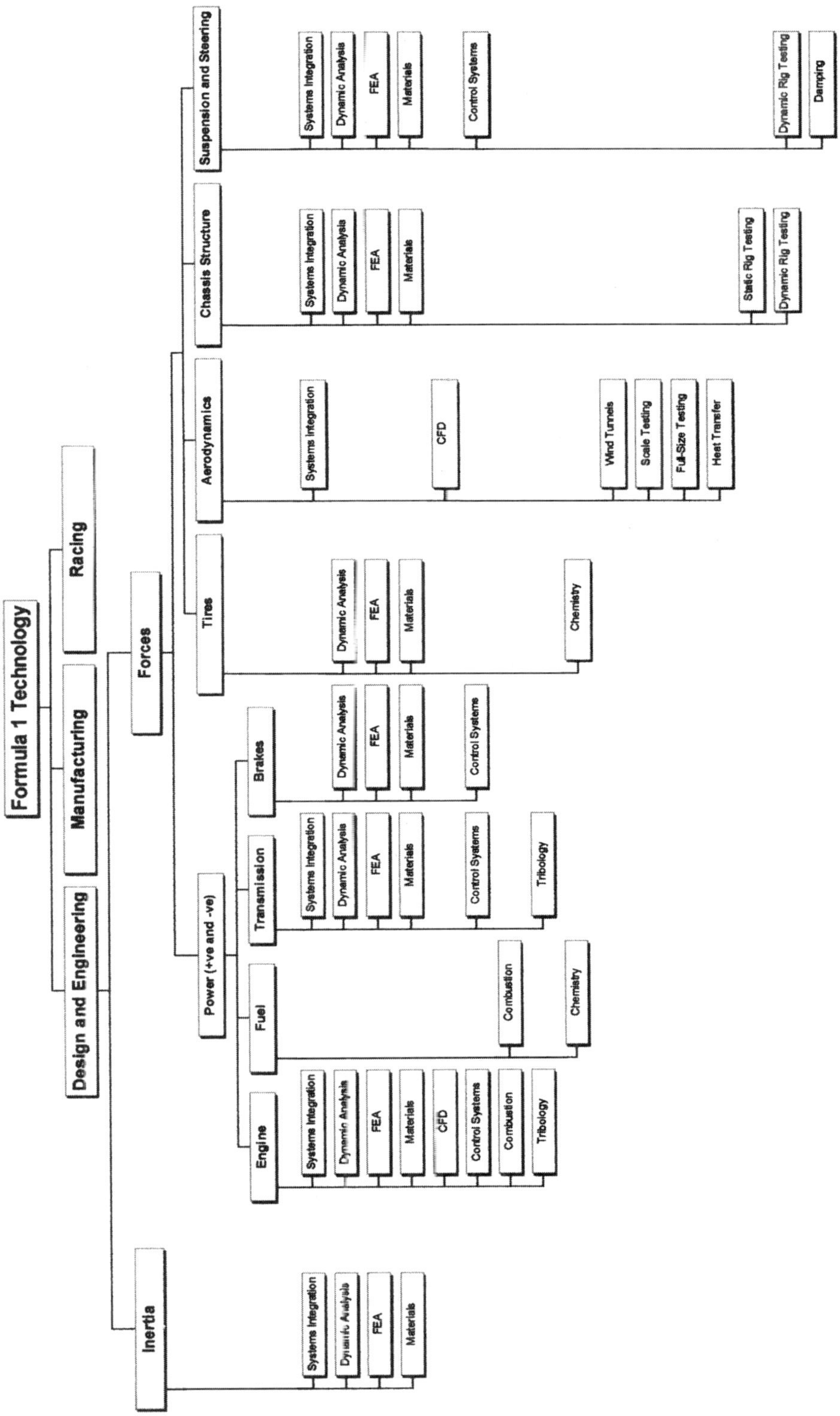

Fig. 3.3 Design and engineering technologies, expanded.

Each of these branches has a number of key technologies, some of which are common across disciplines. Therefore, the layout of the tree can be arranged to align these technologies for clarity (Fig. 3.3).

At the next level of branching, the tree almost explodes into a mass of small branches and twigs, which are difficult to portray clearly on a single piece of paper. The technologies represented at this level are those that make up the intellectual property of their exploiters and, as such, are closely guarded commercial secrets. This book is not large enough to explore all of these sub-technologies, important as they are, because gaining insight into them for publication is almost impossible. Some of them—such as CFRP, ground-effect aerodynamics, and control systems—are covered in greater detail in the appendices of this book. A single technology, CFRP, traced through its branches illustrates the level of complexity into which the Formula 1 technology tree could ultimately be developed (Fig. 3.4).

Manufacturing involves process technologies that are, in the main, commonly available within the aerospace, motor, computer, and electronics industries (Fig. 3.5).

Some teams and their suppliers, however, have developed particular expertise that enables them to manufacture components to an exceptional performance, quality, or speed of manufacture that actually contributes a performance advantage to the car and often a business advantage to the owner. With the life cycle of a Formula 1 car design being approximately 18 months from concept to obsolescence, the speed with which every step of the design, engineering, manufacturing, and testing process can be carried out has a significant bearing on the eventual success of that car. Some technologies, new or new to Formula 1 (e.g., rapid prototyping), also provide an advantage to those having access to them and exploiting them appropriately.

A prototype Formula 1 car, freshly built and standing on its wheels in the workshop for the first time, will not win races or a championship. It is probably not even able to complete a 305-km 190-mile) race at racing speeds. Testing and developing it, adapting it to its driver's characteristics, and racing it successfully involve a separate series of technologies. Included in these, if Formula 1 is to be a successful revenue-earning business, are those technologies that deliver the product that consumers will purchase, thereby providing the ultimate source of funding: TV program and news material in the form of information, images, and sound (Fig. 3.6).

If it could fit onto one page, the assembling of the parts of the technology tree that have been developed so far would illustrate how the main technologies involved in a Formula 1 car fall into vertical groups, which form the subjects of following chapters. Running horizontally through these are a number of technologies that are common to many, such as systems integration, dynamic analysis, and finite element analysis (FEA). Systems integration may not normally be considered a technology in itself, but it is probably the single one that determines the difference between a successful design and an also-ran. It is also the technology that, more than any other, has been a black art for many years and is only now emerging into a science.

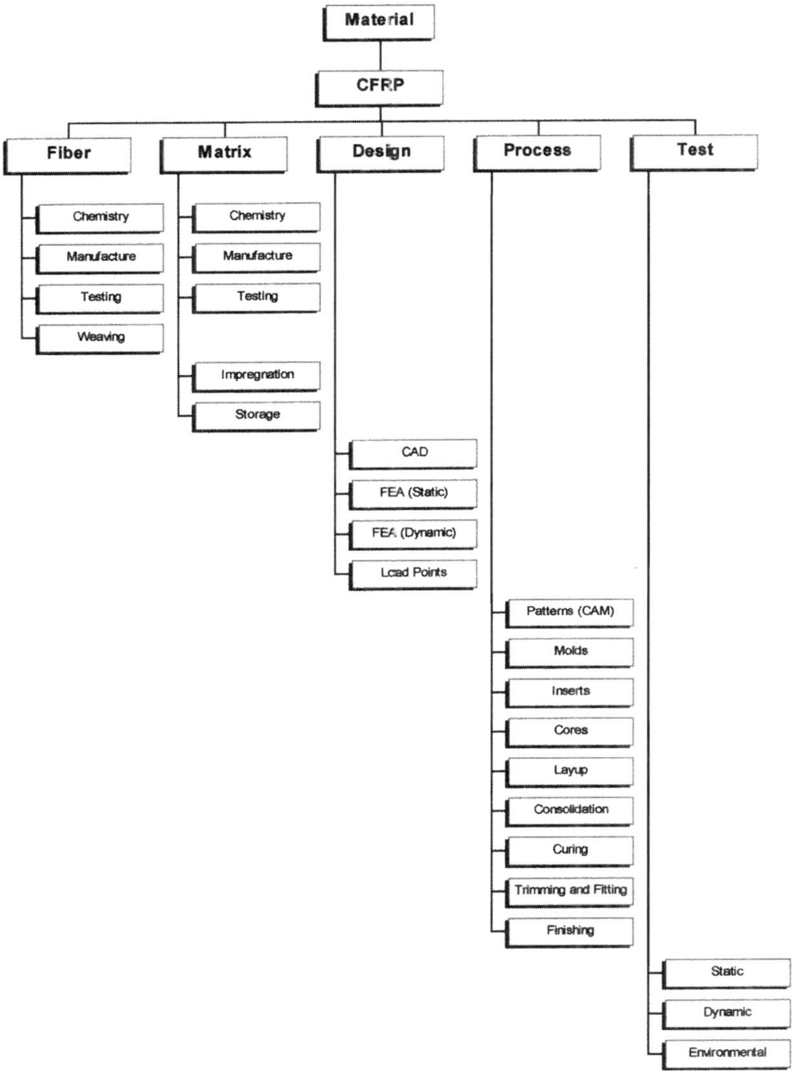

Fig. 3 4 CFRP technology.

Systems integration is the technology of trading off the characteristics of individual components and systems against each other to obtain the optimum performance of the integrated whole. By definition, the characteristics of components and systems must be accurately known for systems integration to be successful, and the optimum performance must be clearly defined. As applied to a Formula 1 car, optimum performance should be obvious but is seldom achieved. Part of the problem is that the combination of straight-line acceleration, top speed, cornering and braking acceleration, change of direction, stability and control, maintainability, reliability, raceability, and driver interface that scores the most championship points will vary from year to year. In motor

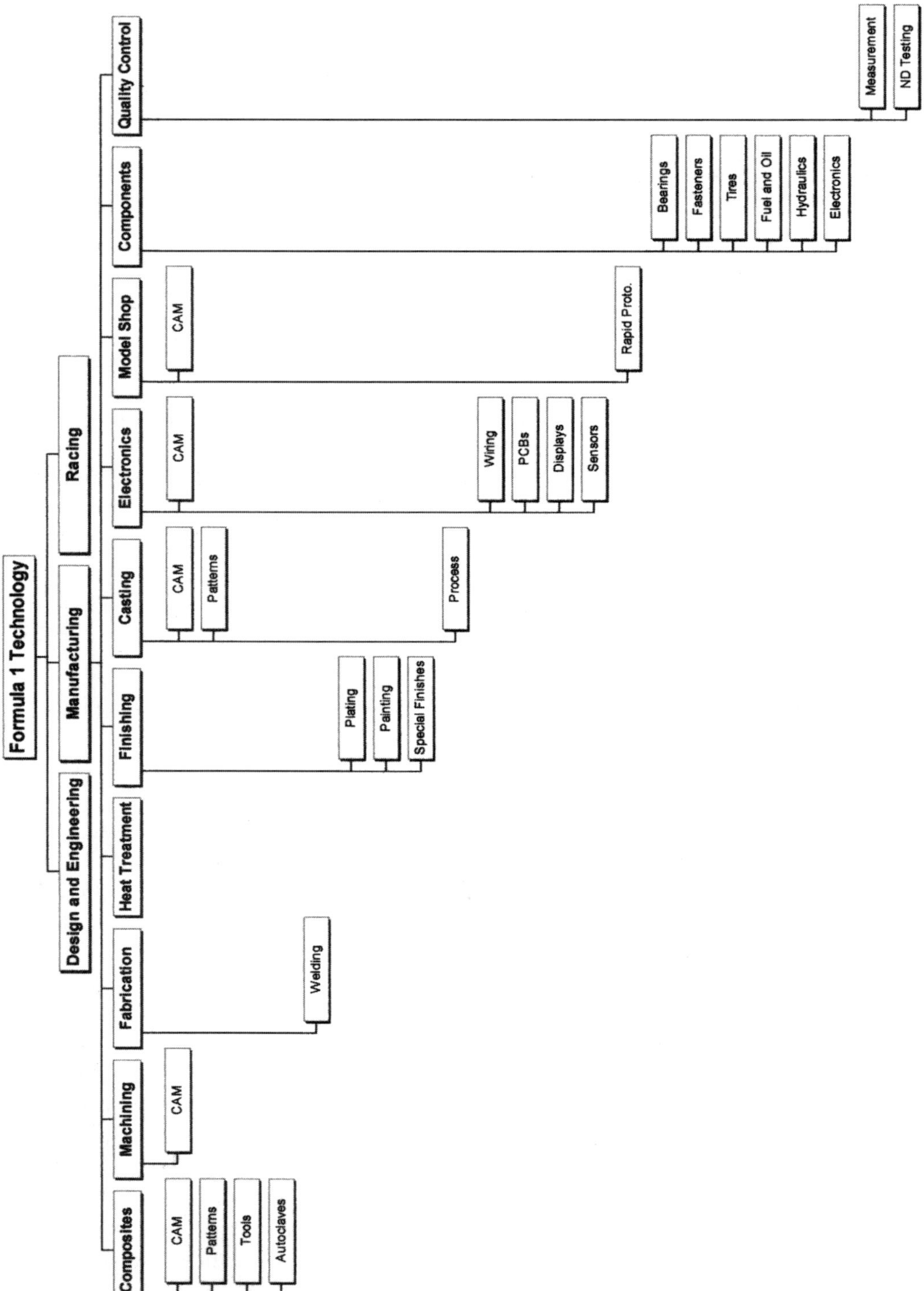

Fig. 3.5 Manufacturing technologies.

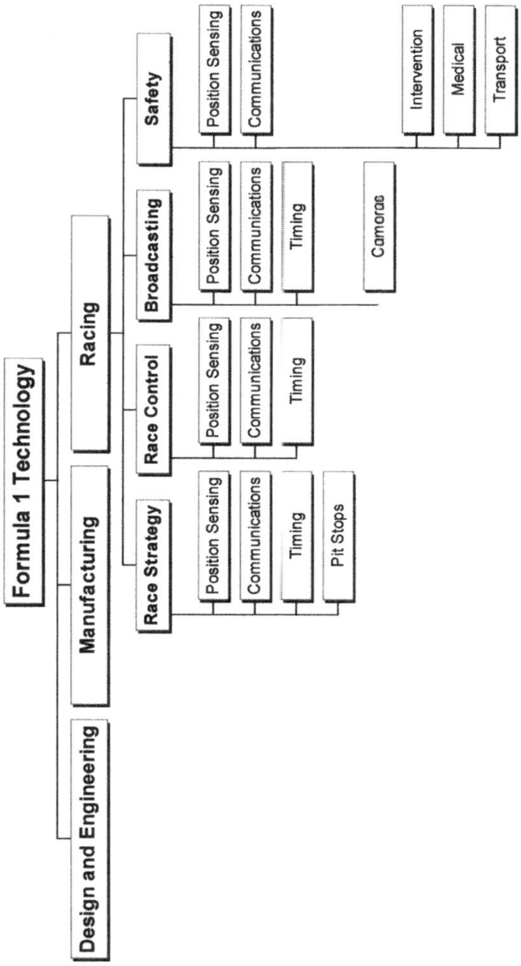

Fig. 3.6 Racing technologies.

racing, as in some businesses and war, optimum performance is a relative term; performance relative to the opposition is the only thing that counts. Determining the right combination of characteristics and choosing the components and systems to yield these characteristics is the key part of the process.

Systems integration is not only about making one component do two or more jobs, although this has been a theme within motor racing for many years and is still applied today. If doing so results in a smaller and lighter package, that is good; however, if it reduces reliability or affects the mechanics' ability to change a failed part, systems integration analysis may determine that the whole system has not benefited from this design approach. The time it takes to change an engine may determine whether the driver makes it into the race or not.

Successful Formula 1 designs were capable of winning World Championships in successive years (e.g., the three World Championships won by the Lotus T72 between 1971 and 1974, which was an example of steady development of a well-integrated design over several years). However, the pace of development today is such that an almost completely new design is required each year to remain competitive. Determining the optimum performance of a design, plus the parts and technologies to be incorporated into it to achieve those characteristics, is influenced by the availability of proven components and systems and then making judgments about those that can be developed effectively in the time available. Formula 1 cars that fail to reach their potential in time to compete successfully are a common cause of lack of success. Those cars may be remembered for their technical wonder, but they do not get into the record books. The history of Formula 1 racing is full of classic examples of successful systems integration and even more full of examples of getting it wrong.

The better systems integration of the light but lower-powered, mid-engined Coopers, Brabhams, and Lotus brought about the extinction of the front-engined Grand Prix cars in 1958. Their lower weight, lower drag, and lower tire use more than made up for less power. Better braking and cornering outweighed the acceleration and the top speed advantage of the more powerful front-engined cars on most of the post-war circuits, to give them a superior performance overall.

Perhaps the classic example of systems integration came about through the cooperation of Lotus's Colin Chapman and Cosworth's Keith Duckworth, with the financial help of Ford—the Ford-Cosworth DFV engine as originally fitted to the Lotus T49 in 1969. The combination of adequate power, light weight, an elegant and light but stiff structure, and reliability through simplicity made the combination a winner from its first race. However, it was partly chance that made this engine the most successful in the history of Formula 1, winning 155 World Championship races between 1969 and 1985. In the same way that the Italian teams, Ferrari and Alfa-Romeo, caught up with the British cars in the mid-1970s, with their more powerful flat-12 engines that also endowed their cars with a lower CG height, the development of ground-effect aerodynamics negated their advantage. The shape of the DFV, a 90° V-8, enabled Colin Chapman and subsequently the other British DFV users to design cars whose monocoque/engine structures exactly suited the new aerodynamics. The V-8 was in cross section a shape very similar to a seated driver: narrow-hipped and broad-shouldered (Figs. 3.7 and 3.8). By placing the fuel tank between the cockpit and engine of the same cross section, the ground-effect tunnels on either side of the monocoque/tank/engine could be maximized. The cylinder heads and induction systems of the flat-12 engines intruded directly into the diffuser, compromising the airflow under the cars. Downforce became more important than power.

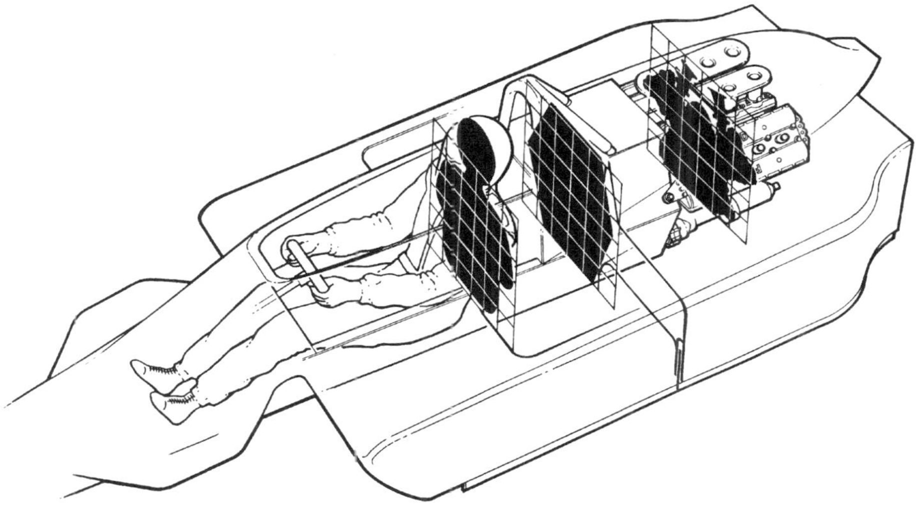

Fig. 3.7 Cross-section comparison among driver, fuel tank, and engine—the Lotus T79-Ford.

Fig. 3.8 The Ford-Cosworth DFV fitted to a Lotus T49. (Courtesy of Phipps/Sutton)

The story of this almost accidentally successful systems integration leads to a classic example of an unsuccessful design that had too much emphasis on one characteristic at the expense of others, such that the whole was less than optimum. Part of Colin Chapman's brief that led to the development of the first ground-effect car, the Lotus T78, was to endow it with aerodynamic characteristics that were more pitch sensitive than the T72. The T72 had suffered in its latter days from understeer on entry to a corner and oversteer on exit. Chapman felt that, if the downforce could be increased on the front under braking, and on the rear under acceleration (or at least reduce the front downforce), the problem would be reduced. The softly sprung cars of that time dived under braking and squatted under acceleration, thereby increasing pitch sensitivity to longitudinal acceleration. The wind tunnel program for the T78 concentrated initially on making the front wings sensitive to their height above the ground, research that contributed significantly to the development of ground effect.

Once the T78 showed its potential, Chapman began to design a new car specifically to exploit the new aerodynamics: the Lotus T79. He specified even more pitch sensitivity. The T79 was successful, but only because its aerodynamics gave it such a performance advantage over its non-ground-effect competitors that its structural, cooling, and braking deficiencies (the first two of which were as a consequence of subjugation to maximizing ground effect) did not quite cripple it. The writing was on the wall.

On the basis that "If some is good, more is better," Chapman specified even greater pitch sensitivity for the successor of the T79—the Lotus T80. The T80 was intended to utilize the entire length of the car for the generation of ground-effect downforce, involving complex bodywork and curved skirts. It had pitch sensitivity in spades! The car was so sensitive that, above a certain critical speed, it became aerodynamically unstable in pitch. One test day at Silverstone, Mario Andretti coined the term "porpoising" to describe the phenomenon when he observed daylight under the front wheels while at speed on the straight. The interaction between the aerodynamic pitch and heave characteristics and the stiffness and damping of the suspension and tires—analogous to flutter in aircraft structures—was not resolved before the end of the season, and the T80 went down in history as a "white elephant," a victim of poor systems integration.

A natural instinct for systems integration is one of the main characteristics that enables certain designers to be so successful in Formula 1 for a period of time. Ron Tauranac (Brabham), Colin Chapman (Lotus), Mauro Forghieri (Ferrari), Gordon Murray (Brabham), Patrick Head (Williams), John Barnard (McLaren), Adrian Newey (Williams and McLaren), and the combination of Ross Brawn and Rory Byrne (Benetton and Ferrari) all were or are skilled at making subjective judgments, balancing in their minds the tradeoffs among weight, power, aerodynamics, and suspension, to arrive at successful combinations without the aid of computers. Even today, the information about component characteristics and the optimization techniques are not sufficiently developed to dispense with that instinctive feel for the right tradeoffs. Adrian Newey emerged from the 1990s as the designer with those attributes best suited to the task. He has been able to balance the critical aerodynamic characteristics—downforce versus pitch sensitivity—to produce a series of successful Williams and McLaren Formula 1 cars. The only designer able to approach his record in the modern era is Rory Byrne (Benetton and Ferrari).

Of course, there is more to systems integration today than aerodynamics. The opportunities for major configuration and layout changes, such as were common in the 1960s and 1970s, are now rare as the combination of Technical Regulations and intense optimization has defined the arrangement of major components in the cars. The driver must be seated such that his feet are behind the front axle, and the fuel tank shape is dictated by the desire to make it as short as possible, without raising the CG height when the tank is full of fuel, the maximum width having been set by the regulations (Fig. 3.9).

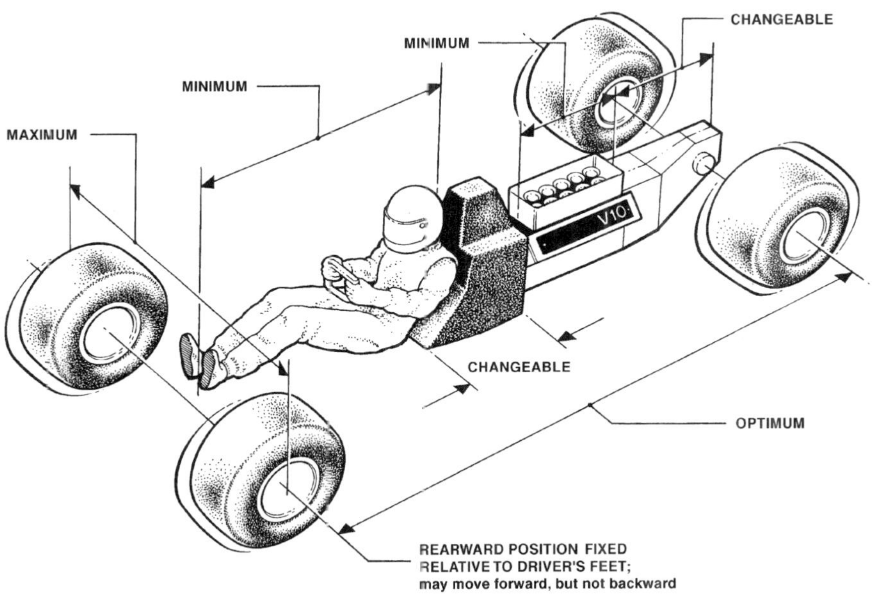

Fig. 3.9 Effect of the Technical Regulations on the layout and dimensions of a Formula 1 car.

Engine layout has converged on the V-10 arrangement, designers having explored V-8 and V-12 layouts since turbocharging was banned. Although the V-8 won World Championships for Ford-Cosworth and Benetton, it lacks ultimate power because of low piston area and inertia limited rpm (Fig. 3.10). Lack of power outweighs its desirably short, structurally stiff, and light arrangement. The V-12, with its high piston area and rpm, gradually lost to the V-10 because of increasing friction and windage losses as rpm rose through development, lower stiffness, and its greater length (Fig. 3.11). The future might have seen engine arrangements such as the W-12, which attempted to combine the higher power potential of twelve cylinders

Fig. 3.10 The Ford-Cosworth V-8. (Courtesy of Sutton Motorsport Images)

Fig. 3.11 The Ferrari V-12. (Courtesy of Sutton Motorsport Images)

with the structural advantages of an engine that is only four cylinders long. The balance of characteristics in making the correct choice between power, driveability, structural integrity, and chassis dynamics would have been a triumph of systems integration. Unfortunately for engine designers, the engine manufacturers have chosen to limit engines to ten cylinders for now (Fig. 3.12).

Gearbox design has slowly changed from longitudinal (gears behind the final drive), via transverse, to longitudinal again (gears ahead of the final drive). The longitudinal arrangement is the slimmest, for good radiator exit air flow and a slim diffuser center-section, but the gears must be positioned between engine and axle to enable the mass of the engine to be moved forward to meet the current requirement for more weight on the front axle. The forward shift of the weight is to rebalance the car in response to larger front tires, developed to compensate for the reduction in overall tread area brought about by the introduction of grooves in the treads and a rear tire limited to 380 mm (15 in.). Positioning the gears ahead of the axle means that the engine and gearbox must be split to change ratios, whereas the older layout allowed quick ratio changes via the rear cover. The transverse arrangement is short and has the gears ahead of the axle, but it requires a gear access aperture in the side of the case, which is structurally undesirable. It is also wider than the longitudinal gearbox. All cars currently have V-10 engines and longitudinal gearboxes, so the current systems integration solution is clear.

Fig. 3.12 The Mercedes-Illmor V-10. (Courtesy of Sutton Motorsport Images)

The scope for distributing other components within the chassis to fine tune the weight distribution is limited. Moving engine systems farther from the engine, such as cooling radiators and the oil tank, must inevitably increase overall weight. Steering and braking systems are inevitably mounted ahead of the driver's feet. The greatest potential for adjusting weight distribution is to eliminate weight from the engine and gearbox assemblies, such that overall weight is below the regulated weight limit, and then positioning the mandatory ballast as desired and as low as possible in the car.

The greatest challenge before the technical director in determining the specification of a new car is deciding what systems to include and how sophisticated or complex he should allow the systems and software engineers to make them. The semi-automatic gear-change system is a classic example. The advantages of abandoning an H-pattern gear-change lever (which the driver operated with his right hand within the tight confines of the cockpit) for switches on the steering wheel are substantial. The largest speed gains came from the freedom given to the driver to concentrate on controlling the car during braking, when it is at its most unstable, and allowing him to flick his fingers to command a gear-change. Instead of taking one hand off the wheel to move the lever precisely across its gate, while synchronizing clutch and throttle, he now simply chooses the moment and commands the gear change, and the system does the remainder of the work. An early down-change used to result in over-revving the engine, as did shifting down instead of up, or selecting the wrong gear during down-changing. Software now refuses a gear-change if it will result in over-revving the engine. The cost savings in rebuilt engines are substantial.

Up-changes can now be performed with the power disconnected for only 20–30 milliseconds. Any attempt to reduce this time will hardly affect the speed of the car but will probably result in more balked gear changes and a possible reduction in gearbox durability. The return on effort to speed up changes is not worth the risk. Down-changes, however, do benefit from development. Speed is not as important during down-changes, but smoothness is. Minimizing the pulse of torque transmitted to the rear wheels as the engine re-engages will result in the least disturbance to the car during the unstable braking phase. Synchronization of engine rpm and good clutch control are essential features of a down-change control algorithm.

Few of the control systems now permitted are going to influence the five key performance parameters or affect the size of the "g-g-V" diagram. They are all concerned with the control and stability of the car and the ability of the driver to use the potential performance. Even the engine control system is as much about driveability characteristics as ultimate power. Assessing the benefits of more complexity, and hence more potential unreliability and driver and engineer confusion, is not easy because the gains may not become apparent on lap simulations. However, the technologies for measuring the precise static and dynamic characteristics of virtually everything on the car are being developed quickly, and the simulation techniques for evaluating the performance tradeoffs are already highly sophisticated. Optimization techniques based on game-theory are under development and will soon provide software for powerful computers to examine and optimize tradeoffs among hundreds of components and systems which will automatically determine the optimum. The complexity and difficulty in measuring the dynamic characteristics of the aerodynamics and the tires are currently the limiting factors in fully automating the systems integration process. It will be a while before the instincts of Adrian Newey and Rory Byrne become redundant.

Chapter 4

The Chassis and Its Dynamics

Nonsense, the car is rubbish; I can only get it to turn-in by using the gearbox.
—Ayrton Senna, in response to a suggestion that his Lotus must be quite good, having attained a front-row grid position at Monaco in 1987

In Chapter 1, the racing vehicle was defined as:

> ...the system by which artificial forces (i.e., non-human forces) are generated in order to accelerate the driver's body while he (the driver) attempts to use the vehicle control systems to maintain the highest possible acceleration level, in the appropriate direction, at all times. For maximum acceleration, the force generation systems must as powerful as possible and their mass, added to that of the driver and including the fuel necessary, as small as possible.

The designer's job is to collect all the components needed for the system so that they perform in a cohesive manner. He also must design it such that it meets certain other practical requirements, namely:

- It must fit the definition of an automobile, according to the FIA Formula 1 Technical Regulations.
- It must conform to dimensional regulations.
- It must meet certain performance-related limitations.
- It must meet safety requirements.

- It must meet certain sporting-related requirements.
- It must be reliable.

The dominant means of generating forces in the plane of the track surface, allowable within the regulations, are the tire contact patches and the drag. (Aerodynamic side force is the other.) How this is achieved is the subject of later chapters on tires and suspension. The tire forces are amplified by aerodynamic downforce, increasing the normal force between the tire and track, at the expense of straight-line speed due to drag. Motive power is produced by the engine, burning stored energy in the form of fuel, transmitted to the tires via the transmission. The brakes provide negative motive power and have a secondary influence on the vehicle control. The assembly of these essential systems and their subsystems into a reliable and structurally adequate entity, with the optimum inertial characteristics, is the subject of this chapter. The dynamic requirements and the regulatory constraints must first be defined to understand the options and techniques that are actually available to the designer for optimizing the chassis layout and construction to achieve his objectives. However, before this can be done, it is necessary to examine what the driver is trying to do with the car as he slows it for a corner, conducts it around the corner as quickly as possible, and accelerates along the subsequent straight. Only then is it possible to start understanding some of the demands placed on the chassis.

For the sake of simplicity, let us consider a 90° bend for which the driver must brake the car down to 150 kph (93 mph) prior to negotiating the corner. At this point, if we define a coordinate system relative to the earth, where the car is pointing in the X direction and the Y axis is at 90° (Fig. 4.1), the initial conditions are

$$V_x = 100$$

$$V_y = 0$$

and the car is pointing along the X axis, i.e., the heading angle,

$$\phi = 0$$

After the corner, when the car is again doing 150 kph,

$$V_x = 0$$

$$V_y = 100$$

$$\phi = 90^\circ$$

The driver must control the car to follow the desired line around the corner and end up with it pointing in the right direction, all conducted at the greatest possible speed. It is a matter of getting the X and Y coordinates of the car, along with the heading angle ϕ, correct at each

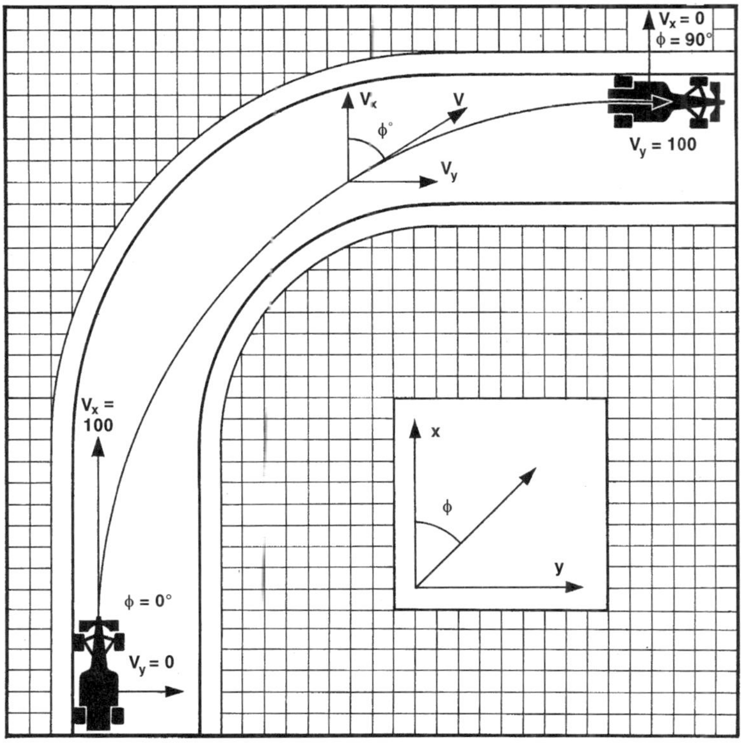

Fig. 4.1 Conditions for traversing a 90° corner.

instant. To do so, the driver must sense and control the first and second differentials (the calculus variety, not the mechanical ones) of X, Y, and ϕ, i.e., accelerations and velocities as well as position. In particular, he must judge the deceleration from 150 kph to 0 in the X direction, and the acceleration from 0 to 150 kph in the Y direction, controlling the magnitude and direction of the force vector by utilizing the longitudinal and lateral force-generating capabilities of the car as it rotates through 90°. Sensing the accelerations, along with the velocities and position, is the key.

X, Y, and ϕ are determined solely by the eyes, as are the rates at which they change, using analysis of optic and retinal flow (Ref. 4.1). A human can process information derived from images quite quickly, so there are only small lags in sensing the position and attitude of the car. Velocities are more difficult. A human being does not really have any speed sensors, at least not any adapted to travelling in wheeled vehicles. When running, he knows the length of stride and the frequency and can estimate speed. He also uses the noise of air rushing past his ears to help estimate speed on a horse or bicycle, for instance, although wind can confuse this data. In a closed car, a speedometer is needed to know speed. However, when driving at the limit, the eyes are too busy to refocus on an instrument. Thus, a driver takes cues from the rate at which his perception of the scenery changes, plus the noise and vibration frequencies from the engine and road and sometimes wind noise, to make a speed estimate. Note how you always drive faster in a quiet car after becoming used to a noisy one, and vice versa. Experience of the conditions and the car count for a lot in determining the accuracy of the estimate.

In particular, quantifying yaw velocity is a very unnatural requirement. A human is not adapted to sense yaw velocity with eyes shut; instead, the rate at which the scenery flows in the field of view gives a sense of rotational velocity. The processing of this visual information by the brain (performing continuous differentiation) is quite slow and must take into account how far away the scenery is. If the head is shaken (i.e., the eyes themselves are moving relative to the car), the noise-to-signal ratio prevents fine resolution of yaw velocity.

Yaw velocity is the prime signal in controlling the force vector of the car. As the car travels around the corner, the heading angle steadily increases until the car points in the new direction. The yaw rate depends on the radius of the corner and the speed. Superimposed on and additional to the heading angle is the slip angle of the car. Any change in heading angle, above or below that predicted by the normal passage of the car around the corner, signals a change in slip angle which the driver uses to control the lateral force via the lateral force characteristics of the tires. The driver must know the yaw angle at all times. To do so, he must sense any variation in rotational velocity relative to the rate of change of heading necessary to rotate the car through the corner.

The acceleration terms in the X, Y, and ϕ senses are less important. The human body, through its acceleration sensors—the semi-circular canals and the forces exerted on the body by seat and straps—is not very good at measuring absolute acceleration. What it is much better at doing is sensing small rates of change in acceleration, sometimes known as "jerk." The longitudinal and lateral accelerations on a car are steady through most of the corner, while yaw acceleration also changes relatively slowly. However, any change will signal a change of grip, an exceeding of the limit of the tires, or a disturbance of the car. The useful jerk signals—those that tell the driver approximately how close the tires are to their limit—tend to be confused by "noise" generated by the disturbed motion of the car over bumps.

What the driver wants to know is any change in what is happening at the tire contact patches, and he needs to know it with the minimum time lag—one essential of any good control system. A Formula 1 car has a natural yawing frequency around 3–4 Hz, determined by its inertia, the cornering stiffness of it tires, their compliance and that in the suspension, and its yaw damping characteristics. By the time the driver receives the motion signals that something has changed, processes those signals, and inputs a control correction, it may be too late. Thus, any means of conveying the essential information more quickly is invaluable. The stiffest and lowest inertia connection between a tire contact patch and the driver is the steering system. If it is free of stiction and compliance, and if the geometry is correct, the steering system will send signals of small changes in the self-aligning torque of the two tires on the front axle, straight to the driver's hands. This information can arrive with less than half the time lag of other information and is the prime sensing system for measuring the front axle grip (Fig. 4.2).

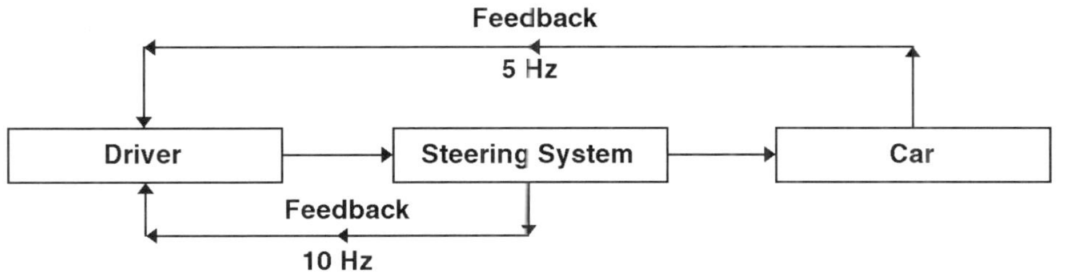

Fig. 4.2 Steering as a control system.

The driver of a car at the limit can often be seen applying small steering inputs. He senses the change in self-aligning torque fed back to him as changes in steering wheel torque, and he gauges how close he is to the limit. An increase in torque for an increase in steer angle indicates less than the limit, zero at the limit, and negative if over the limit (Fig. 4.3).

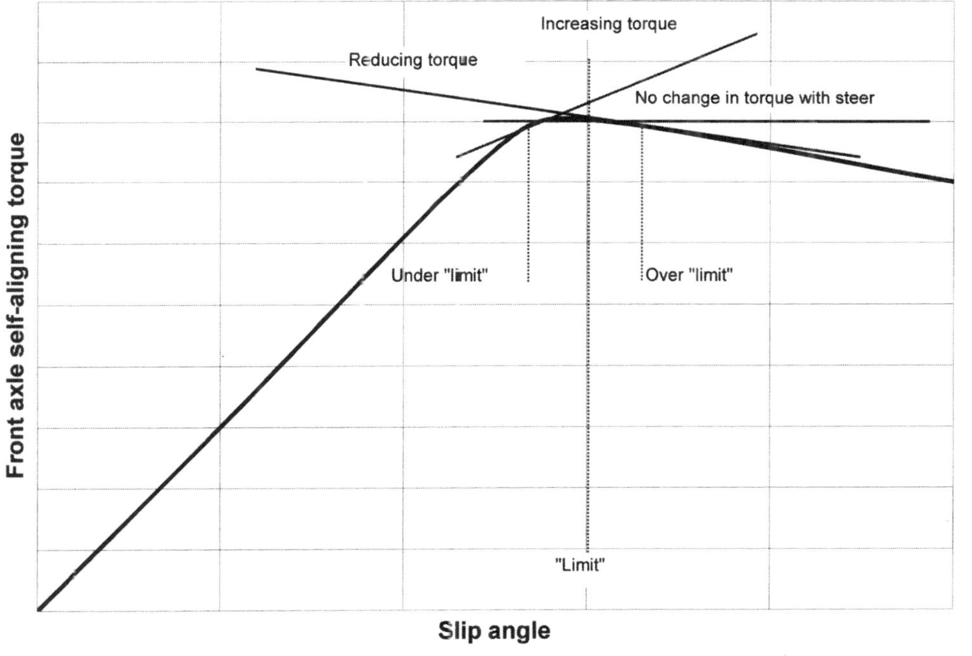

Fig. 4.3 The sign of the change in steering torque (front axle self-aligning torque) with steer angle indicates whether the axle is below, at, or above the limits of the tires. (Author's illustration)

For the rear axle, the driver must rely on any change in yaw velocity to indicate a change in grip. He may also use changes in lateral acceleration, depending where he sits relative to the instantaneous center of rotation of the car. This is why the drivers of power-oversteering dragsters sit as near the rear axle as possible, and why so few drivers of the rear-engined, pre-World War II Auto-Unions could master control of them from the cockpit position so far forward, ahead of the center of rotation.

Thus, the driver receives signals that give him essential information about position, speed, and available grip, which enable him to adjust, closed-loop, the brakes, throttle, and steering to modify the force vector and gain the response he wants from the car. How much feedback he needs and how hard he must work to adjust the response of the car will depend on how accurate the initial, open-loop brake, steering, and throttle inputs were. This is a question of driver confidence in the response of the car.

A fundamental of both aircraft and race car dynamics is that an unstable vehicle will be the quickest to respond to a given input. However, an unstable vehicle will not give confidence to its pilot or driver because it will tend to overshoot and be excited by disturbances. To overcome this problem, the aviation world has put a computer between the pilot and unstable aircraft, using sensors to detect the dynamic motion. Using the artificial stability that this system confers on the aircraft, the pilot can make his demand and the computer will deliver the maximum aircraft response that matches that demand, via the control surfaces, without need for correction by the pilot. Before computers found their way into aircraft control systems, a good approach was to provide the aircraft with good stability and damping, and equip it with powerful control surfaces. This is one of the keys to the legendary appeal of the Spitfire to pilots, and it remains the approach used in competition aerobatic aircraft such as the latest Sukhois and Extras, where computers (i.e., "pilot aids") are inappropriate.

The principles for a race car are the same, but an important difference exists: on a conventional aircraft, the force generators (wings) are separated from the moment generators (tail plane/elevators). Designed correctly, the moments needed for stability and control will still be available when the wing gives up at its limit (i.e., it stalls). On a car, the force and moment generators (tires) are one and the same. At the limit, they all tend to give up roughly at the same time. If the front axle gives up first, there is no control. If the rear axle gives up first, there is no stability. If both reach their limits together, there is neutral stability and no control. It is not possible to provide the driver with both the stability <u>and</u> control he would like to have at the limit. This characteristic of a car enables motor racing to exhibit and match the skills and bravery of drivers. Ayrton Senna's comment, in the quotation at the start of this chapter, was the result of his frustration at not having control at the limit and his skill in artificially destabilizing the car by momentarily locking the rear axle using the gearbox, to persuade the car to turn.

This is the situation to which Stirling Moss referred when he compared driving a race car on full throttle at the limit to a bowler bowling a ball. (See Chapter 1.) All the speed, pitch, spin, and swerve are set before the ball leaves the bowler's hand, and he is unable to influence its trajectory after it has done so. A driver sets up the speed, steer angle, and position of the car on entry to the corner. If it is truly on the limit, he has no further influence on its trajectory. He cannot close the loop in controlling it after he is at the limit in the corner.

In the braking and accelerating phases, when the car is travelling in a straight line, the driver does not require lateral forces or yawing moments unless he wishes to maneuver or the car is disturbed. However, if he does, the lateral force capabilities of the tires may be virtually zero as they are fully utilized propelling or braking the car. Under braking, either or both axles may be at the limit, giving an unstable car if the brake bias is rearward, stable but uncontrollable if it is front biased, and uncontrollably neutral if equal. If the car is in the traction limited phase of accelerating, it will be unstable but controllable.

A top racing driver may be able to live with a marginally stable car. Michael Schumacher can, but his teammates tend to find it more difficult—they lose confidence in the car. The prime control task for the designers and engineers is to provide, by whatever possible means, the maximum control of yaw angle and yaw rate throughout the operating range of longitudinal and lateral accelerations, as well as speed.

Drivers have mastered the control of race cars at the limit ever since cars have been raced, but the necessary skills have changed as the forces involved have increased dramatically. Not only has the physical environment in which the driver works changed, but the control task is different also. In the 1950s, when front-engined cars reigned supreme, the cars had high yaw inertia due to the main masses (engine, fuel, and transmission) being located near the axles, and low force generation (around 1g maximum). They were also inherently stable, although the driver could induce instability with power. The natural frequency of the car in yaw was probably around 1 Hz, well within a human's control bandwidth of approximately 5 Hz. Cornering speeds were relatively low, the tires had large peak slip angles (6–10°), and the fall-off beyond the limit was progressive. Now the cars weigh 15–20% less. They have the main masses located within the axles, as near the CG as possible to lower yaw inertia, and the force generators are amplified by downforce, being nearly four times as powerful. The natural frequency is right up near the driver's limits of control. Cornering speeds are higher, and the tires peak at less than half the slip angle of the narrow tires of the 1950s and fall off more sharply. The cars are unstable under braking and short of control under power. The difference in the frequency responses of the car and those of the driver can be compared to that of a stick and a needle, when trying to balance them upright on the end of a finger. The front-engined cars of the 1950s are more similar to the stick, whereas the driver in a current Formula 1 car must balance a needle on his finger while subjecting himself to up to 4.5g longitudinal and lateral accelerations.

It has long been recognized that it is desirable for all forces on a vehicle to pass through the CG, so that changes in those forces do not result in trim changes that would require control inputs to correct them. On any vehicle that generates its propulsion, braking, and cornering forces through the wheels, that is impossible. Thus, the moments they generate about the CG result in changes to the vertical forces on the wheels. While being undesirable, they have also been exploited by both drivers and designers. Drivers have used longitudinal load transfer to change the vertical loads on each axle and thus change their relative cornering forces, thereby modifying yaw moment and acceleration. The means of generating longitudinal load transfer are the engine and the brakes; the change in thrust on the tires that accompanies this complicates the control mechanism. However, near and at the limit, this mechanism is almost the only one available to the driver for control.

Lateral load transfer has been harnessed by chassis designers to vary the relative loads across the front and rear axles to exploit the load sensitivity of the tires and thereby alter the relative cornering stiffness of the axles. This is possible because cars have four wheels; therefore, the load distribution on them is indeterminate. There is more than one solution for a given longitudinal and lateral acceleration condition. The negative side of this effect is that if the road surface is not perfectly flat, the distribution of vertical loads on each wheel will be affected. The opportunity offered to the designer is that he can engineer the suspension system to determine a desirable load distribution under lateral acceleration by providing a means of altering the relative roll stiffness of each axle.

The designer of a Formula 1 car must do more than bring together the most powerful engine, aerodynamics that create the maximum downforce for the least drag, and the most powerful brakes, and fit the tires with the highest possible coefficient of friction. In addition, he must configure them, along with their support systems, fuel storage, and the driver's control station, in an assembly that dynamically performs in a way that matches the driver's skills, all for the lowest mass. To do this, the designer must decide on certain key parameters:

- Center of gravity
  - Fore and aft position
  - Height
  - Variation with fuel load
- Moment of inertia
  - Yaw (primary)
  - Roll and pitch (secondary)
- Aerodynamic downforce distribution
  - Variation with ground clearance, front and rear
  - Variations with yaw and roll
- Roll stiffness distribution
  - Variation with vertical load

It has been said that four key ratios dominate the vehicle dynamics of a race car: tire cornering stiffness distribution, weight distribution, aerodynamic downforce distribution, and roll stiffness distribution. If these ratios are not correct, it is almost impossible to balance the handling for all conditions of speed and longitudinal acceleration. Balanced handling gives the driver the ability to control yaw angle up to the limit and to minimize adverse yaw responses to disturbances such as bumps on the track.

Unfortunately, these ratios will never remain constant under all dynamic conditions. The ways in which the ratios change relative to each other under the influence of speed and longitudinal, lateral, and vertical acceleration, as well as the attitude of the car relative to the ground, are the real technologies (and in some cases still the black art) of chassis design and development.

Before the designer can draw a line on his CAD screen or enter numbers into his vehicle dynamics simulations, he must read the Technical Regulations to determine the constraints imposed by them, preventing an ideal solution. The foremost of these are the conflict between the tire size regulations and the numerous constraints, mainly for safety reasons, that determine the limited options for disposing of the main masses.

The Technical Regulations set a maximum and minimum tire and wheel width:

	Maximum	Minimum
Front	355 mm	305 mm
Rear	380 mm	365 mm

and a maximum diameter of 660 mm. The FIA Sporting Regulations currently stipulate that both front and rear tires must have four circumferential grooves, 14 mm wide at the tread surface. If, somewhat simplistically, the tread width distribution front to rear is plotted against total tread width (Fig. 4.4), the boundaries are defined.

Fig. 4.4 Tire tread width distribution versus total tread width, as limited by the Technical Regulations. (Author's illustration)

Compared to a number of years ago when Formula 1 cars were permitted to have approximately 1575 mm (62 in.) of available tread width, current cars are under-tired. To use the maximum available width means that the front-to-rear ratio is approximately 46:54. Although tread width does not relate directly to tire cornering stiffness nor the maximum lateral force, it is a reasonable approximation of the relative performance of the tires, front to rear. It is also indicative of the desired weight distribution. Aerodynamic downforce distribution is relatively straightforward to adjust, and it is usually desirable to set it to match the tire and weight distributions, or to set it with a slight rearward bias as speed increases to compensate for the natural reduction in stability that occurs with speed.

Roll stiffness distribution, by which lateral load transfer distribution is affected, is more a means of reducing tire performance on a given axle, with the hope of increasing it at the opposite axle. It is used to modify the effective tire load distribution, especially when the rear axle is transmitting power and its cornering performance is reduced. The range through which it can be adjusted is a function of the axle roll stiffness and the torsional stiffness of the structure joining them. Even with an infinitely stiff structure and rigid suspension, the ratio will be limited by the remaining axle roll stiffness component due to tire vertical stiffness.

Thus, the critical task is to match the weight to the tires. At this point, history intervenes to make it more difficult. The overall layout of the current Formula 1 cars was established when Colin Chapman sketched the Lotus T79 while on holiday in Ibiza in the spring of 1977. For the first time, all the fuel was placed in a single central tank, located virtually on the CG between the driver and the engine. In so doing, Chapman initiated one of the greatest safety features of modern race cars, virtually eliminating fires during accidents. The sides of the car were freed up for ideal positioning of radiators and aerodynamic sidepods. Since then, the aluminum monocoque has given away to CFRP/honeycomb sandwich construction, built to withstand the front, rear, and side impact and static load tests of the FIA, endowing the structure with approximately 20 times the overall torsional stiffness to that of the T79. Radiators have moved rearward to slightly ahead of the engine, and sidepods have shortened, shaped by aerodynamic considerations and refined by wind tunnel testing. For safety reasons, the driver's feet must now be located behind the axle centerline.

Most of the safety and downforce-limiting regulations were formulated during the 1980s and were written, where feasible, to conform to the layout of the cars of that time. This was an era of almost unlimited tire widths, in which it was possible to put sufficient tread on the car for the power at that time. The tire balance hovered around 40:60, and the weight distribution matched it, giving excellent traction. Thus, the layout of

Front Axle—Driver—Fuel Tank—Engine—Gearbox—Rear axle

became virtually enshrined in the regulations, with little other disposable mass available to adjust the weight distribution, as shown in Fig. 3.9. The radiators could be moved fore and aft on the car. However, the farther they were from the engine, the heavier the installation became, and distribution and sensitivity to pitch of the aerodynamic forces generated by the flat underneath of the car dictated short sidepods. The fuel tank could be shortened only by making it taller because the width is constrained to 80 cm (31.5 in.), but this goes against the desire for a low CG height. The introduction of pit stops allowed the capacity, and hence the length, to be reduced, enabling the engine and its mass to be moved forward with an extension to the gear case between engine and gearbox. Apart from minor adjustments—for instance, by relocating some of the black boxes and electrical components—the CG was effectively fixed for a given wheelbase. A longer wheelbase does give some freedom to adjust CG—approximately 1.5–2.0% per 100 mm (3.9 in.) (Fig. 4.5). A longer wheelbase increases overall weight and yaw inertia and has other effects on vehicle dynamics. With 120-liter (31.7-gal) fuel tanks to give the potential of one-stop races, a weight distribution of approximately 43:57 is possible in a 3.0-m (9.8-ft) wheelbase. This is still several percentage points rear-biased compared to the tires and can lead to an unstable car.

Attempting to use the other ratios to compensate (e.g., a rear-biased aerodynamic distribution and a very stiff-in-roll front axle) leads to excessive understeer under certain conditions such as turning-in under power. Using excessively biased ratios to compensate for other unbalanced ratios is a patch that does not cure the problem.

The solution lies in lightening the car, particularly components behind the CG. The minimum weight limit of 600 kg (1322 lb), including driver, effectively means that the cars weigh in at 530 kg (1168 lb) dry. With extensive use of CFRP, metal-matrix composite (MMC), titanium, and the latest aluminum alloys, plus refined design, stress analysis, and precise CNC manufacturing techniques, it is possible to build a car that weighs less than 450 kg (992 lb). More than 70 kg (154 lb) of depleted uranium, compressed silicon carbide, or tungsten ballast necessary to increase the weight to the minimum is available for strategic placement in the car. Mounted as low as possible, it can all be placed ahead of the CG, although too far ahead will start to raise the yaw inertia (Fig. 4.6). If we assume that part of the weight reduction (say, 30 kg [66 lb]) comes from lightening the engine and transmission, and this mass is now placed as far ahead of the CG as it was behind, the weight distribution will increase by 2% to approximately 45:55. From this, it is possible to conclude that those chassis designers who are blessed with under 100 kg (220 lb) engines, instead of the 125 kg (275 lb) engines that were more common in the late 1990s, are able to achieve not only weight distributions that match the characteristics of the tires but are also equipped with the most powerful means of tuning the balance of the car. Engine design is about more than power.

The techniques for achieving the aerodynamic and roll stiffness distributions are covered in later chapters. Both have significant effect on the detail design of the chassis. The designer must provide a structure that joins together the suitably positioned masses and that is stiff and strong enough to cater to all loads fed into it, while having the smallest possible cross section and the lowest possible mass. The principle loads include those induced by the engine, the aerodynamics, brakes, road irregularities, and inertia loads due to the masses under accelerations and vibrations. In addition, the structure must withstand impact loads, having absorbed part of the crash energy

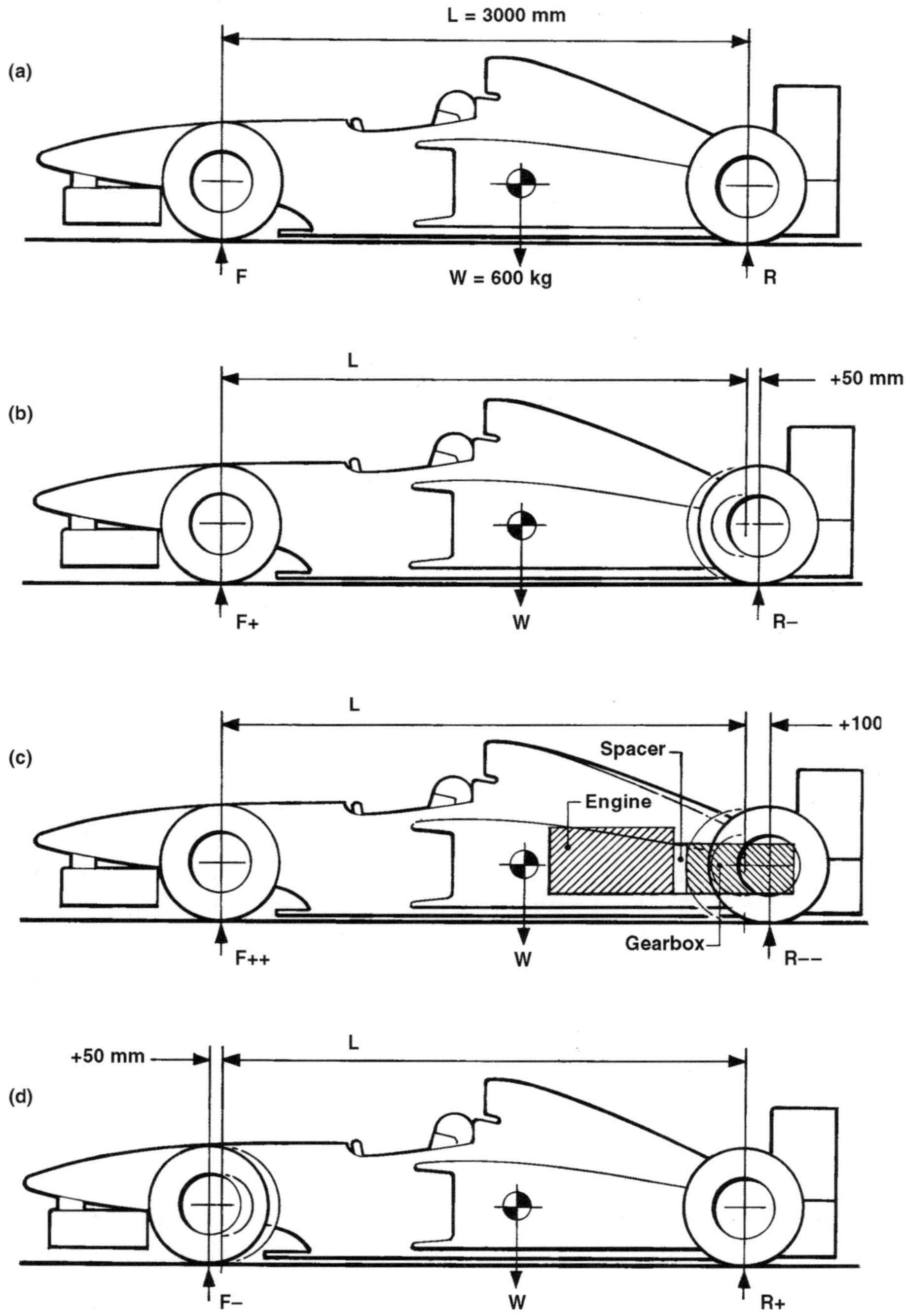

Fig. 4.5 Effect of wheelbase changes on longitudinal CG position. (a) Standard car, 43% front. (b) Rear wheels moved backward, 43.9% front. (c) Rear wheels and gearbox moved backward, 44.8% front. (d) Front wheels moved forward, 42.3% front.

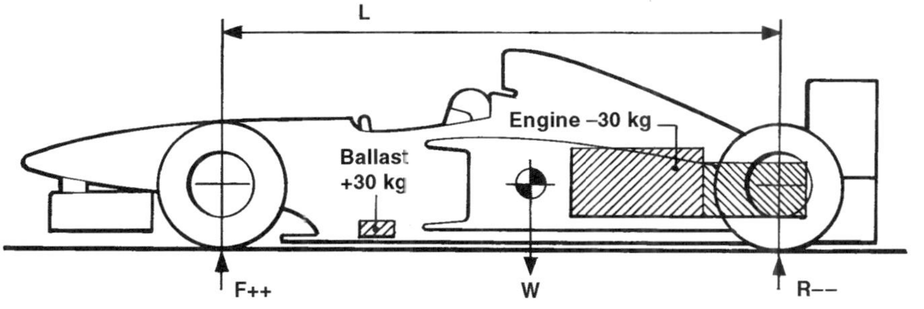

Fig. 4.6 Effect of engine weight reduction on longitudinal CG position; light engine and ballast forward, 44.7%.

by deformation. Generally, the resulting structure is so strong that stiffness is the design case rather than strength. Not only must the primary structure provide a stable platform such that the suspension springs and dampers can control wheel loads, but its stiffness will determine the natural frequency of its modes. The dominant forcing frequencies are the car on its suspension (approximately 5–7 Hz), wheel hop (approximately 15 Hz), and the engine (more than 250 Hz). The bending and torsion natural frequencies of the chassis need to be kept well above the suspension frequencies but will inevitably be below those of the engine, except at very low rpm. The axle-to-axle torsion frequency is more than 30 Hz. The only mechanisms for damping these modes are the suspension dampers themselves (which do not work well at 30 Hz, if at all) or the movement in joints between components at higher frequencies (e.g., engine mounts, engine-gearbox interface, and joints between individual structural engine components), all of which are extremely undesirable.

Composite honeycomb construction, using CFRP skins, has transformed monocoque construction, creating immensely stiff flat panels and eliminating joint movement. Finite element analysis (FEA) software for composite structures has enabled optimized lay-ups to be specified, using high-modulus carbon fiber where stiffness is critical, and high-strength fibers where impact loads and load concentrations, such as suspension loads, must be fed into the structure. The cross section of the monocoque is now regulated such that the construction needed to withstand the FIA safety tests affords high torsional stiffness, even in the cockpit area with its large cutout. The engine and gearbox, with their ever diminishing cross sections and their joints to each other and the monocoque, have become the critical items in determining overall axle-to-axle torsional stiffness. The monocoque is very stiff in bending; however, again it is the engine-to-monocoque joint that is critical.

In designing the overall layout of the car, the designer must take into account servicing and maintenance of the vehicle. As components become more integrated in the quest for low weight and reliability in the face of increasing complexity, maintenance and servicing of the cars has become a major issue. The ability to perform routine servicing tasks at the race track, such as engine changes, tire changes, and refueling, and non-routine tasks such as diagnosing and replacing failed components or repairing accident damage, can be critical to meeting the event schedule and hence maximizing the time available for practice and setting up of the car. It may even determine qualifying and race results.

Computerized diagnostic tools and scheduled component replacement parallel similar requirements in the aviation world. "Line replaceable units" are becoming the norm, with new and rebuilt assemblies available at the track. Engine changes can be accomplished in less than one hour, making it feasible to replace a unit between morning and afternoon practice sessions. Engineering for the shortest possible pit stops has taken on a new urgency because overtaking now frequently occurs during these stops. Tires can easily be changed within the time it takes to refuel, and with the standardized fuel rig and its prescribed flow rate of 12 liters (3.2 gal) per second, only the time taken to engage and disengage the nozzle can be influenced by the team. These seconds, gained or lost, can affect race results.

Chassis design, as far as it affects the performance of the car, has reached a level of sophistication where values can be assigned to all key relevant parameters. Whether they are front-to-rear ratios or individual values, they are required if the simulation engineers are to model the dynamics of the car. Much of the R&D relating to the chassis is spent in careful measurements on rigs and on extensively instrumented test cars on the track. It is essential to establish accurate values for all key parameters and to validate models by calibrating the sensitivity to changes in those values. The problems associated with vehicle dynamics modeling and simulation of the motion of the car are discussed in more depth in Chapter 12, particularly those aspects involved in transient motion. The key to success in this endeavor is the fidelity of tire data. Tires, being the force generators, are somewhat analogous to the wings and control surfaces of an aircraft, components whose precise characteristics are researched and tested in wind tunnels and in flight tests in the greatest possible detail. Of particular interest are their characteristics near the stall and post-stall. Tires have similar linear ranges and also similar nonlinear characteristics at the limit, the details of which determine the performance, stability, and control of the car. Tire modeling and the measurement of quasi-static characteristics are in their infancy, even in Formula 1, and this lack of detailed information is imposing a limit on how far the science of race car dynamics can progress. It is possible to measure the steady-state characteristics of a racing tire, but transient tire characteristics remain an unknown. To understand how cars are affected by road irregularities, gusts, and driver inputs, tire transients and the transient aerodynamics, damper, and structural dynamic characteristics must be researched by modeling and validation. The potential to design and build a Formula 1 car that brings a big smile to its driver's face the first time it runs and thereafter lies in these developments. Meanwhile, black art exerts a big influence, and those designers who have knowledge of the secret numbers, or luck, are both rare and much sought after.

Chapter 5

Engines

A horse, a horse, my kingdom for a horse.
—Shakespeare, Richard III

Of all the parts that make up the Formula 1 car, the engine produces the most outward manifestation of performance and technology. The deafening sound of thousands of explosions per second and the sheer speed of the cars convey straight to the senses the impression of phenomenal performance by automotive standards. There is no need to make mental calculations to appreciate the effect of the engine, as there is with the cornering and braking performance, nor to attempt the impossible task of deducing how much downforce the shape of the car is generating. Would motor racing be so popular without the senses being assailed by the racing piston engine? Neither the electric motor nor the gas turbine engine, the only potential alternatives for racing in automobiles, could ever create the same emotion in those who follow the sport as that produced by the great racing engines of the past and present.

Yet the Otto-cycle piston engine is old technology compared to the prime movers used in other fields. The gas turbine rules where high power to weight and efficiency dominate the requirements. The electric motor is acquiring better performance and efficiency, while providing the ideal means of computer controlling the output. Formula 1 regulations stipulate closely the type of engine permitted: it must be a four-stroke, normally aspirated (NA) engine of 3.0 liters swept volume, with no more than ten cylinders and no more than five valves per cylinder. Lotus tried its gas turbine-engined T56B in Formula 1 in 1971, having found the turbine to be competitive at Indianapolis. It did not prove successful for several reasons: lack of engine braking, high fuel consumption, poor throttle response, installation compromises, and possibly some gyroscopic effects. Whether all these problems could be overcome today—with the use of (full authority digital engine control) FADEC-controlled, small, lightweight, fuel-efficient turbo-shafts developed for helicopters, with carbon-carbon brakes to slow it down—will never be known. However, the power/weight advantage is there:

Gas turbine	10.0 hp/kg	4.5 hp/lb
Normally aspirated four-stroke	5.5 hp/kg	2.5 hp/lb
	(including cooling system and gears)	

Ultimately, the deciding factors would be throttle response and fuel consumption.

While road cars are still fitted with piston engines, it is inevitable that, for commercial reasons relating to automobile manufacturer involvement in Formula 1 and for entertainment reasons (i.e., the sound), piston engines will be mandatory for a long time.

To understand why the piston engine has clung to the motorcar and will continue to do so for the foreseeable future, it is necessary to examine the strengths and weaknesses of the piston engine and their relevance to its use in motor racing. All heat engines, including the piston and gas turbine engines, obey certain physical laws, and these laws guide us toward understanding the pros and cons of a particular cycle or type of engine while providing a measure by which the numerous technologies involved can be assessed.

The theoretical maximum rate at which energy can be generated by a heat engine (i.e., power) is a product of the oxygen flow rate through it and the mass and calorific value of the fuel that is burnt in combination with the oxygen; that is,

$$\text{Maximum theoretical power} = K_1 \times \left[\text{theoretical flow rate of } O_2\right] \times \left[\text{theoretical calorific value of fuel burnt}\right]$$

In reality, the actual heat energy released by the combustion process depends on numerous factors, the most important of which are the compression ratio and the heat losses to the exhaust and to the cooling systems of the engine. The latter are necessary to prevent components from exceeding their material limits. These all add up to a loss of energy that is quantified as the thermal efficiency—the ratio of actual heat released to the theoretical heat released.

All heat engines are a form of pump in which air, with its fixed proportion of oxygen, is drawn in from the atmosphere, compressed by a certain ratio in one or more stages, heated by the combustion of fuel with the oxygen, and expanded again to atmosphere in one or more stages. The expansion of the hot gases liberates more power than needed to compress the air, and the balance is available to do work. The theoretical amount of air, and hence oxygen, that can pass through the engine with each revolution can be calculated from the physical details of the pump. However, the actual amount will depend on the losses incurred as the air and combustion gases flow through it. The actual volume compared to the theoretical volume is quantified as the volumetric efficiency. Due to intake and exhaust resonance effects, this exceeds 1.0.

Not all the net work available from the process it useful. Some is used in overcoming friction, pumping the air, and driving accessories necessary for the functioning of the engine (e.g., fuel pump, lubrication and cooling pumps, and generator). The ratio of the actual work available tc the theoretical work is the mechanical efficiency.

The actual power generated is now

$$\begin{aligned}\text{Actual power} = K_1 &\times [\text{theoretical flow rate of } O_2] \\ &\times [\text{theoretical calorific value of fuel burnt}] \\ &\times [\text{thermal efficiency}] \times [\text{volumetric efficiency}] \\ &\times [\text{mechanical efficiency}]\end{aligned}$$

The theoretical flow of oxygen is a function of the physical volume of air pumped by the engine per revolution (swept volume) times the number of revolutions in a given time (rpm). The equation that describes the performance of an engine now becomes

$$\begin{aligned}\text{Actual power} = K_1 &\times [\text{swept volume}] \times [\text{rpm}] \\ &\times [\text{theoretical calorific value of fuel burnt}] \\ &\times [\text{thermal efficiency}] \times [\text{volumetric efficiency}] \\ &\times [\text{mechanical efficiency}]\end{aligned}$$

Air pumps are either rotary or reciprocating, and engines are one or the other of these, or a combination of the two, in one or more compression and expansion stages, as shown in Table 5.1.

TABLE 5.1
RACING ENGINE CYCLES

Engine Cycle	Compression	Combustion	Expansion
NA piston, four-stroke	Reciprocating (single stage)	Intermittent (1 per 2 rev.)	Reciprocating (single stage)
NA piston, two-stroke	Reciprocating (two-stage)	Intermittent (1 per rev.)	Reciprocating (single stage)
Supercharged, four-stroke	First stage rotating Second stage reciprocating	Intermittent (1 per 2 rev.)	Reciprocating (single stage)
Supercharged, two-stroke	First stage rotating Second stage reciprocating	Intermittent (1 per rev.)	Reciprocating (single stage)
Turbocharged, four-stroke	First stage rotating Second stage reciprocating	Intermittent (1 per 2 rev.)	First stage reciprocating Second stage rotating
Turbocharged, two-stroke	First stage rotating Second stage reciprocating	Intermittent (1 per rev.)	First stage reciprocating Second stage rotating
Gas turbine	Rotary (multi-stage)	Continuous	Rotary (multi-stage)

Variations are possible within these cycles as to the point at which power from the expansion stage is taken to drive a given compression stage, and from where the power output is taken. Additional combustion is possible between expansion stages, if free oxygen remains in the gas stream.

Each cycle or engine type achieves different fundamental thermal, volumetric, and mechanical efficiencies and, perhaps more importantly, different characteristics of how these efficiencies vary with rpm and load.

The piston engine has lower thermal and volumetric efficiencies than the gas turbine at its optimum design point because of its reciprocating motion, intermittent combustion, and tortuous airflow paths past the valves. However, it performs well away from the design point through a reasonably wide range of rpm, giving good control and operating characteristics well matched to an automobile. Balanced against these advantages, it requires a multi-speed gearbox to provide its peak torque through the speed range of the car; it vibrates due to its reciprocating motion; the intermittent combustion requires silencing; it needs a cooling system; it is inefficient at part load; and it produces noxious emissions. The motor industry has invested vast sums in producing effective solutions to these problems and even more in facilities for producing piston engines cheaply. It is likely to be the prime mover in automobiles for several decades, probably integrated with energy storage and electric drive as a variety of hybrid power-train systems.

For Formula 1, which requires some stable level at which competitors can technically compete with each other, the four-stroke NA piston engine is the chosen cycle today because it represents the most common one currently used in road cars. However, to further limit the maximum power developed, it can be seen from the previous equation that one or more of the swept volume, rpm, and fuel calorific value must also be regulated. Traditionally, swept volume and, more recently, calorific value have been selected, leaving rpm as the final technical challenge. Furthermore, the engine designers' and developers' task is to achieve the highest possible thermal, volumetric, and mechanical efficiencies throughout the required rpm operating range.

Figure 2.11 in Chapter 2 illustrates the change in the maximum power of engines used in Formula 1 since 1965. As such, it does not tell us much about the development of engine technology except during periods when the formula was held constant. Instead, it shows how the FIA has had to change the formula to control power because power has escalated due to development within a selected set of regulated parameters.

In the 1960s, the 1.5-liter NA formula was deemed to be not powerful enough, and the swept volume was doubled in 1966, thereby doubling power. The rate of progress was generally even until 1981, when a different cycle was finally made to work and turbocharging, combined with a relatively free definition of the fuel, caused power to double in approximately five years—without a change of formula! Turbocharging was phased out by applying ever more restrictive pressure ratio limits until a 3.5-liter NA was a better option and Formula 1 was regulated to this capacity and cycle. However, fuel was not further regulated at the same time. Greater funding in Formula 1, particularly via the technical involvement of motor manufacturers in engine development and by the energy industry in fuel development, maintained the rate of power increase. The prospect of power reaching similar levels to that of the turbo-engines, plus an increasing environmental conscience, led to fuel finally being regulated to closely represent publicly available pump petrol. Stimulated by the safety issues raised in the wake of the fatal accident of Ayrton Senna at Imola in 1994, the FIA and the engine manufacturers agreed on a return to a 3.0-liter NA formula.

A better measure of how engine technology has developed, within a given engine cycle and fuel specification, is to rearrange the previous equation to

$$\frac{\text{Actual power}}{\text{Swept volume}} = K_3 \times [\text{rpm}] \times [\text{thermal efficiency}]$$
$$\times \; [\text{volumetric efficiency}] \times [\text{mechanical efficiency}]$$

If the three efficiencies are maintained as rpm is increased through development, then the relationship between actual power/swept volume and rpm is linear. In Fig. 5.1, the performances of the 1967 Ford-Cosworth DFV, the 1983 Ford-Cosworth DFY, and a current engine, typified by the current Mercedes-Illmor V-10, are plotted in this way.

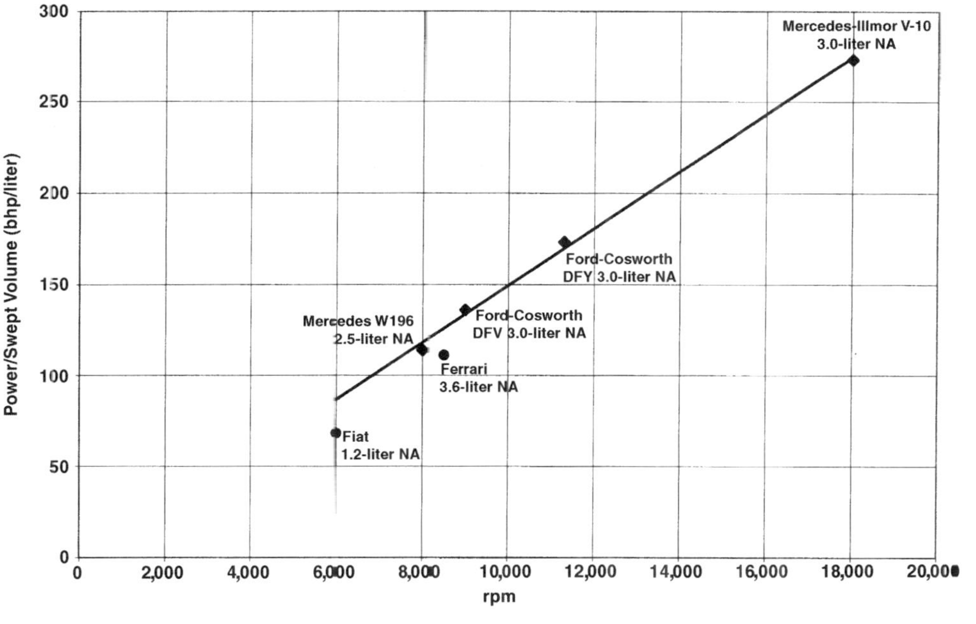

Fig. 5.1 Specific power versus rpm. (Author's illustration)

These three engines—all 3.0-liter NA engines running on fuel that is close to pump fuel, and designed 16 and 15 years apart, respectively—lie on a straight line passing through zero. Plotted also are the 1955 Mercedes M196 2.5-liter NA engine and two 1999 high-performance road car engines: a Fiat 1.2-liter four-cylinder, and a Ferrari 3.6-liter V-8, both NA engines. Although the Mercedes GP engine ran on a specially formulated fuel, it too lies on the line, indicating that it may have had a lower efficiency than the later engines. The road car engines lie only slightly below the line of the racing engines due to the effect on volumetric efficiency of intake filters, exhaust catalyst, and silencer, while the additional accessories needed on a road car (e.g., large alternator, power steering) reduce the mechanical efficiency. Forty-four years lie between the Mercedes racing engine and the similarly performing Ferrari road engine.

The fact that the four racing engines lie closely on a straight line passing through zero means that the product of the three efficiencies has been kept constant over the 44 years, while rpm has been more than doubled. The technologies of the Formula 1 engine center on these four main technologies: increasing rpm, while maintaining the thermal, volumetric, and mechanical efficiencies.

Figure 5.2 plots rpm against calendar years and shows that rpm is increasing at an ever greater rate; therefore, the same is true for real development in the three efficiencies. Plotted against dollars expended instead of years, the progress might well be more linear.

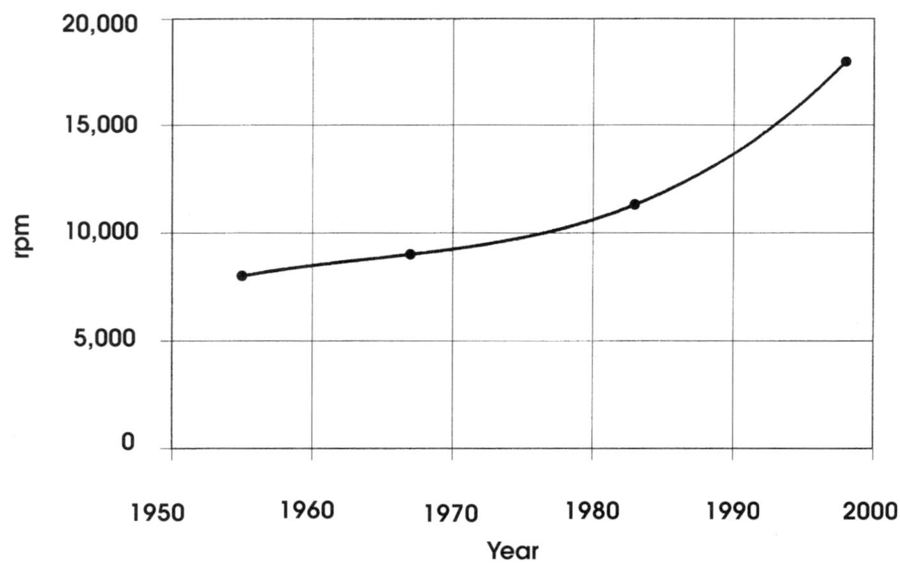

Fig. 5.2 Plot of rpm versus calendar year. (Author's illustration)

Apart from potential losses in efficiency, which will be discussed later, other factors mitigate against increasing the rpm of a piston engine, mostly related to the reciprocating motion of some of its components. Inertia loads in the pistons, connecting rods, crank pins, and webs, as well as the resulting loads in their bearings and the crankshaft, all increase as the square of rpm. The same is true for the valves and their actuation system. The only remedies are to minimize the reciprocating masses and reduce the amplitude of the reciprocating motion (i.e., stroke and valve lift, respectively). These considerations, among others, determine the optimum layout of the engine with respect of the number of cylinders, the arrangement of the cylinders, and the overall dimensions. There have been successful 3.0-liter NA engines with eight, ten, and twelve cylinders, all in V arrangements, apart from a brief period in the early 1970s when the Ferrari flat-12 was dominant. Currently, the V-10 layout is used exclusively in Formula 1. Honda pioneered the V-10 in 1989 as a 3.5-liter engine, while Ferrari (V-12 until 1995) and Ford-Cosworth (V-8 until 1996) were the last to convert to the layout that is currently the best compromise among rpm (i.e., power), weight, size, shape, structural integrity, driveability, and fuel consumption. The V-8 is limited to approximately 15,000 rpm by secondary torsional vibrations in its crankshaft, while the V-10 will not reach its crankshaft limitation until other factors allow rpm to rise to approximately 19,000. The V-10 has 405 possible firing orders, each of which produces a different combination of primary and secondary out-of-balance forces and couples, with their associated bearing, crankshaft excitation, and engine mount loads. Only three or four of these are practical solutions for a racing engine, and analysis of the dynamic response of the rotating components requires a multi-degree-of-freedom computer model. Honda has used a balance shaft on its first- and second-generation V-10's, choosing to balance the secondary forces and use the balance shaft to counteract the out-of-balance primary forces. Others have followed different solutions, with weight being the ultimate goal. However, a light but harsh engine will shake a chassis to such an extent that chassis weight must increase to ensure durability.

The smaller cylinders of the V-12's compared to those of the V-10's (250 cc compared to 300 cc), with their associated shorter stroke and smaller bore, permit higher rpm due to lower inertia loads and piston speeds. The smaller cylinders also have lower heat losses, and thus the twelve-cylinder should run at higher rpm and produce more power than the ten-cylinder if the extra friction and windage losses of more cylinders are kept under control. Against this, the V-12 is longer and has a smaller cross section which, when used as a section of the chassis structure of the car, must be less stiff for a given weight. The extra length also has weight distribution consequences. The long crankshaft presents its own torsional stiffness problems and, if each crank pin is supported by main bearings, more friction. The V-10 has six main bearings against the seven of the V-12, although Ferrari is reported to have experimented with V-12's having four main bearings.

With rpm continuing to rise steadily and the 19,000 rpm theoretical limit of the V-10's within sight, twelve cylinders (the maximum permitted by the regulations) were inevitably being considered again by engine designers, but perhaps not as V-12's. Audi has built a number of experimental W-12's, a configuration whose length is shorter than that of a V-8 of the same capacity and potentially stiffer. If the packaging problems of intake and exhaust systems could have been solved and the bulky cross section did not inhibit rear-end aerodynamics, the advantages of twelve cylinders might have been realized without compromises. However, in 1999, the engine manufacturers agreed to a ban on engines having more than ten cylinders until at least 2007.

As rpm has risen, stroke has fallen (Fig. 5.3). The bore-to-stroke ratio has risen from 1.3:1 of the Ford-Cosworth DFV to approximately 2.25+:1 of today's V-10's.

Even so, piston speed has increased by more than 30% and piston acceleration by 75% (Fig. 5.4). The implications of the higher piston speed on mechanical efficiency and the high piston acceleration on mechanical design will be discussed later in this chapter.

One much sought-after advantage of a shorter stroke is that the shorter crank throw allows the crankshaft and hence the whole engine to be lower. The CG height reduction has a significant effect on handling through reduced load transfer across the car during cornering. The sensitivity of this parameter has been increased in the last few years due to the regulated narrowing of the track by 200 mm (7.9 in.) and a reduction in effective tread width of the tires, leading to increased load sensitivity.

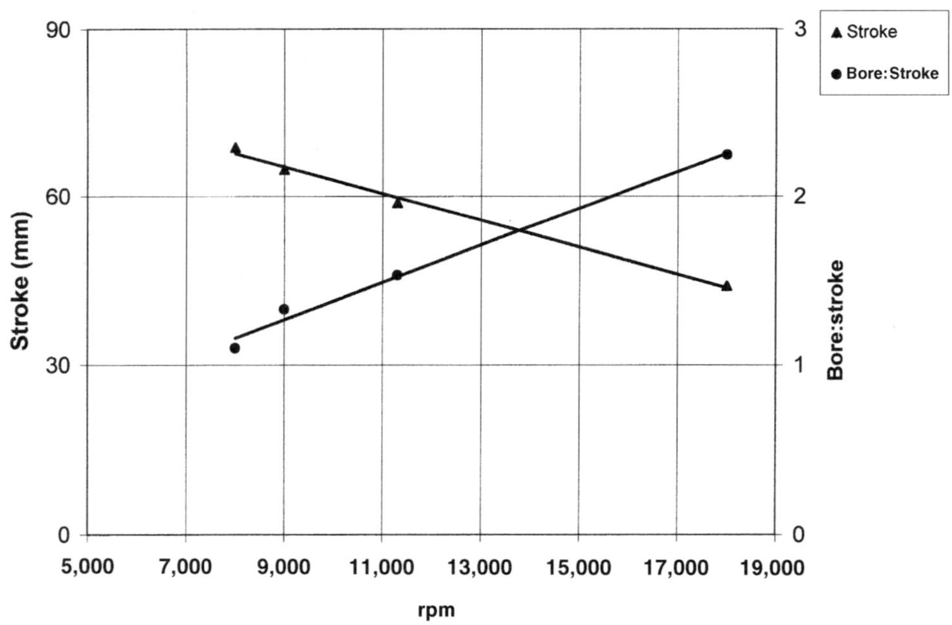

Fig. 5.3 Stroke and bore-to-stroke ratio versus rpm. (Author's illustration)

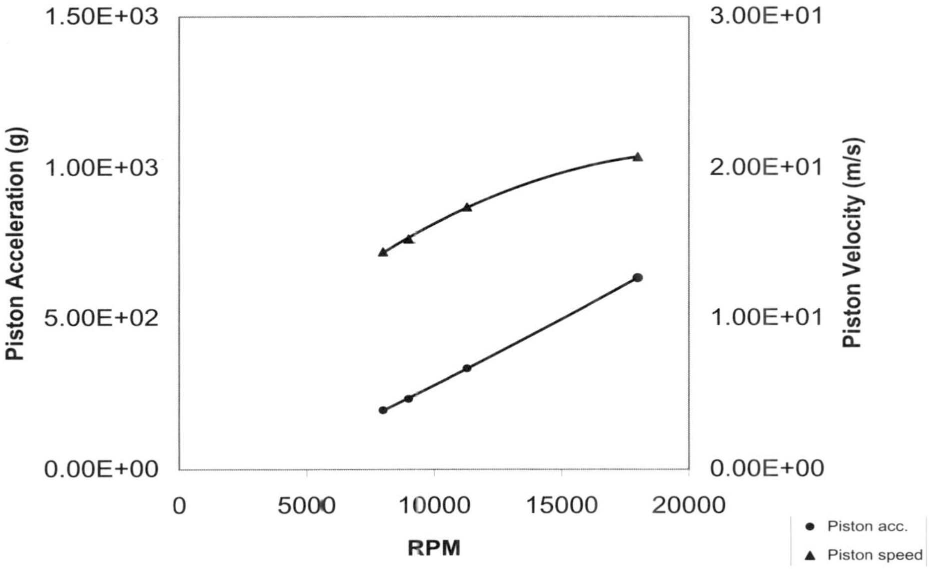

Fig. 5.4 Piston acceleration and piston velocity versus rpm. (Author's illustration)

The limit to rpm is currently set by mechanical considerations—the bottom end of the engine, not the valves. For many years, the valve gear set the maximum rpm, the natural frequency of the mechanical springs determining the rpm at which they were no longer able to control the closing of the valves. Reduction of valve mass and titanium springs allowed peak rpm to rise to approximately 15,000 rpm. However, it was not until Renault developed a pneumatic valve closing system for its 1986 1.5-liter turbocharged V-6 engine, and subsequently applied the technology to its successful 3.5-liter NA V-10 engines, that the valve gear ceased to be a limitation. The system provided a massless spring, avoiding spring surge, with a high pre-load and a rate that was low enough to lower the natural frequency outside the operating range. More aggressive cam accelerations were possible with this system, increasing the possibilities for valve timing (Fig. 5.5).

At the moment, burn rate is not a limitation on rpm either, in spite of the disc-shaped combustion chamber that results from a bore-to-stroke ratio in excess of 2.25:1. However, the detailed design of the combustion chamber, both in the cylinder head and piston crown, is critical. The valve pockets constitute a significant part of the combustion volume, and the flat disc that is formed around top dead center effectively quashes the carefully engineered tumble and turbulence of the intake charge. To increase rpm requires a shorter stroke, which creates a thinner disk for the combustion chamber. Where the limit may lie is closely guarded information and the subject of intense research in computer simulations, rigs, and single-cylinder engines by race engine manufacturers, their motor manufacturing partners,

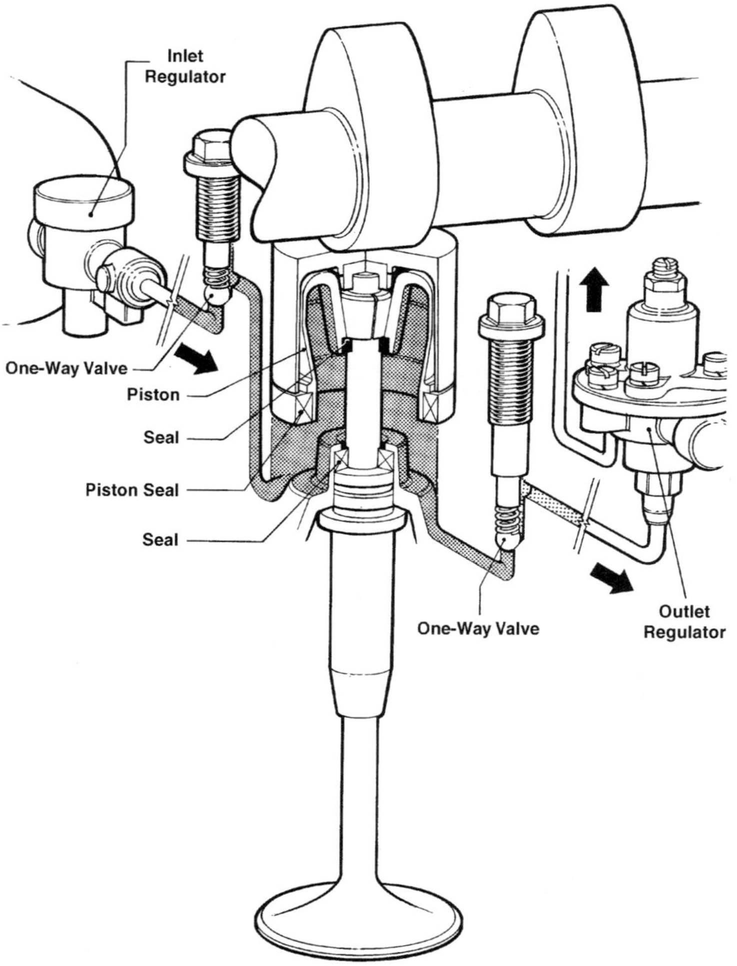

Fig. 5.5 Pneumatic valve closure system.

and fuel companies. When fuel chemistry was free, combustion problems could be solved by various "rocket fuel" recipes, brewed by the chemists. With only pump-specification fuel permitted, exact constituents and the quality with which the fuel is blended are crucial to good combustion.

The components that dominate development for greater rpm are the piston, connecting rod, crankshaft, and bearings. The design of the piston for maximum strength and minimum weight, as well as crown design, makes this component the most closely guarded part of the engine. In addition to the titanium connecting rods, aluminum alloy pistons are the structural Achilles' heel. As rpm increases, the stroke must decrease and the bore increase for a given piston speed. In reality, piston speed and inertia loads are traded off, depending on available solutions for both issues. It is not ultimate load that breaks pistons and connecting rods, but the using up of a component's fatigue life. At higher rpm, the rate at which fatigue cycles are consumed increases, with both loads and cycle frequency increased. Racing engines are not designed for an infinite fatigue life, but rather one of approximately 1.5 to 2 times racing distance, or 300 to 500 miles. The detailed design, material specifications, manufacturing processes, and quality control of the bottom-end components are all critical. It is not surprising that engines sometimes explode so spectacularly, as the fatigue life of a single component terminates prematurely at maximum rpm.

As rpm rises, main and big-end bearing surface speeds can be reduced only by making journal diameters smaller, compromising the bending and torsional structure of the crankshaft, reducing its critical natural frequency, and putting up bearing loads. Cavitation in the oil of a plain bearing occurs if the speed is too high and causes an immediate breakdown of the oil film with catastrophic results. Any interruption in the pressurized supply of oil to the bearings (e.g., due to air inclusion) has a similar effect.

Freed of the valve system constraints by Renault's development of the pneumatic valve spring system, Formula 1 engine rpm continues to rise and shows no sign of leveling off, although the imposition of a ten-cylinder limit may ultimately cause a slowdown in progress. The development of technologies that enable the engines to operate at these high speeds has been accelerated by CAD, CAE, and CAM—tools to assist in the detailed design, analysis, and precision manufacture of components. Allied to material technologies released by the aerospace industry as it adjusts its business to peace time, perhaps the greatest accomplishments lie in the research and development to maintain overall efficiency as speed has risen.

In 1967, the BMEP of the Ford-Cosworth DFV was approximately 14.0 bar at 8,500 rpm. This figure has fallen to only around 13.8 bar at 17,500 rpm in today's engines—a reduction of less than 2% as rpm has doubled. The change in BMEP with rpm is indicative of how the product of the key efficiencies

$$[\text{thermal efficiency}]\times[\text{volumetric efficiency}]\times[\text{mechanical efficiency}]$$

has fallen with increasing rpm. Many of the factors that affect the first two of these (thermal and volumetric efficiencies) are common, and it is virtually impossible to separate and analyze them independently. Factors such as valve position, area, and timing not only affect the losses incurred by the intake and exhaust gases, but also the turbulence level and the shape of the combustion chamber. A gain in volumetric efficiency due to a change may result in a loss in thermal efficiency due to poor combustion. The development of the optimized design of the intake system, valve gear, cylinder size, combustion chamber shape, exhaust system, and fuel injection and ignition systems is a major candidate for systems integration. Understanding the tradeoffs among all the details of these components that contribute to the thermal and volumetric efficiencies is a big part of racing engine R&D and is where the major auto manufacturers and fuel companies can contribute their expertise, facilities, and dollars. The results of this research are regarded by everyone involved as some of the most secret intellectual property emerging from their investment in motorsport.

The technical literature is full of papers on simulation and modeling techniques for gas flows in piston engines. The fact that the flow is non-steady, involves heat flow and combustion, and the chemistry of the gas changes during the cycle has meant that little can be borrowed from other sources. The computational fluid dynamics (CFD) techniques can be applied for the turbulent flow in the combustion chamber but are currently of limited use for gas flow analysis right through the engine. Steady-state rigs likewise have limited potential, and the single-cylinder engine is the main tool for research. Up to a dozen of these research engines may be in use at any one time for the development of a given engine, each applied to a particular component or system research program, such as

- Valve timing and lift characteristics (variable valve timing is not used due to in-cylinder geometry constraints).
- Variable intake length systems and throttles.
- Ignition—Firing the charge on part throttle is a key driveability issue.
- Injector positioning—At least two injectors per cylinder are used to provide precise control over a wide range of fuel flow rates, with timing being critical and derived experimentally.
- Injection and ignition mapping.
- Combustion chamber design.
- Bore-to-stroke ratio and compression ratio.
- Coolant temperatures—Chassis designers want higher water temperatures (130–140°C [266–284°F] maximum) for smaller radiators, but higher water temperatures present the engine with a problem as more heat goes into the oil.
- Sensor development.
- Port design.
- Lubrication.
- Fuel.

The exhaust system cannot be simulated on a single-cylinder engine because coupling between cylinders is a significant effect. At least as many full V-10's are needed for development, although as much of their testing deals with durability as with producing power.

The overall configuration of a cylinder in competing engines does not vary greatly, but the details do. All engines have four valves per cylinder, set at a narrow angle which is a compromise between valve area and combustion chamber shape, and a single central spark plug. Variable-length intakes assist in spreading the torque curve enough to limit the required number of gears to the maximum seven permitted by the regulations. Injectors are placed both in the ports and above the intake trumpets to provide maximum mixing and even distribution. Barrel throttles are the most common type used.

Short exhaust ports, for minimum heat rejection to the coolant, lead to a five-into-one exhaust system for each bank. Ferrari has experimented with variable-length components in the exhaust system, but similar developments are no longer permitted. The exhaust system performance is sensitive to the firing order, but so too is harshness and vibration which affect durability. Exhaust pipe length has been influenced by the underbody design of the car, but Ferrari finally broke from this constraint by exiting its exhausts through the top of the body, on the basis of a better compromise between optimum exhaust tuning and aerodynamics.

The fueling and ignition computer, which also controls the drive-by-wire throttle and variable-length intakes, has multiple sensor inputs. Intake charge temperature and pressure, as well as exhaust temperature and free oxygen, are monitored, along with temperatures and pressures throughout the engine and its cooling, lubrication, fuel, and hydraulic systems. Individual cylinder control is not necessary because intake air distribution is good. Protection from detonation due to lean mixtures and over-revving are both programmed into the control. Pistons do not last long during detonation, nor do bottom-end components during over-revving. The ability to monitor these critical conditions and prevent the engine from entering danger zones has allowed the engine to be operated much closer to the failure threshold.

Generally, as peak engine speed increases, both volumetric and thermal efficiencies decrease. Therefore, for the overall efficiency to remain constant, mechanical efficiency must increase in compensation. The success in this area of development is the factor that has enabled power to increase in proportion to rpm. Mechanical efficiency is based on the losses of useful work incurred in

- Friction
- Pumping
- Windage within the engine, including oil churning
- Accessories

Friction losses have some components proportional to speed, and others proportional to the square of speed, associated with the viscosity of the oil film. The development of reduced friction is a three-way program among the engine designer, who attempts to reduce loads and speeds; the surface finish engineers, who determine the finish and coatings applied to rubbing surfaces through heat treatment, vapor deposition, and applied coatings; and lubrication chemists, who develop additives for the oil. Rigs, and ultimately the motor-dynamometer, are used to prove the techniques, while quality control of machined finishes and component tolerances ensure that the fruits of these programs turn into horsepower. One of the benefits of reducing the inherent friction between the surfaces of moving parts is that lower viscosity oils can be used, with a reduction in the friction due to velocity squared and lower heat input to the oil. More oil is pumped into the engine than is required for lubrication because it is also used to cool the pistons by being squirted onto their undersides. After use, the oil is churned inside the crankcase which, along with the air moved by the pistons, rods, and crank, creates losses—again, an rpm squared effect. To minimize these windage losses, each pair of cylinders is isolated in a sealed chamber, effectively dividing the engine into five V-twins. The pressure in each chamber is reduced to a near vacuum to reduce the air density and hence the aerodynamic losses. The scavenge pumps (up to three per bay and fifteen total) are tiny Roots-type air pumps to evacuate each chamber and also remove the oil under any condition of horizontal acceleration. The air/oil mixture is centrifugally separated, and the oil is further de-aerated in the oil tank.

The efficiency of these pumps, along with the oil pressure pump, water pumps, fuel pump, hydraulic pump, and generator (all of which are a burden on the net power output of the engine) are individually maximized by careful detailed design and rig testing. Rotating inertia must also be minimized because it adds to the mass of the car to be accelerated, particularly in the low gears.

The reduction of friction as the engine speed has risen, to compensate for the inevitable fall-off of thermal and volumetric efficiencies, has enabled the power potential of raising rpm to be realized. The technologies developed in all areas are of relevance to production cars in the development of cleaner and more efficient engines.

There is no inherent reason why a 3.0-liter racing engine should become heavier as rpm and hence power increase. Strengthening components to meet the increased inertia loads by adding material is not a solution, due to the negative effect of increased mass. The design objective is to increase strength to weight and stiffness to weight, the latter raising natural frequency of components as the forcing frequencies become higher. Gas pressure loads and torque are no higher, but the systems must be able to service the higher heat throughput. If engine weight is plotted against rpm, on the same graph as power/liter versus rpm, for the same four racing engines (Fig. 5.6), and against the calendar year, on the same graph as rpm versus year (Fig. 5.7), a steady reduction in engine weight can be seen. However, that reduction is diminishing toward a minimum weight, barring new technology. It is not clear what that is, but it will not be much below 100 kg (220 lb) without the introduction of new materials. Most of the materials with the greatest potential for significant weight reductions, composites, and aerospace alloys in particular are now banned from use in the highly stressed areas.

Engines are still built of the same basic materials used 30 years ago when the Ford-Cosworth DFV was built: cast aluminum for the block and crankcase, aluminum alloy pistons, and steel for the crankshaft, camshafts, gears, and exhaust system. Connecting rods are now titanium, as are valves. Lightly stressed, cool parts (e.g., covers, intakes, ducts, and brackets) are now

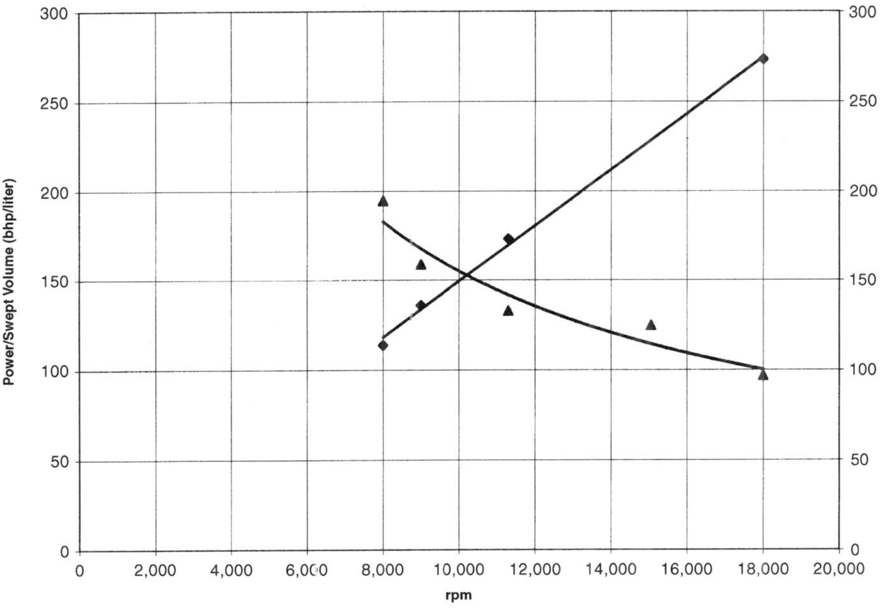

Fig. 5.6 Specific power and weight versus rpm. (Author's illustration)

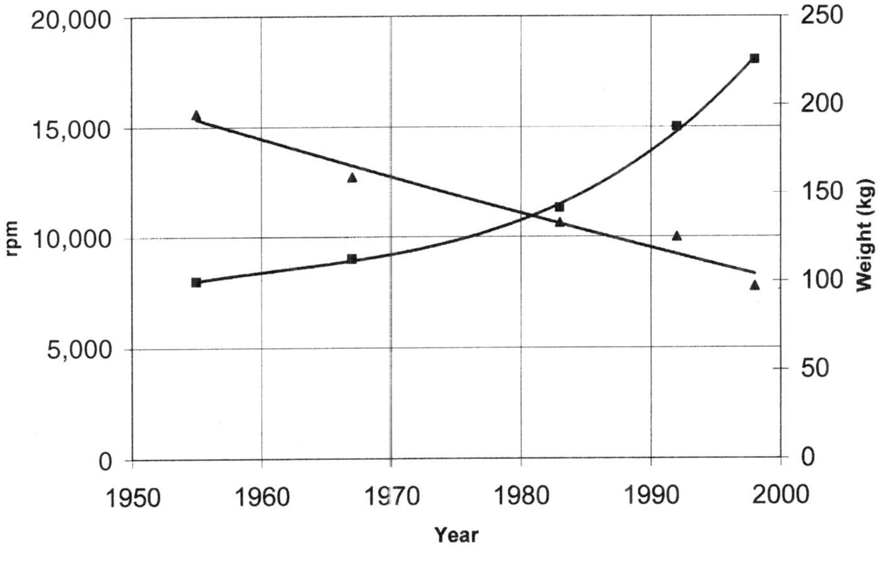

Fig. 5.7 rpm and weight versus calendar year. (Author's illustration)

CFRP instead of cast magnesium or aluminum and steel fabrications. Modern engines carry many sensors and their associated heavy wiring looms and connectors. The scope to use composites in crankshafts, camshafts, pistons, heads, and blocks is limited by the regulations, as is the use of metal alloys with a specific modulus greater than 40 GPa/(g/cm^3). The weight reduction has resulted chiefly from better design and manufacturing processes, particularly the CAC machining of all the surfaces of castings.

CAD and CAE have provided the tools to analyze and optimize every component to the actual load conditions they experience, as measured by multi-sensor, high-speed data systems on rigs and dynamometers and in the car. Mario Illien, the designer of the Mercedes-Illmor engine, has stated that the secret of the modern racing engine is attention to detail of every part. He has always calculated the weight of every component at the design stage and claims he has never been more than 1 kg (2.2 lb) in error in his estimation of the final engine weight. The precision of design and stress analysis and the precision and quality of manufacture that computer-aided techniques have brought about have yielded significant weight reductions. The opportunity to reduce the contingencies built into components to take account of a lack of knowledge or the inability to make exactly what the designer desires brings the weight of all parts down to a minimum. Only better materials and more accurate measurement of service loads and conditions will now yield significant weight reductions of individual parts. Rationalizing the design, reducing the number of parts, and better systems integration continue to offer opportunities to reduce overall car weight and will surely be the benefit of closer cooperation among chassis and engine makers, especially as they come under single corporate umbrellas.

For a long time, the engine has been used to stiffen the chassis in race cars, where vibration is not a comfort issue, by bolting it rigidly to the chassis structure. However, it was not until 1969 that the engine replaced part of the chassis, dispensing with any other structure behind the front engine mounts. As suspension frequencies rose, driven upward by ground-effect aerodynamics, and the significance of chassis torsional stiffness became more important, the stiffness of the hot engine and the integrity of the mounting points have received greater attention. Systems integration has brought engine and chassis designers closer to cooperate on integrating the engine, its accessories, and its interfaces with the chassis. As the cross section of the engine has shrunk, with reducing stroke and greater design refinement, it no longer matches the cross section of the fuel tank as the Ford-Cosworth DFV did (Fig. 5.8). Attaching the engine to the rear of the fuel tank has become more of a problem because the mounting points on the crankcase and heads no longer coincide with stiff areas on the fuel tank bulkhead. The smaller cross section also reduces the torsional and bending stiffness of the engine, and the joint faces of the major structural components (heads, block, and sump) also become significant. Stiffness is not only essential for the chassis handling performance, but deflections in the engine will distort parts and may easily cause additional friction that leads to overheating and failure. Hot rig testing and FEA are key processes used by designers and development engineers to analyze structural performance.

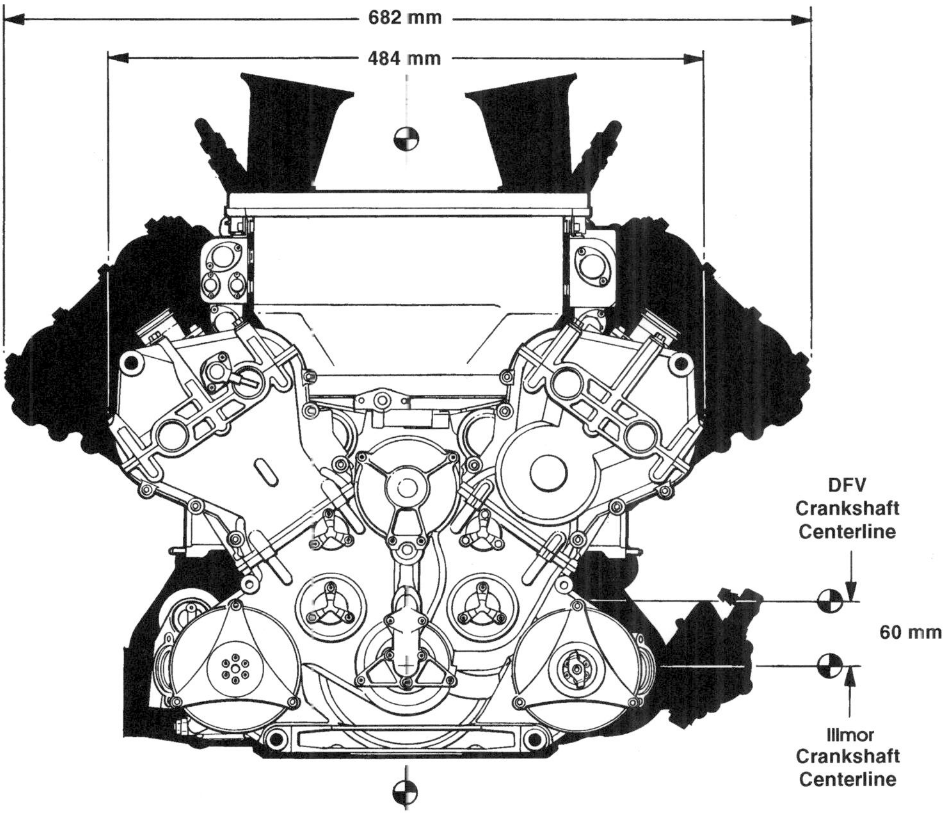

Fig. 5.8 Comparison of cross sections and crankshaft heights of the 1967 Ford-Cosworth DFV and the 1998 Mercedes-Illmor 3-liter NA engines.

There is no part of a race car to which the adage "To finish first, you must first finish" is more relevant than the engine. It cannot be replaced, and faults can seldom be corrected during a race. Terminal failures are there for all to see and cannot be explained as "an electrical problem." The hardest parts of any machine to make reliable are the hot, rotating ones, and a modern Formula 1 engine has plenty of these, plus even more highly stressed, hot, reciprocating ones. Attention to detail, as with weight reduction, lays the foundations of reliability. However, it is rig, dynamometer, and track testing, as well as actual racing, that prove it. Only under racing conditions do the combinations of rpm, throttle, air temperature, cooling, g-level, and wear occur, that find the limits of all the parts individually and combined as systems. Rig testing, however, is only as good as the parameters used to set the test conditions. Sensors and high-speed data systems have allowed the track conditions to be measured and reproduced much more faithfully, but the phrase "we never experienced this failure before," used to explain an early retirement, remains common. As testing becomes more restrictive, the ability to carry out race distance tests is limited. Thus, engine dynamometers capable of reproducing the true duty cycle are used more frequently for proving durability of new components and systems.

All aspects of the engine in a car can be reproduced except for the effect of acceleration experienced by the engine and its systems. To reproduce the effect of 4g braking or cornering would require the entire dynamometer to be mounted in centrifuge. There is no other way. The mounting of the engine on the test bed is unlikely to reproduce exactly the vibration characteristics of the engine in the car; therefore, these destructive accelerations will not be simulated. The engine is mated to a gearbox, with the semi-automatic gear-change under the control of the dynamometer's computer, to simulate the rpm transients that occur during gear-changes. The gearbox output drives either a servo-controlled, eddy-current dynamometer, whose dynamic performance allows the load to be changed rapidly enough to follow the throttle and gear-changes, or to a flywheel, whose inertia represents the mass of the car, between the gearbox and the dynamometer. All ambient conditions (i.e., atmospheric temperature and pressure and ram-air conditions from the air box) are controlled. The engine cooling, lubrication, fuel, and generator systems all function representatively. Also, the cooling radiators of the car may actually be used; temperatures will be controlled precisely. The duty cycle of any circuit can be programmed, including the sequence for the start of the race and pit stops. Everything that can be measured is measured and a race-distance number of cycles simulated to weed out any weaknesses.

Formula 1 engine development is based on a high level of long-term research and development, rigorous analysis made possible by computers, attention to design detail, and CNC manufacturing processes. Technologies that revolutionize racing engines are rare. Although the Ford-Cosworth DFV introduced the concept of systems integration to the chassis and engine, it was not truly revolutionary for all its brilliance. Turbocharging represented a change in engine cycle and temporarily produced a revolution in Formula 1 racing. Electronic fuel injection, which enabled turbocharged engines to fulfill their potential, brought about a slow revolution as it took over from mechanical injection. Perhaps the key technology that has prevented a flattening of the performance curve is the pneumatic valve spring, which has removed a major mechanical constraint to increased rpm. To gain the benefits, combustion and friction are the two areas that have received the most attention, in addition to the mechanical design of rotating and reciprocating parts.

With the goal of 300 hp/liter in sight, the 20,000 rpm required should be achieved in 2001, at current rates of progress, unless limited by the V-10 configuration. The R&D effort to maintain combustion efficiency at a bore-to-stroke ratio of more than 2.5:1 (i.e., a stroke of less than 40 mm and a bore in excess of 100 mm) and the further reduction in losses necessary should not be underestimated. Nine hundred horsepower from a piston engine weighing less than 100 kg (220 lb) will be no mean accomplishment, and it should sound good, too!

Chapter 6

Transmissions

All memorable quotations referring to gearboxes involve expletives.
—Peter Wright

The mechanical arrangement by which the power of the engine is transmitted to the driven wheels and the mechanism by which the appropriate ratio between the engine and the wheels is engaged has not changed significantly since the earliest race cars. Power is transmitted via a multi-plate clutch between the engine and the gearbox to enable the connection to be disengaged during gear selection, to smooth the torque pulse as the engine and gearbox input speeds are re-matched after a gear change, and to facilitate standing starts. Spur gears are used for the various ratios, with face dogs to engage the selected ratio to its shaft. Bevel gears are used to turn the drive through 90° where appropriate, either before or after the gear cluster. Final drive to the driven wheels is via a differential and articulated half-shafts.

However, major changes have been made to the implementation and detail of transmissions over the last 50 years. Moving the engine to a location behind the driver enabled integration of the engine with the gearbox, which in most cases was located on the rear, driven axle. In the mid-1960s, the engine and transmission became part of the chassis structure, taking all suspension and driving loads from the rear axle into the chassis monocoque, increasing the level of integration with the engine such that they became a unified structure. The detail layout of the gears has changed to suit the latest thinking on aerodynamics, both internal and external, and weight distribution. The development of computer-controlled systems and hydraulic power supplies for Formula 1 cars in the 1980s enabled the gear-changing process to be semi- or fully automated and provided the driver with "carefree" gear changing while keeping his hands on the steering wheel. The development of carbon-reinforced composites has influenced clutch design (the reduced-diameter friction plates are now made from carbon-carbon material) and is hovering on the verge of taking over the gear-case from magnesium, aluminum, or titanium.

In the aviation world, it is said that the parts of an aircraft that usually give trouble are the "hot, rotating bits." The same is true with race cars, the culprits being engines, gearboxes, drive shafts, brakes, wheel bearings, and tires. This truism survives in spite of these being the components that receive the greatest attention to detail at the design stage, are manufactured to the closest tolerances and material specifications, and are most rigorously inspected. The perpetual quest that is Formula 1—more performance from less weight—pushes these parts hardest. Thus, it is not surprising that they keep appearing on the list of reasons for DNF's (did not finish).

After the engine, the gearbox can cause the most trouble if the principles and detail engineering are not right. The engineering disciplines needed to design and develop a racing gearbox are aligned more closely to those needed for the engine than the chassis. Manufacturing tolerances, heat treatment specifications, and manufacturing processes are similar to those employed in making the engine; few Formula 1 teams have the resources and specialist knowledge to design and manufacture the critical gearbox components (e.g., gears, shafts, dogs, and change mechanism) that determine the reliability of the unit. Therefore, they turn to specialist firms. History shows that those racing teams that have successfully made their own engines tend to have the least trouble with gearboxes (e.g., Ferrari, Porsche, and BRM).

The addition of computer-controlled, electro-hydraulic systems to operate the clutch, gear selection, and the differential has had a negative effect on transmission reliability, adding an additional complex layer of sensors, electronics, hydraulics, and software to the system.

Layout

Since the 1965 Ferrari 1512 pioneered the concept and the Lotus T49/Cosworth DFV consolidated it, the accepted practice has been to use the engine and gearbox as extensions of the primary structure, carrying rear suspension, drive, and wing loads via the engine into the monocoque. Although this function tends to determine the external layout of the gear case, the internal design is dictated by the arrangement of gears and final drive.

Until 1975, when Ferrari first ran a transverse gearbox and, in doing so, turned the drive through 90° before the gear cluster instead of after it, the standard layout positioned the gears behind the final drive. The space between it and the rear face of the engine, around the clutch, was used to store engine oil in a tank integral with the magnesium gearbox case. This layout afforded the ability to change any of the five or six ratios in approximately 30 minutes. The disadvantage of this layout was the mass of the gears behind the rear axle.

When the regulations moved the driver backward to prevent his feet being forward of the front axle, it became difficult to achieve the forward CG the tires demanded. This, coupled with an increasing awareness of the influence of the yaw moment of inertia, caused other designers to follow Ferrari's lead, placing the transverse gear cluster forward of the final drive.

In the last five years, again led by Ferrari, all teams have reverted to longitudinal gearboxes, with the gears located forward of the final drive. This arrangement provides the current optimum tradeoff among underbody aerodynamics, internal airflow, mass distribution and weight, time to change ratios, structural stiffness, overall length, transmission efficiency, and reliability.

The original longitudinal layout, with gears placed behind the final drive, lent itself naturally to a two-stage speed reduction from the 13,000 rpm of the engines in the 1970s to the approximate 2500 rpm necessary at the wheels, equivalent to 320 kph (199 mph). One reason for the change to a transverse layout was the need to introduce an additional speed reduction stage, to accommodate the ever increasing engine rpm—up to 18,000+. In this arrangement (Fig. 6.1), the input shaft drives a bevel gear (first stage and 90° direction change) which then drives the input shaft of the gear cluster (second stage). The output shaft of the cluster has a spur gear on it that drives the final drive ring gear (third stage).

The new generation of longitudinal boxes also have three stages, with the extra gear enabling the mass of the gear cluster to be positioned as low as possible (Fig. 6.2). Here the input shaft drives the gears directly (first stage), which incidentally results in high gear tooth speeds but lower torque and hence tooth loads, and the output shaft drives a bevel (second stage and 90° direction change) which then drives a spur gear final drive (third stage). An alternative arrangement swaps the second and third stages, with the output shaft of the gears driving a step-down spur gear, which drives the bevel final drive. This arrangement restricts the options in placing the final drive relative to the output shaft of the gears.

High rpm, narrow power bands, and almost instantaneous carefree gear changing have led to seven-speed options being built into most gearboxes. This is not so much for pure acceleration as to give the driver a wider choice of ratios for a given range of corner speeds on a given track. The ratio selected for a corner (usually two possible ratios) influences the progression and gain of the throttle response by determining the part of the power curve in which the engine is operating when the driver starts opening the throttle.

The biggest drawback to the longitudinal layout, which due to its narrowness theoretically impedes the internal and external airflow the least, is the necessity to split the gearbox from the engine to change gear ratios through the removable front bulkhead. However, the lack of side covers and the apertures they close results in a sounder structure. Covers and removable bulkheads are spigoted for optimum structural integrity, a practice that CNC machining has made affordable.

To speed the operation of ratio changes, the engine oil tank, while remaining in the same position between the engine and gearbox until the late 1990s, is not integral with the gearbox case. A CFRP tank is mounted to the engine (within the bell housing or, more recently, in front of it) as designers strive to move weight forward. The tank remains attached to the engine during ratio changes, thus permitting all oil lines to remain unbroken.

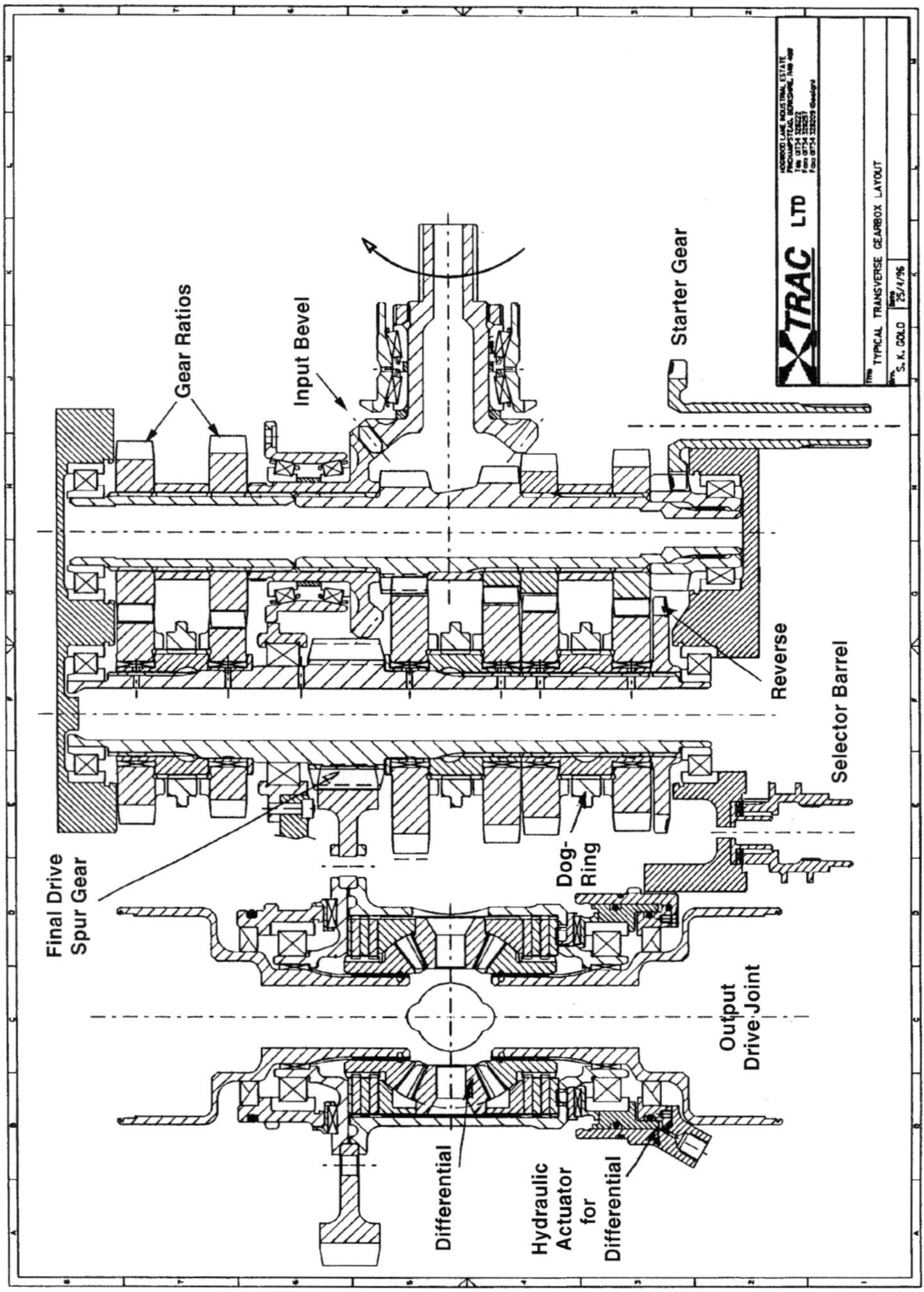

Fig. 6.1 Transverse gearbox layout. (Courtesy of Xtrac Ltd.)

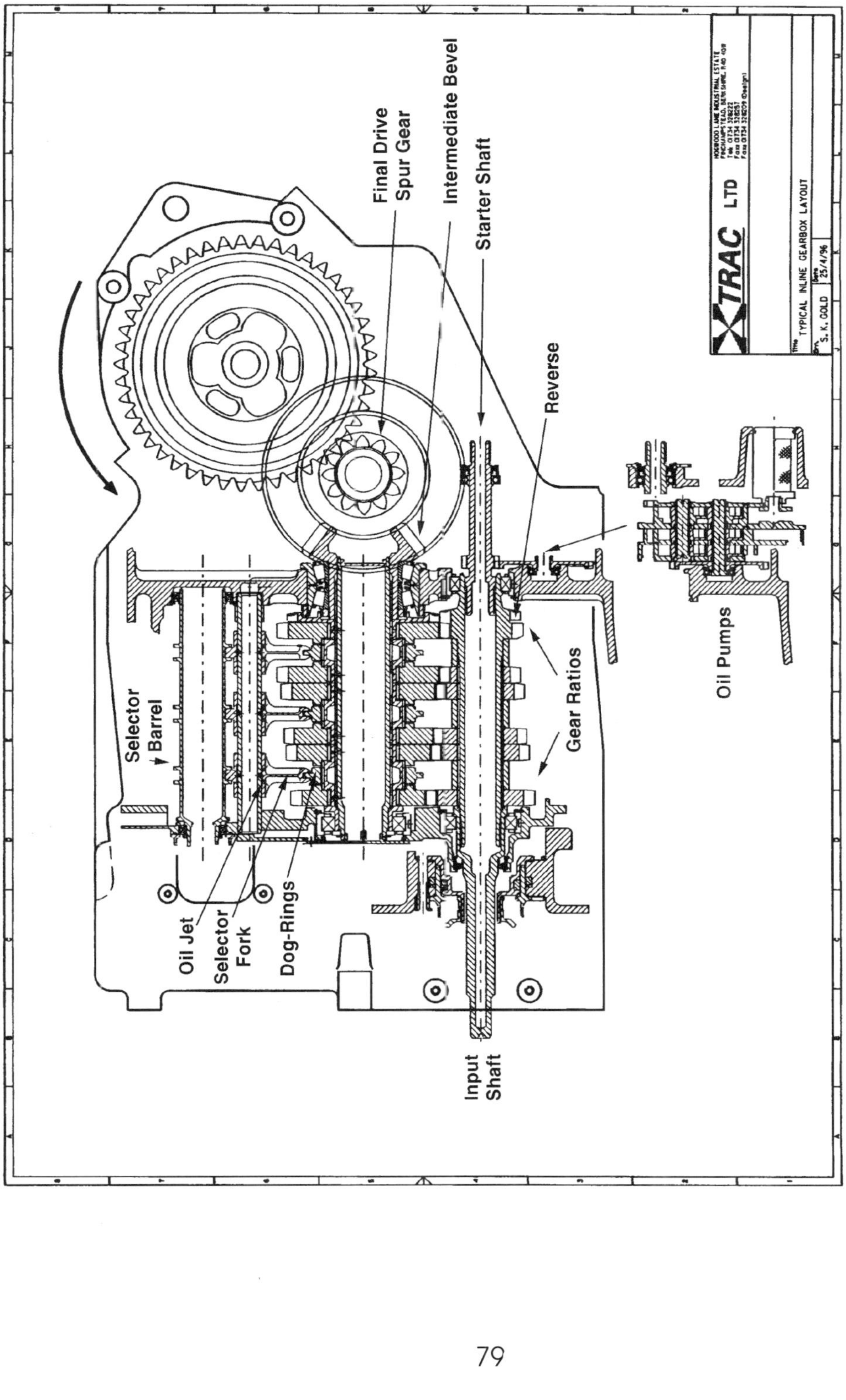

Fig. 6.2 Longitudinal gearbox layout, with the gears located ahead of the axle. (Courtesy of Xtrac Ltd.)

Casing

Not only must the gearbox case provide precise location of all shafts under gear parting loads, it is also the primary structure of the car. The need for an adequate cross section for torsional and bending stiffness and to pick up suspension mounting points efficiently has meant that the volume of the case is considerably greater than is strictly required to enclose the gears. The additional volume is used for oil tanks for one or more of the engine, gearbox, and hydraulic systems, as well as the catch tank.

Choice of material for the casing is governed by two fundamentals: stiffness at elevated temperatures (100+°C [212°F]) and manufacturing process. The majority use either MSR magnesium (high silver content, good stiffness with temperature, and high notch sensitivity) or precision-cast thin-wall aluminum. Sauber was the first to go to aluminum in 1993, employing the Zeus foundry that made the castings for the Illmor engine it used that year. In theory, there is no reason to prevent magnesium from being cast with the same degree of precision, and we may see a return to this material as the precision pattern-making techniques are adapted by foundries that deal with magnesium.

Both die casting and investment casting are processes that, at a price, offer potential to reduce case wall thickness and hence weight. The Minardi team built an investment cast titanium case in 2000.

Stimulated by John Barnard, Ferrari explored the use of machined-all-over steel for the gearbox case, using the CNC machining technology developed to manufacture the steel crankcases of Ferrari engines at that time. In 1995, Barnard substituted titanium for steel, desperate to reduce the weight at the back of the car behind the long V-12 engine. The latest Ferrari gearbox is a combination of titanium and CFRP. The bell housing and rear structure are CFRP, while the center section, which locates all the shafts and carries all the gear loads, is fabricated from NC-machined segments in titanium. Suspension loads are fed into both CFRP and titanium sections. The cost per kilogram saved by this stunningly expensive approach is probably beyond the pockets of most teams. The more common approach is to integrate the bell housing and the gear case in magnesium (thereby eliminating a heavy and stiffness-compromising joint at the change in material interface) and to integrate the final drive bearing caps and rear case with the CFRP rear impact structure. Suspension loads are more easily fed into the magnesium case than they would be into a CFRP case.

In 1998, Barnard (then working for Arrows and forever the innovative designer) and Stewart Grand Prix's designer Alan Jenkins both opted for all-CFRP gearboxes. The 10+ kg (22+ lb) potential weight savings and greater hot stiffness were the real potential performance gains. It is more difficult to design load points, such as suspension mountings and bearing housings, in CFRP than in cast alloy, and therefore aluminum or titanium bulkheads incorporating bearing housings are used. Differential thermal expansion between the metallic and composite parts presents real problems. Choosing the right mix of materials and joining them in an integrated structure is the key to weight reduction in the gearbox. The fact that Stewart dropped the

CFRP gearbox for 1999 and no other new ones have emerged is due to the inevitability of any new gearbox causing sleepless nights, while radically new gearboxes cause nightmares. (The mechanics and engineers suffer the sleepless nights; the team principals have the nightmares.) The effort of getting on top of new gearbox problems during the all too short pre-season test period after new cars are launched has a compounding effect: gearbox faults are not usually fixed quickly, and it is impossible to run the car until they are. New parts, particularly castings, CFRP cases and shafts and gears take time to manufacture; therefore, the test program cannot continue while the gearbox is not functioning. As an inevitable consequence, the development of the rest of the car and the opportunity for the race engineer and driver to learn how to set it up is severely compromised, with ensuing poor race results. Teams developing radical gearbox concepts are wise to make them parallel programs and not to commit the race program to them until those concepts are fully proven.

Gears and Shafts

During the era of transverse gearboxes, the quest to shorten the gear cluster and hence narrow the box was dominant. One millimeter saved in the arrangement for a single gear factored up to give a 6-mm (0.24 in.) narrower case. Prior to semi-automatic gear changing, gear width and tooth profile were dictated by the need to survive inadvertent multiple downshifts, during which the inertia load of accelerating the engine bent the gear teeth backward, ready to snap off on the next application of power. Software has eliminated the miss-shift. Development of the ratio gears, bevels, and final drives is centered around profile design, manufacturing tolerances, and material and heat treatment specifications.

The rate at which the life of a gear is used is a function of the deviation from the design tooth profile. As the gear wears, the rate at which it approaches the end of its life increases. If the manufacturing process is imperfect, the teeth start their working life already part of the way down the path to failure. Manufacturing imperfections are similar to wear in terms of their effect on gear life—greater precision in the tooth-forming process has a large effect on that life. Computer-generated profiles use many more increments than hand-calculated profiles (effectively a series of flats), and CNC grinding machines reproduce that profile to far greater accuracy than hard cutting and final grinding. The application of FEA is used to explore optimum profiles under load.

Computer-controlled heat treatment processes have brought a level of precision to material properties that is limited only by the specification tolerances of the base material from which gears and shafts are manufactured. This material is usually 300M steel. Current emphasis is to control more tightly the process of manufacturing the steel billets.

Lubrication

Gearbox lubrication is as much about cooling bearings and gear tooth surfaces as pure friction reduction. If surface temperature exceeds 150°C (302°F), material properties degrade. High rpm make it increasingly difficult to extract the oil/air mixture thrown out by the gears from the walls of the gear case, separate the air from the oil, cool it, and pump it to the high-load high-temperature moving parts of the gearbox.

Multi-lobe pumps have replaced gear pumps for their ability to pump air, self-prime, and avoid cavitation. Pressures have risen to approximately 0.2 MPa, and, while solid oil is fed to bearings, a mist is directed into the mesh of gears. There is a neat arrangement whereby the selector fork in use opens a port in the selector fork rail, and oil is fed through the fork itself to the gear selected. The resulting oil spray lubricates free-running gears. Scrapers remove the oil from the rotating gears and direct it to the pump pickups, before feeding it to a swirl pot for centrifugal air separation.

Gearbox lubrication is now universally dry sump. Twin pumps scavenge the casing, and a single pressure pump is gravity fed from the integral tank. Careful design enables oil quantities to be as low as 1.5 liters (0.4 gal) (every additional liter adds 0.85 kg [1.9 lb]). The use of Neo gearbox oil, with its low-viscosity characteristic, has permitted the gear-change hydraulic system to share the gearbox oil after suitable filtration to avoid contamination of the Moog servo valves, thereby saving the weight of an additional tank, cooler, and fluid.

Drive Shaft Joints

The quest for a narrow gearbox and the high underbody air temperature in the exhaust from the radiators have caused the GKN Tripod constant velocity joints (GSJ's) to be moved inside the final drive assembly where they can be cooled and lubricated by the gearbox oil, instead of relying on their own grease. The solution was pioneered in off-road racing, in particular on Paris-Dakar cars where the required substantial wheel movements take the joints to extremes of angularity and cause them to run at high temperatures. The arrangement illustrated in Fig. 6.3 shows oil-cooled and lubricated GSJ joints, as close together as possible inside the differential.

Differential

The role of the differential has come under much scrutiny during the last decade, as vehicle dynamics modeling has enabled engineers to appreciate its strong influence on vehicle stability and control. It was not long after man decided to propel a vehicle by driving two, laterally spaced wheels mounted on a common axle, that he discovered the need for a mechanism to allow the wheels to rotate at different speeds as they followed different paths around a corner, and yet both transmit torque to the ground. Pecquer invented the differential in 1827 for use on steam-powered traction engines.

The open differential delivers equal torque to each wheel, irrespective of the relative rpm of the two wheels. In a corner, when load is transferred from the inner to the outer wheel, the slip ratio of the inner tire must increase to compensate for decreasing vertical load to maintain

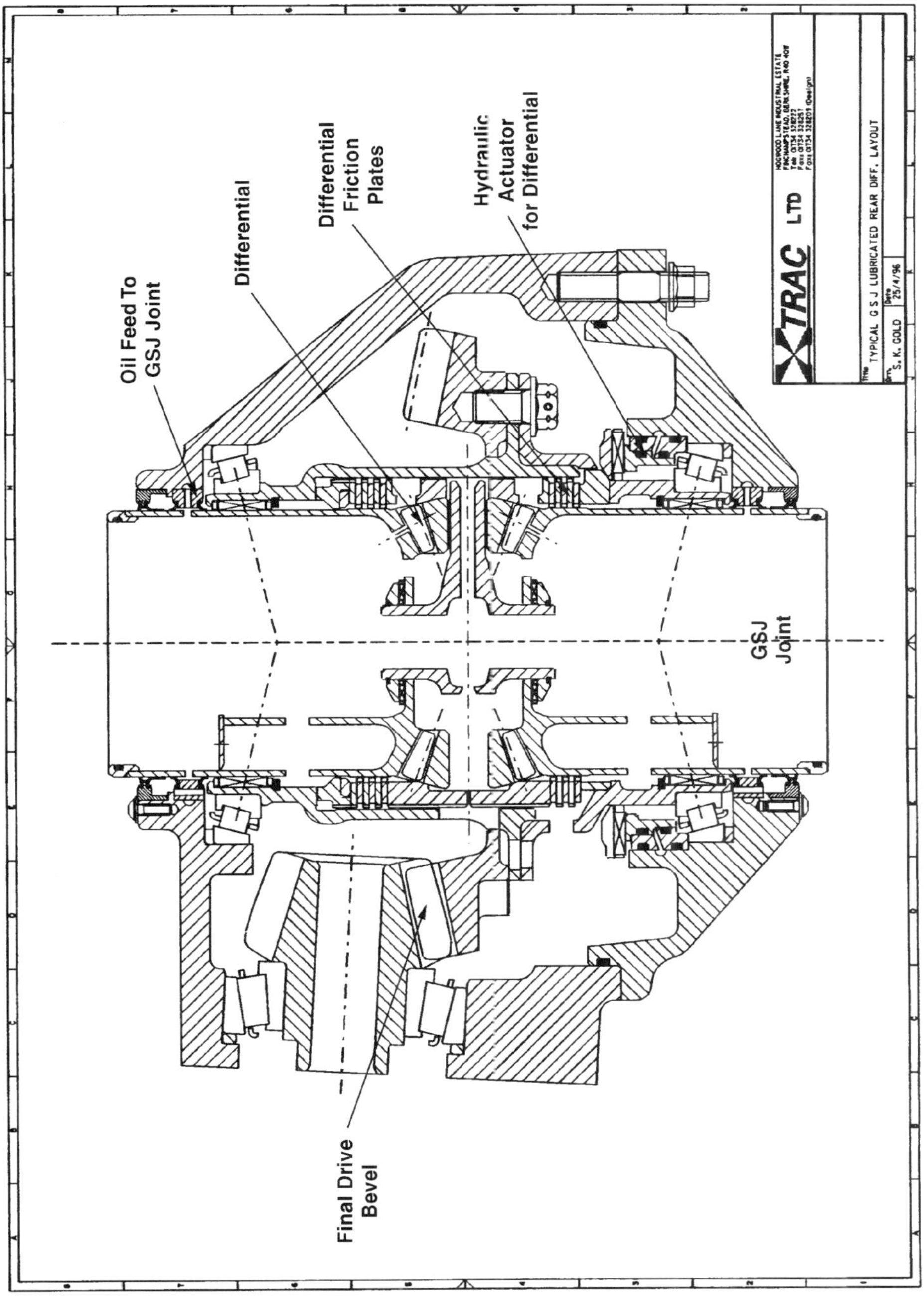

Fig. 6.3 Typical lubricated GSJ and controllable differential unit. (Courtesy of Xtrac Ltd.)

equal torque. Eventually, the slip ratio exceeds that for peak longitudinal forces, and the wheel spins. At that point, the differential ensures that the torque transmitted to the outer wheel equals that of the inner (i.e., low), and any increase in acceleration becomes limited.

If a resistive torque is applied to the inner limiting wheel, the differential must transmit more torque to that wheel and hence will also transmit more to the outer wheel. The inner tire is not able to transmit more thrust, but the outer one can. Thus, the sum of the two will be greater. The resistive torque can be applied in either of two ways: (1) via the brakes, as in "fiddle brakes" on trials cars and the McLaren of late 1997 and early 1998 (prior to their ban), and the latest types of traction control on road cars that use the ABS system to differentially apply the brakes; or (2) as a friction torque, generated between the two output shafts of the differential. All differentials have some degree of internal friction and therefore are not truly "open." However, the intentional application of friction is the basis of all limited-slip differentials.

The characteristics of a limited-slip differential depend on how the friction is generated and how it changes with factors such as differential wheel speed, total torque, differential torque, and temperature. A wide variety of differentials have been developed over the years, each with a different set of characteristics that are determined by the mechanical arrangement. As such, they tend to have fixed characteristics, to be sensitive to manufacturing tolerances and wear, and sometimes to have undesirable side effects.

I would refer the reader again to *Race Car Vehicle Dynamics* by Milliken and Milliken (Ref. 6.1) for descriptions and analyses of the small number of differential types that have stood the test of time and remain in popular use for race cars and high-performance road cars. It is necessary to understand these devices and the differences and limitations of their characteristics prior to considering what electronic control can do for them. This is particularly important in Formula 1, where the limitation is imposed that the differential cannot be active and must conform to tight regulations concerning its characteristics. (See Article 9.7 of the Technical Regulations, found in Appendix D of this book.) No such restriction exists for World Rally Cars, which are free to use complex, active-feedback differentials in the front, rear, and middle of their four-wheel drive systems. Much of the pioneering work on differentials has taken place in this field.

Current mechanical limited-slip differentials do a good job of controlling single wheel spin on a race car, even in the wet. So why go to all the complication and potential reliability of an electro-hydraulically controlled device? The answer lies in the ability of the differential to steer the car by controlling differential torque across the axle. The difference in thrust at the wheel, side to side, generates a moment about the CG that can be large enough to significantly affect the stability and control of the car. Among the first to recognize and exploit this were the cinder track and midget racers in the United States. Because of the limited grip, and the fact that the corners all go one way, the cars are fitted with locked differentials or "spools" (actually, no differential at all) and different-diameter rear tires (stagger). This generates a turning moment into the corner and overcomes the inherent power understeer of a short-wheelbase, front-engined car. The technique has been further refined by oval track cars and was even used in Formula 1 in the 1970s, when cross-ply tires that varied in static and dynamic radius were used, albeit with limited-slip differentials.

Depending on the torque distribution, either under power or on the overrun during braking, a limited-slip differential will either resist yawing or introduce a turning moment in addition to any steering input. Under hard braking when the car is unstable, a yawing resistance is a stabilizing moment and hence is desirable. Salisbury differentials can achieve this with suitable ramp angles and pre-load, but the effect is quite small. Strong lock-up under power will have the same effect but in this case can be undesirable because it adds to power-on understeer. What is desired is a system that provides the right torque distribution and hence yawing moment to meet the requirements of the driver and the response of the car at any given moment.

In an unregulated world, a fully active system would be used, sensing and feeding back all driver inputs (throttle, brakes, and steering), car response (speed, lateral and longitudinal acceleration, and yaw acceleration), and rear wheel response (slip ratio, vertical load, and torque) to determine the optimum torque distribution for the rear wheels. The differential mechanism alone, however complex the internal friction control was, would be insufficient to ensure that the torque could be varied from a 0/100% to 100/0% split across the axle. The rear brakes, differentially applied, would also be needed. Of course, in Formula 1, none of this is now permitted.

What is permitted is an electronically controlled differential having characteristics that adhere to Article 9.7 of the Technical Regulations. Making it electro-hydraulic allows the desired characteristics to be closely controlled and easily set up, and it permits the engineer or driver to select from a library of characteristics. In practice, a Salisbury-type limited-slip differential is fitted with hydraulic actuators, which load the clutch packs, modifying the internal friction between the two output shafts. The hydraulic pressure is modulated by a Moog valve, which is controlled by the chassis computer. Only the individual teams know exactly what the control algorithms are, but the options are limited by the regulations. Using differential wheel speed is straightforward, providing it can be shown to the FIA software inspectorate that it is *differential* wheel speed that is being used, not *mean* wheel speed. That would appear too similar to an element of a traction control system.*

Torque measurement is much more difficult. A reliable, race-worthy system for torque measurement on a race car has not yet been developed, although systems are used in testing on both the gearbox input-shaft and the half-shafts. That leaves the necessity to calculate engine output torque from all engine parameters that influence it, to derive the torque from a map of engine performance that has been calibrated on the dynamometer, and then to multiply it by the gear ratio to determine final drive input torque.

* See author's note on page 290.

The differential clutch pack load is calculated according to a function of input torque, differential wheel speed, or both. The functions must be fixed around the circuit, variable only when the car is stationary. As such, they will be less than ideal at all times but should be more optimized than a purely mechanical setup.

Electronically controlled differentials in Formula 1 will not dramatically reduce lap times nor produce a discernible difference in the way the cars are driven. More stability under braking and an increased ability to turn in under power are the potential benefits. The need to program their characteristics into software has promoted from being part of the black art of racing.

Figures 6.3 and 6.4 also show a hydraulically controlled, friction plate differential, of the type being used in Formula 1. The friction across the output shafts of the unit is set by the axial load on the friction plates (as in a Salisbury limited-slip differential), and this load is controlled by the annular hydraulic actuator.

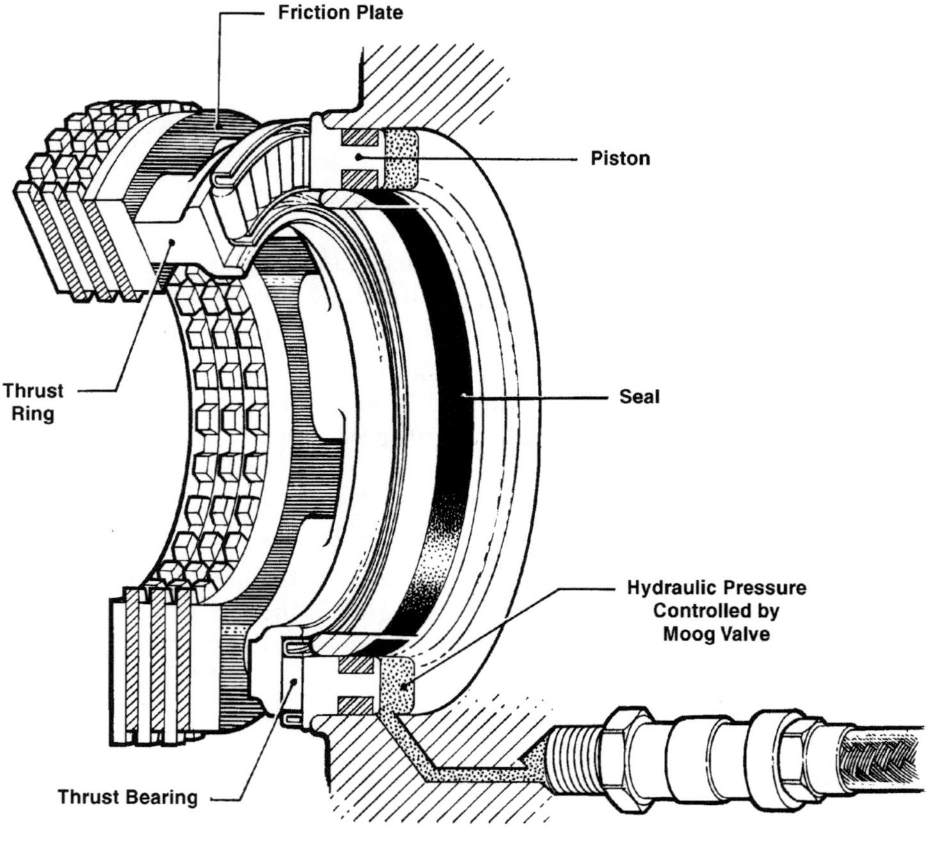

Fig. 6.4 Hydraulic actuator for controllable differential.

Future Developments in Differentials

McLaren achieved the ideal torque distribution system with its fiddle brakes and was able to achieve manually what the most sophisticated anti-skid or active stability road car systems achieve automatically. For McLaren to have used automatic control would have clearly been against the regulations. These road car systems apply the rear brakes differentially to impart a yawing moment to the car to either steer it or stabilize it. They also control single or double wheel spin as needed. McLaren provided its drivers with a switch to select either rear brake, according to the direction of the corner, which they were then able to apply to either turn the car in under power (by braking the inner wheel) or prevent inner wheel spin when accelerating out of a corner. Braking of one rear wheel under power will reduce the torque delivered to the wheel on that side and increase it to the opposite side via the open differential. The net result is a turning moment imparted to the car, which steers it toward the side of the braked wheel.

In the aftermath of Ferrari's protest of the McLaren system, the FIA modified the Technical Regulations relating to both brakes and differentials, adding two relevant clauses:

- The brake system must be designed in order that brake caliper pressures in each circuit [i.e., front and rear circuits] are the same at all times.
- The design and control of the differential may not permit a greater ratio of torque distribution than the ratio of grip between the inner and outer driven wheels.

Delivering a torque ratio from the differential and then modifying it by absorbing part of it in the brakes was no longer permitted.

However, the advantages of directional control and stability enhancement through the use of differential torque control across the car was now as obvious to Formula 1 designers as to road car engineers. Benetton sought to use its front axle differential to achieve stability under braking and turn in, and probably something similar under power, with its twin-clutch gearbox. If use of the brakes were not permitted, could the differential be made to do something similar? As usual in such circumstances, others had been there before Benetton.

In a corner, the inner wheels follow smaller path radii than the outer wheels and will therefore rotate more slowly if permitted. An open differential allows them to do that. A locked differential forces them to rotate at the same speed, and a limited slip attempts to equalize them to a degree dependent on its characteristics. The inner wheel therefore runs at a locally higher slip ratio than that of the outer wheel. It is impossible to make the outer wheel run at a higher slip ratio than the inner wheel with these devices and thereby transmit more torque to the outer wheel to generate a destabilizing, turning-in moment. To do this requires a totally different approach, or the ability to brake the inner wheel. As any student of military vehicle engineering will know, these are the two techniques used to steer tracked vehicles such as tanks.

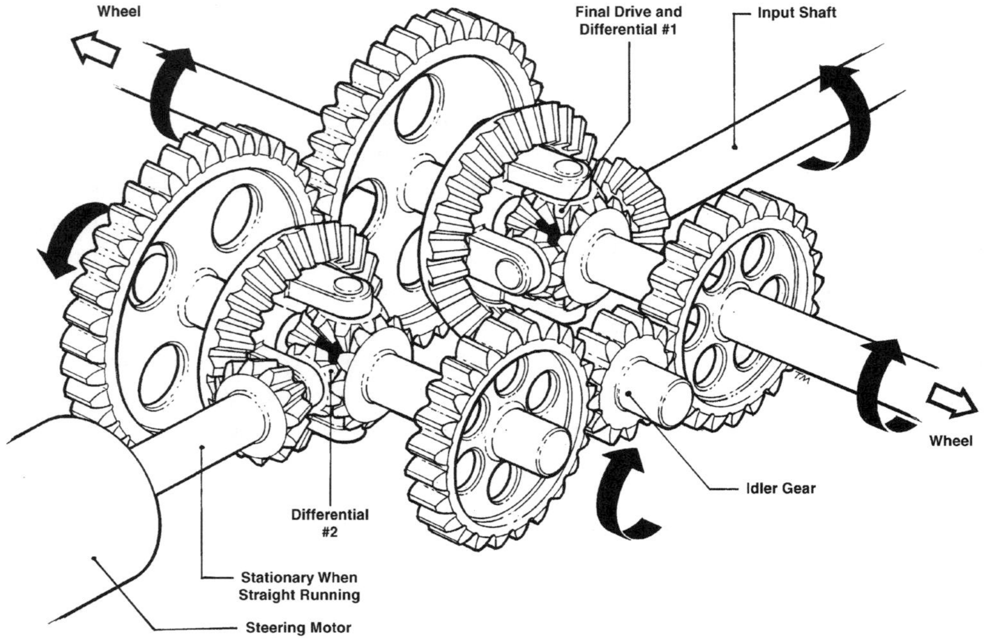

Fig. 6.5 "Tank steer" controlled differential.

Differential wheel speed can be achieved using double, back-to-back differentials (Fig. 6.5). A conventional open differential provides drive to the output shafts. The output shafts of a second differential are connected by gears to those of the first differential, but with an intermediate gear on one side to reverse the rotation. In straight line running, the output shafts of the second differential rotate in opposite directions but at identical speeds. As a result, the input shaft does not rotate. However, if the input is made to rotate, one output shaft will increase in speed and the other will slow down, depending on the speed and direction of rotation. If a controllable and reversible electric or hydraulic motor is connected to the input shaft of the second differential, the vehicle can be steered precisely by controlling the speed and direction of this motor.

Such a system was developed by Lotus Engineering for the active fore and aft torque split on a Chevrolet Active Concept Demonstrator in the 1980s. It also was fitted to a four-wheel drive Peugeot 205 Turbo16 Rally car as a development mule. Active axle differentials of a similar design were considered, but the active-suspension, four-wheel steer vehicle was already complex enough. The system worked, and the gear whine could be heard in the office when it was being tested on the test track at Lotus. The noise would change as the torque was shifted back and forth for automatic correction of understeer or oversteer at the limit.

Such a system is fine for steering tanks and is interesting as a research tool, but it is complex and inefficient. Honda has developed a different system called a direct yaw control system, which is fitted to some of its road cars including the Honda Prelude R. In this system, the speed ratio between the inner and outer driven wheels (front-wheel drive, in the case of Honda) is fixed, with the outer wheel rotating faster. Two epicyclic gear sets determine the ratio, and the appropriate set is engaged by a clutch, according to the direction of turning. In a straight line, both clutches are open. A computer controls the whole system automatically. It could definitely be termed a "twin-clutch" gearbox because there are two additional clutches, over and above the normal clutch for taking up the drive from the engine. Figure 6.6 shows schematically the functioning of the differential, although the implementation is somewhat different due to the use of epicyclic gears. The ratio difference between left and right wheels, defined by the gear ratios, is exaggerated in Fig. 6.6 for clarity. Honda also developed a four-wheel drive version, in which the ratio between the front and rear wheels is fixed when cornering in a similar manner (the speed of the rear wheels is increased to deliver more torque), as well the ratio between the two rear wheels.

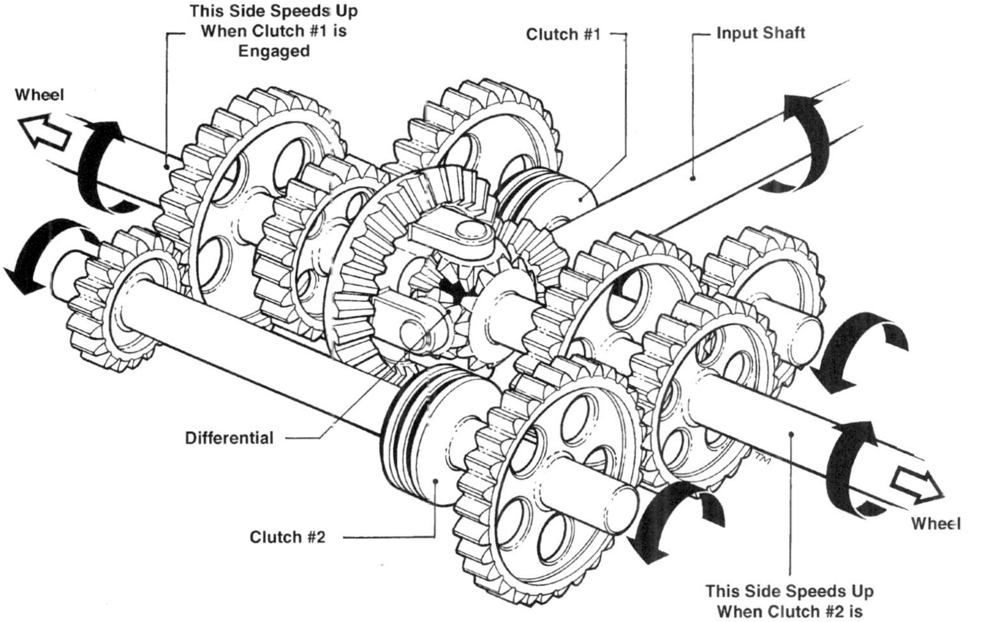

Fig. 6.6 "Steering" differential—Honda concept.

This system can be optimized for only one radius of corner and would not provide the degree of control and optimization desired in Formula 1. However, it does point to a potential method of implementation. If the clutches are allowed to slip, a degree of control over the speed difference is possible. Now only one gear shaft is required, but it must be driven at the mean wheel speed and geared identically to both half shafts at a slightly different ratio to provide slip at the clutch.

Figure 6.7 shows a conventional final drive and open differential. An additional shaft is driven by the final drive gear and is geared to the output shafts of the differential such that it tries to drive both shafts at a slightly different speed from the differential. Two clutches disengage the gears from driving the output shafts, preventing seizure of the whole system. However, if one of the clutches is partially engaged, torque is transmitted to that output shaft, increasing its speed, with the opposite shaft slowed proportionately by the differential. Modulation of the clutches provides control of torque distribution, and the gears and clutch controls are designed to enable the speed of the outer wheel to increase, while the inner is slowed. The lay shaft can either increase the speed or slow the output shafts, with the clutch engagement strategy designed accordingly. The ratio and clutch characteristics must be chosen to provide the degree of control required. The result is a twin-clutch system simpler than that of Honda, which provides a means of controlling the output shaft speed differential.

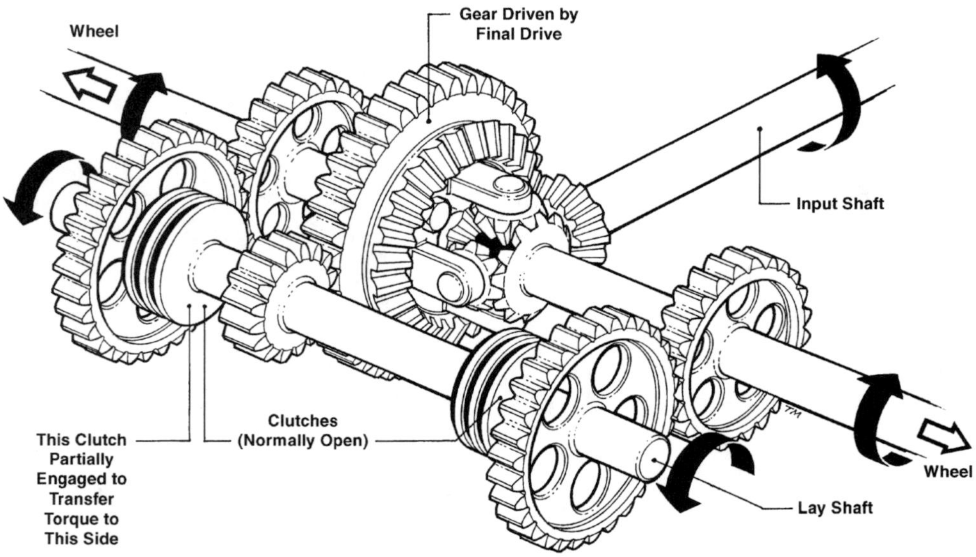

Fig. 6.7 Formula 1 "steering" differential.

Such a system needs lateral and longitudinal acceleration inputs to determine the control of the clutches. These are not permitted inputs to an electronically controlled differential in Formula 1; therefore, they would have to be purely hydro-mechanical in operation. The Automotive Products Ltd. semi-active suspension system that Williams initially adapted to its Formula 1 car was a purely hydro-mechanical system, before Williams converted it to computer-controlled electro-hydraulics. It employed pendulums driving hydraulic valves to signal lateral and longitudinal acceleration. Such a system is perfectly feasible, even if it is not ideal in this age of electronics.

Slipping clutches that transmit significant torque for long periods would require cooling and thus would have to run in oil. Porsche used such a system on the four-wheel drive Porsche 959 to control torque distribution fore and aft. Multiple sintered-metal clutch plates running in oil did the job.

Thus, all the technology for total control of torque distribution across the rear axle of a Formula 1 car within the Technical Regulations is available. By calling the individually operated rear brakes of McLaren's fiddle-brake system "clutches," putting them inside the gearbox mixed with the differential and operating them automatically, the regulations concerning brakes, driver aids, and rear steer are circumvented. The effect is almost exactly the same. The understanding of how it can be used beneficially is well developed by both the road car manufacturers and the leading Formula 1 teams.*

Selection

The last attempt to change the concept of selecting a gear in a racing gearbox, away from the system of face-dogs was, to the best of my knowledge, the Lotus "queerbox" fitted to the T78 on occasion in 1977. Based on a principle used in the Gogomobile bubble car, the gears were selected by balls being forced out through holes in the shaft, into grooves in the internal bore of the gear to be driven. The arrangement resulted in a very short gear cluster, having removed the need for gear face-dogs and dog-rings and probably would have worked well with a hydraulic gear change system to provide the selection load. As it was, the feedback through the gear lever gave the drivers tennis elbow, and I distinctly remember the day at Snetterton when the gearbox selected a gear by itself while the engine was being warmed up. The car wandered off across the paddock, chased by the mechanics.

The revolution from a manual H-pattern selection mechanism to a semi-automatic motorcycle-type sequential selection was accomplished in two stages. Again, Ferrari pioneered the semi-automatic system but did so originally applied to three individual hydraulic actuators driving the selector rods, relying on software (and classic mechanical interlocks) to avoid the wrong gear being selected and multiple gear selection. The sequential system of a ratchet-driven shaft that drives the rods via cams ensures that only one gear can be selected at a time and in sequence. The system lends itself ideally to a semi-automatic control system and a single actuator.

The development of computer-controlled, power-assisted gear-changing is probably the single most significant technology to come from Formula 1 in decades. As it spreads to other racing formulae, it is being actively developed for road cars as the solution to the problem of how to combine automatic gear-changing with high transmission efficiency. Power trains of the future will require a fully integrated and optimized engine and transmission to meet emissions and fuel

* Since this section was written, the FIA has issued a clarification of Article 9.7 of the Technical Regulations that effectively prohibits the use of torque-steer differentials, as described here.

consumption targets that can be achieved only by divorcing the driver from the control of the system. Because the technology is right for racing, as well as the wider context, it is one of the few control technologies to have survived changes made in the 1994 Formula 1 Technical Regulations that banned "driver aides."

Whenever we look back at the development of any successful technology, we may wonder why it was not thought of in the past and why it took so long to occur. The technologies needed to implement electrically switched gear-changing have been available since the 1950s, even if all of the sophisticated enhancements and checks, performed by a digital computer, only became practical on a race car in the 1980s. We now see all of the advantages of hands-free gear-changing, 0.02 second upshifts (approximately five times faster than manual gear-changes), carefree downshifts thanks to over-rev protection (which pays for the whole thing by itself, by the avoidance of over-revs at more than $150,000 per enforced engine rebuild), and precision gear-changes that minimize the disturbance of the car when carried out while cornering at the limit. However, these were not the reasons that John Barnard sought to initiate it when he first joined Ferrari in 1986.

Traditionally, a significant part of the car was affected by the need to transfer motion from the driver's hand, back past the fuel tank and the engine, to the gearbox. Cockpit dimensions, monocoque design and structure, fuel tank, engine and exhaust layout, and rear suspension all had to take into account the need for an articulated rod, with linear and rotary motion, to pass through or by them. At Ferrari in the 1980s, the engine came first, second, and third; the gear-change had to share the remaining space with everything else. Offended by such compromises, Barnard set Ferrari and the associated R&D resources of Fiat to the task of devising a system of electrically signaled gear-changes, commanded via steering-wheel mounted switches and actuated by hydraulic cylinders attached to the selector shafts.

A small gear pump, driven off the end of a camshaft, supplied oil to the hydraulic system. Solenoid valves controlled high-pressure oil to three double-acting actuators that drove the selector shafts. A conventional, mechanical interlock system ensured that only one shaft was operated at a time. An adaptation of the 1987 gearbox was tested at the end of that year, utilizing fully automatic upshifts from the first runs. This convinced Barnard to press forward with a system for the 1988 3.5-liter V-12 engine Ferrari 639. The car was never raced, Ferrari persevering with a 2.5 bar, turbo V-6. It was not until mid-1989 that the system raced in the Ferrari 640.

As the advantages of such a system became apparent when allied to the full capabilities of a computer, the speed of development increased as other top teams strove to catch up to Ferrari's lead. High-speed automatic changes, which were permitted at that time, obviated the need to skip-shift when changing down through a large number of ratios. Suddenly, the motorcycle system of sequential gear changing appeared very attractive as a system for automatic or semi-automatic control. Developed for operation by a foot, which cannot easily control motion in two planes as a hand can, all it required was a single actuator to provide rotary motion of the selector barrel or drum. With inherent mechanical sequencing preventing two or more gears from being engaged simultaneously, the simplification was obvious and Williams quickly spotted the potential of this approach, which became the standard for a time.

Selection Mechanisms

The selector forks (three in the case of a six-speed gearbox, four for a seven-speed) require linear motion to slide the dog-rings in and out of engagement with the gears. They must be selected in turn and mechanically linked, such that one can be moved into gear only when the others are all in their neutral positions. With a conventional H-pattern gear-change, across-gate motion rotates the gear-linkage rod to swing the selector finger through slots in the rods on which the selector forks are mounted. When the finger is aligned with the appropriate rod, fore and aft motion of the gear lever is transmitted, via linear motion of the linkage, finger, and selector fork, to move the dog-ring into mesh with the selected gear. Detents hold the selector forks in the required position.

With a sequential change system, pegs on the selector forks or their rods engage in a circumferential groove in a cylinder (the barrel or drum) (Figs. 6.8 and 6.9). The groove is zigzagged for approximately one-quarter of its circumference. As the cylinder rotates and the peg engages the zigzag portion, the selector fork is driven first one way, meshing the dog-ring with one gear, and then the other to engage the next gear. Further rotation returns the fork to the neutral position. There can be either three grooves spaced along the cylinder (one for each selector fork, with the forks sliding along a shaft) or one groove (with the selector shafts' pegs disposed around the circumference). In the first case, the zigzags in the three grooves are arranged such that the selector forks engage their gears in the appropriate sequence. In the second case, the zigzag in the single groove engages each of the three pegs in sequence. The cylinder is detented into the engaged-gear and neutral positions (a neutral exists between each gear), and only one gear can be engaged at a time.

With a manually operated sequential gear-change, barrel rotation is driven by a ratchet system. In this system, forward strokes drive the barrel one way (sequencing up through the gears) and rearward strokes rotate it the other way (sequencing down through the gears). It is impossible to determine at the gear lever which gear is engaged; thus, some form of indicator is required. The easiest way to understand such a system is to strip down an old motorcycle gearbox and inspect the mechanism closely.

Actuator Systems

The motion and forces needed to operate the selector forks can be supplied by electro-hydraulic, electro-pneumatic, or electrical means. A comparison of the power densities of the systems weighs in favor of the electro-hydraulic approach, with pneumatics not far behind it. Although modern ceramic magnets have helped the size and weight of the electrical system, I know of no

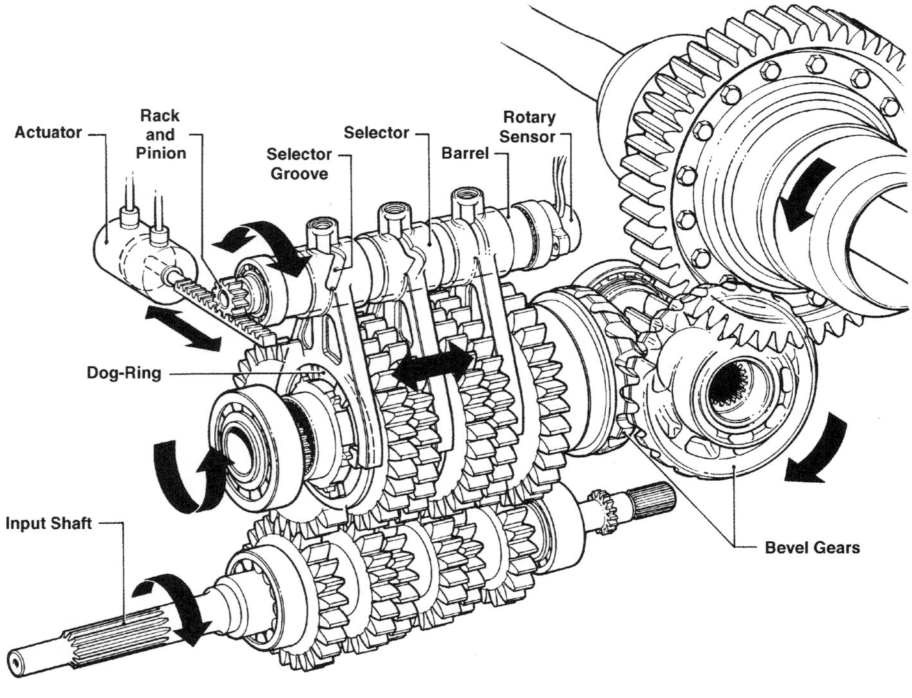

Fig. 6.8 Semi-automatic, sequential selection mechanism.

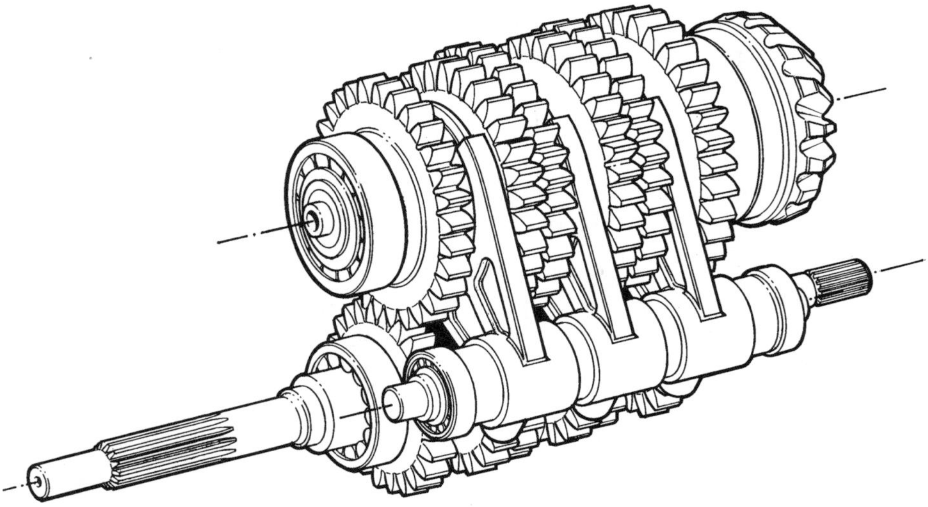

Fig. 6.9 Typical gear-change barrel arrangement relative to the two gear shafts.

applications in Formula 1, although Ferrari did consider it. Pneumatics avoids the need for fluid tanks and coolers. Likewise, leaks are not a disaster, and the compliance of the working fluid (air) has a significant effect on the characteristics of the change sequence—it is similar to a manual change. Although Tyrell pursued this approach, the lower pressure involved means that all components are larger than when hydraulics are used.

Originally, Ferrari fitted a double-acting hydraulic actuator to each selector shaft, each with its associated hydraulic solenoid valve(s) and position sensor. Sequential mechanisms require only one double-acting actuator, one valve, and one position sensor to control and sense the selection of all gears. If a ratchet system is used to convert linear motion into rotary motion of the barrel, the ratchet lever actuator can be driven open-loop. This allows a simple solenoid valve to be used; however, it is more typical to fit a Moog electro-hydraulic servo-valve because it permits precise control over the acceleration and velocity of the actuator. Alternatively, the barrel can be driven by a rotary actuator—either a linear actuator driving a rack and pinion, or a vane type. This must be controlled closed-loop, with the exact position of the barrel for each selection movement controlled by the servo-system. This approach requires much greater precision (e.g., sensor drift with temperature would be a disaster), but it gives greater control over the selector forks and is ultimately probably faster.

To watch a gear cluster, set up out of the case and in a rig, being cycled through the gears from first to seventh gear and back to first gear in less than one-fifth of a second, is an impressive sight. The selectors blur, and the process sounds like a quick burst from a Gatling gun.

Dog-Rrings

Detail design of the face-dog clutches that lock the selected gear to its shaft when slid into mesh with the matching dogs on the face of the gear wheel has undergone significant change to meet the requirements of high-speed engagement. The number of dogs has been reduced from generally eight to four. The dogs themselves have not increased in size; the freed space has been left open to increase the ease of engagement. Undercuts have reduced to speed disengagement under load; the pent-roof detail, used as a crude form of syncromesh in manual changes, is gone. The slow speed of a manual change did not allow the dog-ring to move far enough into engagement when there was significant differential speed between gear and dog-ring. The hydraulics move the dog-ring so quickly that quite high miss-synchronization can be accommodated, provided the space between the dogs is large enough.

Clutch Control

Controlling the clutch requires another Moog valve and a position sensor. If the clutch pedal is retained, concentric hydraulic cylinders operating in parallel are used as the slave cylinder. If the driver prefers, the pedal and master cylinder are dispensed with, and a steering wheel lever controls the clutch position, proportional to the lever position.

Without tactile feed back through the lever, clutch takeup during a standing start is difficult to sense by the driver. Wear in the clutch also will vary the takeup point, and the control system must sense and compensate for it.

Hydraulic System Requirements

The gear-change is only one demand on the hydraulic system, which may include fly-by-wire throttle, clutch, power steering, and differential. Although the peak hydraulic flow during a gear-change may be quite large, the mean flow is small and is determined mainly by the leakage flows inherent in Moog valves. However, the system must be capable of working when the engine is only ticking over (approximately 3000 rpm) so that gears can be engaged in the pits or during pit stops. The system requires a small pump (to meet mean flow) and sufficient accumulator capacity to meet peak flow, which is determined by the likely number of retries (subsequent attempts to engage a gear after a balked change—see the following discussion).

A hydraulic system model—calculating pump input, system leakage flows, and actuator demands to determine the system pressure based on the accumulator status—allows optimization of tradeoffs among pump size, gearing, and accumulator capacity. Data from track testing (engine rpm and actuator velocities—throttle, clutch, steering, and gear-change) are input into the model, and the calculated pressure and accumulator status are output.

Sensors

To perform the gear-change sequence, the rpm of the two shafts of the gear cluster, as well as the position of the selector barrel, must be known. In addition, to control the clutch slippage, the values of the two rpm's across it are required, as well as its position. These requirements are met by measuring engine rpm, gearbox input shaft speed, and the final drive speed (mean of the speeds of the rear wheels), and factoring them according to the intermediate, step-up/step-down gear ratios.

Rotation and position sensing uses the same technology sensors found elsewhere on the engine and chassis, although the temperature requirements are high (greater than 150°C [302°F]), especially for sensors mounted near the clutch.

The barrel position sensor is used, via the computer, to display to the driver which gear is selected.

Switches

Twin rocking paddles, mounted behind the steering wheel and which can be pulled by the driver using his fingertips, have become the standard method of commanding gear-changes. The paddles operate twin micro-switches, or no-moving part proximity switches, to command either an upshift or a downshift. Springs and snubbers, as well as the geometry of the rocking pivot, provide the right tactile qualities—as important as those of a computer keyboard, which is also a set of manually operated, high-speed micro switches. Software is used to ignore switch bounce, again as with keyboards.

A button or buttons on the steering wheel may be allocated to command skipped downshifts (e.g., sixth to second), as allowed by the regulations. (See the following discussion.)

The clutch lever drives a position sensor. Again, tactile qualities are all important, and many drivers have difficulty controlling the clutch during starts because no force feedback is possible. The driver does not receive the force cues telling him when the clutch is about to bite, which he is used to receiving through the clutch pedal.

Regulatory Restrictions

Given a free hand by the regulations, the control computer would automatically change up when the rev limit was reached and change down to a gear, either selected by the driver or preprogrammed into the computer, based on where the car was located on the circuit, as the speed reduced. However, the regulations insist that the instigation of an individual gear-change must come from the driver, as occurs with a manual change, and the computer can control only the execution. Thus, each upshift and downshift must be as a direct result of a switch signal. *

If the computer calculates that a downshift will cause the engine to exceed a set rpm limit, it can reject it; however, the driver must repeat the command for the change. The computer cannot wait until the speed has dropped sufficiently and then make the change. Holding down the paddle switch is not permitted (i.e., continuously asking for a downshift until it is within rev limits) and can be detected in the software as well as monitored by the overhead, onboard television camera.

Skip-shifting using a button is permitted, in the same way that a driver can shift with a manual change. Multiple downshifts, using this method, must be carefully judged because if the rev limit rejects them, the driver may arrive at a second gear corner while still in sixth or some intermediate gear.

Gear Change Strategy

Upshifts

A manual upshift takes approximately 0.1 second, during which time the car will decelerate at up to 1g due to high-speed drag, losing approximately 3.5 kph (2.2 mph) in the process. The earliest attempts at semi-automatic gear-changing reduced this time to 0.04 second, and now 0.02 second is possible.

---

* See author's note on page 290.

A seven-speed gearbox must have its ratios distributed to cover a range of speeds from 150 kph (93 mph) (first) to 350 kph (217 mph) (seventh), i.e., an average of 33 kph (21 mph) per step. Connected to an engine with a useful peak rpm of 18,000, the rpm drop at each change will be approximately 1500–2500 rpm, with the smaller drops likely to occur in the higher gears.

Upshifts are carried out clutch-less and at full throttle. To slow the engine, the ignition is cut, the start of which is timed to coincide with disengagement of the gear. The small residual undercut on the dogs would prevent the dog-ring from being pulled out of mesh if full torque were still applied to it. With the throttle open, the engine slows at the rate of more than 30,000 rpm per second and during the 0.02 second time that power is off will drop around 600 rpm. Thus, when the next gear is engaged, the gears are not synchronized and the residual energy in the rotating engine is transferred to the rear wheels. This is good for acceleration, but it excites the transmission system. The resulting disturbance of the tire contact patch can upset the car. How quickly the upshift is made is a tradeoff between acceleration performance and car stability. In the lower gears, generally upshifted while still in a corner, the balance may tend toward a slower, smoother shift; the changes from fifth to sixth and sixth to seventh, generally made in a straight line, will be as fast as possible. All this can be programmed in the software. It is rumored that some engine manufacturers have fitted exhaust backpressure valves—possibly to apply braking to the engine during upshifts and so increase the rpm drop during the gear-change.

Downshifts

Downshifts are generally performed with the throttle closed, and the clutch is used to unload the dogs from the overrun torque of the engine. When a down change is commanded, the software first calculates the rpm at which the engine would be turning if the new gear were engaged. If this exceeds the set rpm limit, the gear-change is rejected and the driver is informed. If it is acceptable, the sequence proceeds. While in neutral, the drive-by-wire throttle is "blipped" (separate "blippers" are used in the absence of drive-by-wire) to increase the engine speed and to synchronize the gear speeds prior to engagement. There is generally no great hurry to execute down changes during braking; if they require 0.04 or 0.05 second, car performance is unaffected. The engine speeds up at more than 50,000 rpm per second, and thus 0.04 second will provide a 2000 rpm increase. The last thing the driver wants while braking hard is for the rear wheels to momentarily lock due to a miss-synchronized downshift.

Nigel Mansell, driving the early Williams semi-automatic gearbox, discovered the benefits of full-power downshifts. Entering a flat-out corner in, say, fifth gear, he would select fourth as the speed was scrubbed off, without lifting the throttle. The change was performed smoothly enough to avoid disturbing the car and gave him the right gear to accelerate out of the corner. The control algorithm must be able to deal with this case, particularly the period when the gearbox is in neutral, with the engine at full throttle.

The only time the regulations permit the engine ignition, throttle, and clutch to be under the sole control of the computer is while actually executing driver-commanded gear-changes.

Retries

The timing and phasing of the various operations that make up a successful gear-change are determined experimentally and depend on the dynamics of the sensors and electronics, the actuation system, the engine, and the transmission. Once set, they will permit repeated perfect gear-changes until some part of the system changes and one or more of its characteristics goes out of limit. Then, miss-shifts will begin to occur.

Occasionally, dog hits dog square on, and the gear does not engage. The rhythm of the sequence is lost, and a "retry" algorithm is instigated. This involves withdrawing to neutral, adjusting gear speeds with throttle and ignition cuts, and retrying to engage the gear. More than one retry tends to signal something going wrong with the actuator, the mechanism, or the state of the dogs. Statistics for retries are logged and provide a good indicator of the state of health of each gear, upshifting, and downshifting.

Neutral

With a neutral between each gear, plenty of potential exists for the driver to lose his way in the gearbox when coming to a halt, especially if he is spinning. A button on the dashboard or steering wheel commands the barrel to the neutral position between first and second gears.

The regulations demand that the track marshals are able to put the car into neutral. This is achieved by opening an accumulator that pressurizes and disengages the clutch.

Reverse

A separate actuator and solenoid valve are generally used to engage reverse, with software ensuring that it is not actuated unless the other gears are in neutral. Alternatively, a cable operated reverse is used for simplicity.

Data Logging

Diagnosing a failure in a process that takes only a few hundredths of a second and which is reported by a driver who is concentrating on driving the car when it occurs would be almost impossible without modern data logging. Apart from the statistics of retries, gear-change sequences that result in one or more retries lasting up to 0.05 second or so are often logged with sensor and computer outputs, recorded at a 1 kHz sampling rate. The information this record provides to the engineers will, with experience of the system, allow them to diagnose where the fault lies—mechanical, control system hardware, sensors, or software—and to monitor the health of the gear-change system.

Clutch

The ability of the carbon clutch to continue to shrink in diameter and weight, while transmitting ever greater power, is one of the lesser known phenomena of Formula 1. The current 4.0-inch-diameter AP Formula 1 unit (Fig. 6.10) must be close to the limit, and it is unlikely that engine designers will be able to lower the crankshaft centerline much more. The tiny clutch would not appear out of place on a 250-cc motorcycle engine. Clutch duty cycles have been made slightly easier with the advent of computer-controlled gear-changing, where the clutch is not operated at all during up-changes and is controlled precisely by the computer on down-changes. The start takes clutches to their limits of temperature, and they are still expected to accommodate the bulk of the slipping needed to maintain engine rpm when leaving the start line. Tires lose grip if slipped too much, while the carbon plates do not, unless their surface temperature becomes too high. Ultimately, the thermal capacity of the clutch determines where the size limit lies.

Figures 6.11 through 6.13 show three types of gearboxes.

Future Developments

There are three main objectives in future developments of the gearbox: (1) smaller/lighter/stiffer, (2) more reliable, and (3) more efficient. As with most things in engineering today, progress can

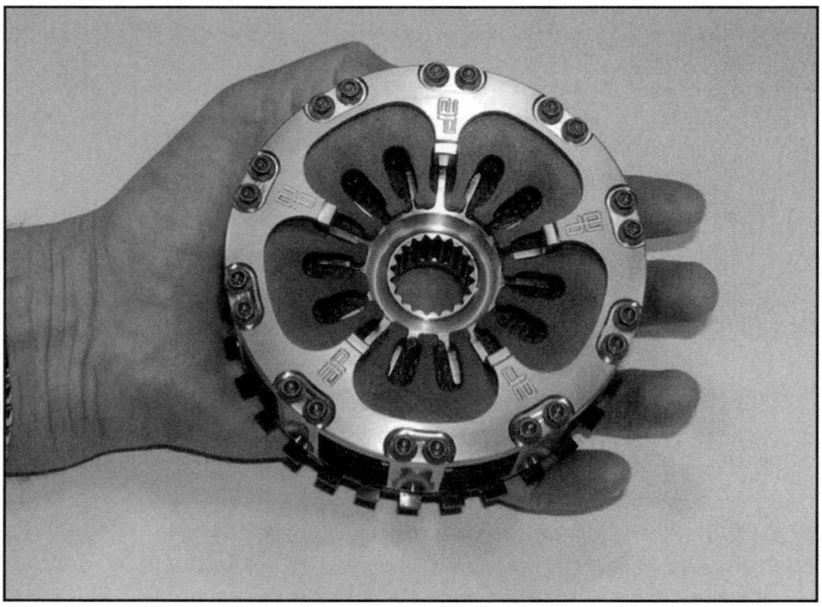

Fig. 6.10 A 4.0-inch-diameter, carbon-carbon clutch. (Courtesy of Automotive Products Ltd.)

Fig. 6.11 Jordan gearbox, complete with rear-impact absorbing structure. (Courtesy of Sutton Motorsport Images)

Fig. 6.12 Williams FW21 gearbox, installed in the car with rear suspension fitted.

Fig. 6.13 Ferrari F300 gearbox. Construction is a combination of titanium and CFRP.

be made only if a sound database of factors exists, such as speeds, loads, temperatures, and pressures involved in the system. Speeds, temperatures, and pressures have been measured for some time, but the measurement of loads in a gearbox has not. The engine companies also want to know loads, particularly the output torque from the engine, and gearbox input shafts are now fitted with torque sensors. With the addition of the torque in the drive shafts, the gearbox efficiency at any combination of speed, load, ratio, and temperature can be established. This is the raw data needed to start real development to improve efficiency. Bearings, gear tooth profiles, lubrication systems, and oils all will be affected.

To establish the loads seen by the components and to develop them for lower weight and longer life is somewhat more difficult. The engine generates torsional oscillations, superimposed on the steady-state torque, at a variety of frequencies. The gearbox teeth impose their own torsional oscillations. Coupled with the torsional flexibility of the shafts and tires, the backlash in gears and dogs, and the slip characteristics of the clutch and tires, these oscillations excite a multi-degree-of-freedom torsional system. Modeling it and validating the model with real data are the keys to analyzing the potential for reducing the size and weight of components and evaluating alternative layouts.

This technology exists today but is by no means trivial. The transmission of a Formula 1 car is as complex and highly loaded as the transmission of a helicopter. Even given the resources of the aviation industry, the hot rotating bits fail most often.

Chapter 7

Tires

Oh, eye of newt and leg of toad—things like that.
—Goodyear compound chemist, on being asked what ingredients go into a race tire compound

They may appear to be merely round and black, but tires are arguably the most sophisticated component on a race car. They provide the only points of contact between the car and the road, and all forces generated by the car, except aerodynamic drag and side force, are reacted at the four small contact patches where the rubber meets the surface of the track. Tires are one of those components of a race car that, regardless of how long and hard you look at them, you cannot possibly figure out how they work. With only the width, diameter, and wheel size (regulated to 330 mm [13 in.] in Formula 1) to compare, even an experienced engineer really cannot determine the difference between one design and another. Without the serial number molded into the sidewall, the tire engineers themselves are unlikely to be able to identify the many differences.

How a Tire Generates Forces

Any attempt to describe how a tire works, supporting the car and transmitting the accelerating, braking, and cornering forces to the track and providing the means by which the driver can control its motion, quickly resorts to diagrams, graphs, and complex equations that are beyond the scope of this book. The serious student of vehicle dynamics must turn to the technical literature for a complete analysis of tire mechanics. A good place to start is *Race Car Vehicle Dynamics* by William F. and Douglas L. Milliken (Ref. 7.1). However, to understand the fundamental principles of how tires transmit these forces, only two mechanical concepts must be understood. The first concept is that any structure, whatever it is made of, deflects when a load is applied to it (i.e., everything is a spring). The converse of this is that if a structure is deflected or deformed, it will generate a resisting force. A tire is a complex structure, which if deformed appropriately, will generate a force either vertically, longitudinally, or laterally, or in some combination of all these and moments about the three axes.

The second concept is friction. When two surfaces are pushed together, they will withstand a force that tries to make them slip over each other, up to some peak load, above which they will slide relative to each other. Conventional, mechanical theory states that the peak sliding force is proportional to the load holding the surfaces together and, for most materials, does not exceed that load (i.e., the coefficient of friction is less than 1). However, tires do not obey that law because they possess very complex "friction" properties, more of which will be discussed here later. However, for purposes of analyzing tire mechanics, it is easiest initially to assume they obey the laws of friction.

Let us first examine how the tire supports the vertical load. It is similar to a balloon—a pneumatic spring with a flexible membrane container. Increase the load, and the pressure will increase due to a reduction in the volume, partially compensated for by the deformation of the elastic skin. The "skin" of the tire is much stiffer than that of a balloon, but the principle is the same. If the load is applied to a flat surface, the area over which it acts also increases until again the pressure × area = load, and equilibrium is restored. The near-vertical sidewalls of a tire also contribute to the load carrying and behave similarly to a pair of vertical springs, complicating the analysis.

Next is the longitudinal load. The vertical load has provided a contact patch between the tire and the road, rather than a point or line contact, which would occur if the tire and the road were both infinitely rigid. If the wheel did not rotate, the tire would generate a longitudinal (or lateral) force reaction in the same way any stationary object sitting on a surface would do. It would resist movement until the friction force was overcome and then slide. This is what happens when a driver locks up a wheel under braking. The rotating wheel and tire behave rather differently.

Figure 7.1 shows an enlarged section through the tread, at the contact patch of a tire generating a tractive force. For the longitudinal load to be carried by the tread (tangential to the surface of the tire), it must deform—in this case, in shear. However, at the leading and trailing edges of the contact patch, it carries no load because it does not contact the road. Therefore, at these points, no shear deformation can occur. The deformation develops progressively back from the leading edge and in doing so creates a relative movement between the top and bottom edges of the tread. This means that the top and bottom of the tread are moving at different speeds where the tread is in contact with the road.

At the trailing edge, as the tread starts to lift off the road and the load on it decreases, the shear reduces and the bottom of the tread must readjust to the speed of the top. This cannot happen instantaneously at the trailing edge; therefore, a progressive slip between the tread and the road occurs, facilitated by the reducing contact patch load as the tread peels off the road. As the tractive load increases, the shear deformation increases, the relative speed between top and bottom of the tread increases, and the area of slip increases, moving forward in the contact patch. When it reaches the leading edge, no further increase in tractive force can occur and the wheel starts to spin. The speed difference in the tread is a measure of the

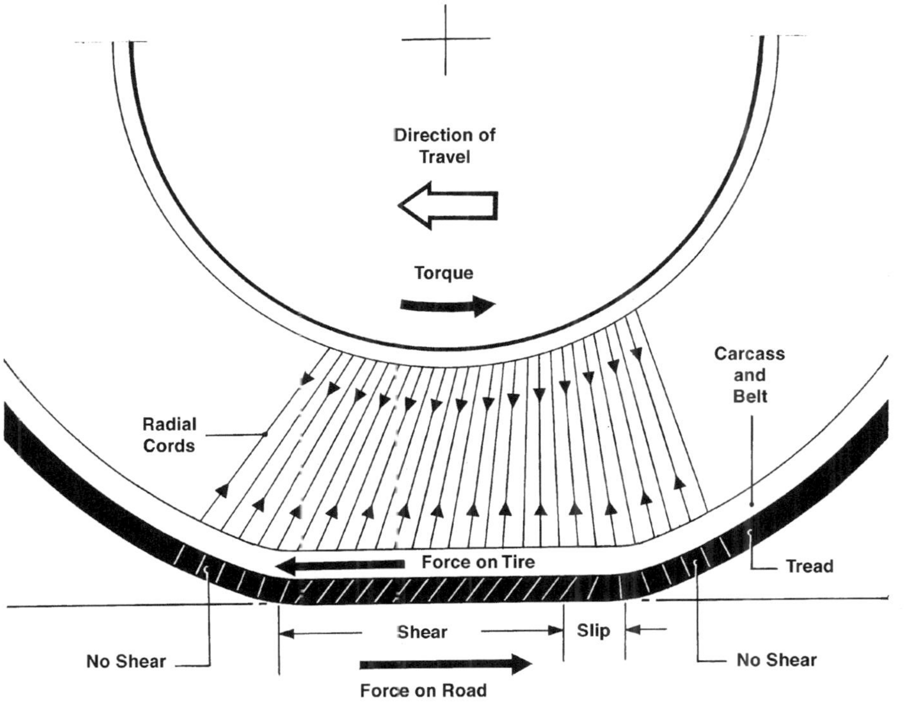

Fig. 7.1 Tire longitudinal force generation.

tractive force and contributes to the slip ratio, which is the ratio between the speed difference and the actual road speed. In the case of a braking force, the system is reversed.

The loads in the tread are transferred to the wheel via the sidewalls of the tire, which also deform to carry the loads. How they do so depends on the tire construction, which will be discussed later. However, this further deformation also adds to the slip ratio. It is slip ratio that is measured by ABS and traction control systems to determine when the wheel is about to lock up or spin.

The mechanism for generating lateral force is similar, with the tread being sheared across the contact patch, starting progressively at the leading edge and slipping on the road toward the trailing edge back to the undeformed state as the tread leaves the road again (Fig. 7.2). To generate this shear deformation, the direction in which the tire rolls must differ slightly from the direction in which the car is moving. The angular difference is the slip angle. As the wheel rolls, any given point of contact between tread and road is displaced sideways relative to the tire centerline, shearing the tread. As with the longitudinal case, the load in the tread is transferred to the wheel via the sidewalls, which likewise deform laterally. As the slip angle increases, the lateral shear load increases in the tread, and the area of lateral slip moves forward in the contact patch. Again, when it reaches the leading edge, no further load can be carried, and the tire starts to skid.

Lateral load can be generated in one other way. If the tire is cambered, camber thrust is generated by a distortion of the contact patch, caused by tilting the tire. When the tire rolls, the distorted patch is forced to "straighten," thus generating a lateral force. This effect becomes less useful the wider the tire. Sufficient camber to generate a useful force would overload one edge of the tire. On motorcycles, however, it is the primary mechanism for generating lateral force.

The tread of the tire transfers longitudinal and lateral loads to the road surface by a combination of mechanical interlocking and molecular bonding. Because the road surface is irregular, the rubber of the tread deforms into the irregularities and "grips" it. How hard it grips depends on the detail of the road surface, the hardness of the rubber (which itself depends on temperature), and its ability to avoid being sheared off, causing accelerated wear. Molecular bonding is akin to the way a Post-It® Note sticks to various surfaces but can be removed again. The bond is weaker than the adhesive, the paper material, and the surface material to which it is stuck. Similar to the Post-It Note, the tread rubber sticks to some surfaces better than others. The compound engineer must study the detail of the road surface, at both the macro and micro levels, to choose the best compound, taking temperature into account.

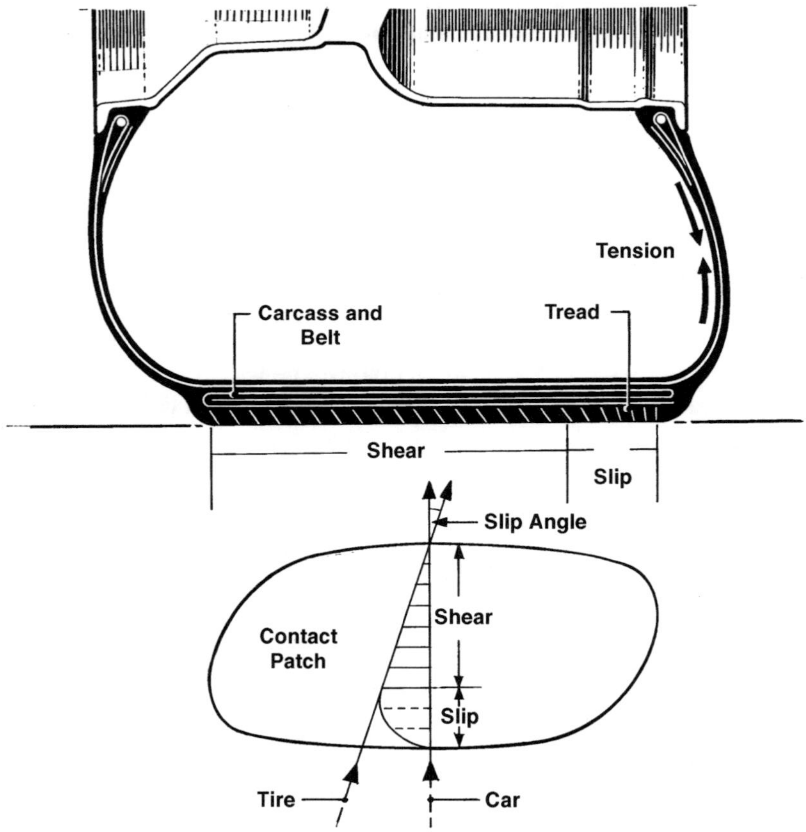

Fig. 7.2 Tire lateral force generation.

Typical Formula 1 race tires have a ratio of lateral (or longitudinal) load to vertical load of approximately 1.7. The qualifying tires of the early 1980s were as high as 2.2. That is much more than the typical coefficient of friction values for most engineering materials. For instance, the coefficient of friction for the carbon-carbon material used in racing brakes and clutches is typically around 0.6.

Tire Characteristics

Because race car drivers try to drive their cars in such a way that the tires are generating the maximum possible force in the appropriate direction for as much of the time as possible, the most important tire characteristics are the slip ratio at which maximum longitudinal force is generated and the slip angle at which the maximum lateral force occurs.

The slip ratio is the parameter that describes the magnitude of the slip between tire and road (10% slip ratio at 100 kph [62 mph] road speed means that the wheel and tire are rotating at 110 kph [68 mph] when accelerating or 90 kph [56 mph] when braking). The slip angle is equivalent to the ratio of the lateral velocity of the tire to the velocity in the rolling direction. Because force × speed = power, the greater the slip ratio or slip angle for a given force, the greater the power absorbed by the tire. Lost power is inefficient, slowing the car. Furthermore, although putting energy into the tire helps heat the tread rubber to its optimum temperature, too much heat will degrade the compound, causing loss of grip and eventual failure due to blistering.

As accelerating, braking, and cornering levels have increased with the development of aerodynamic download, the forces the tires must generate also have increased. A doubling of the peak force at any given speed must be accompanied by a halving of the peak slip ratio and slip angle to maintain the power absorbed and heat input to the tire at the same level.

Table 7.1 shows examples of this:

TABLE 7.1
POWER ABSORBED BY TIRES DURING ACCELERATION, BRAKING, AND CORNERING

Speed	g-level	Slip	Power Absorbed
100 kph (62 mph)	1.2g acceleration	3% slip ratio	9 hp
100 kph (62 mph)	1.2g acceleration	6% slip ratio	18 hp
300 kph (186 mph)	3.5g braking	3% slip ratio	78 hp
300 kph (186 mph)	3.5g braking	6% slip ratio	156 hp
240 kph (149 mph)	3.5g cornering	3° slip angle	108 hp
240 kph (149 mph)	3.5g cornering	6° slip angle	215 hp

As both slip ratio and slip angle increase, the forces increase proportionally with them until the limit is reached. At the limit, the force may continue to increase slightly, remain level, or decrease, depending on the tire characteristics. The slope of the linear section is the cornering stiffness of the tire and determines the sensitivity to the steering and the slip angle of the car. How smoothly and suddenly it flattens or falls off in the nonlinear region around the peak will have a great bearing on the feel of the car and how easy the car is to drive at or near the limit (Fig. 7.3).

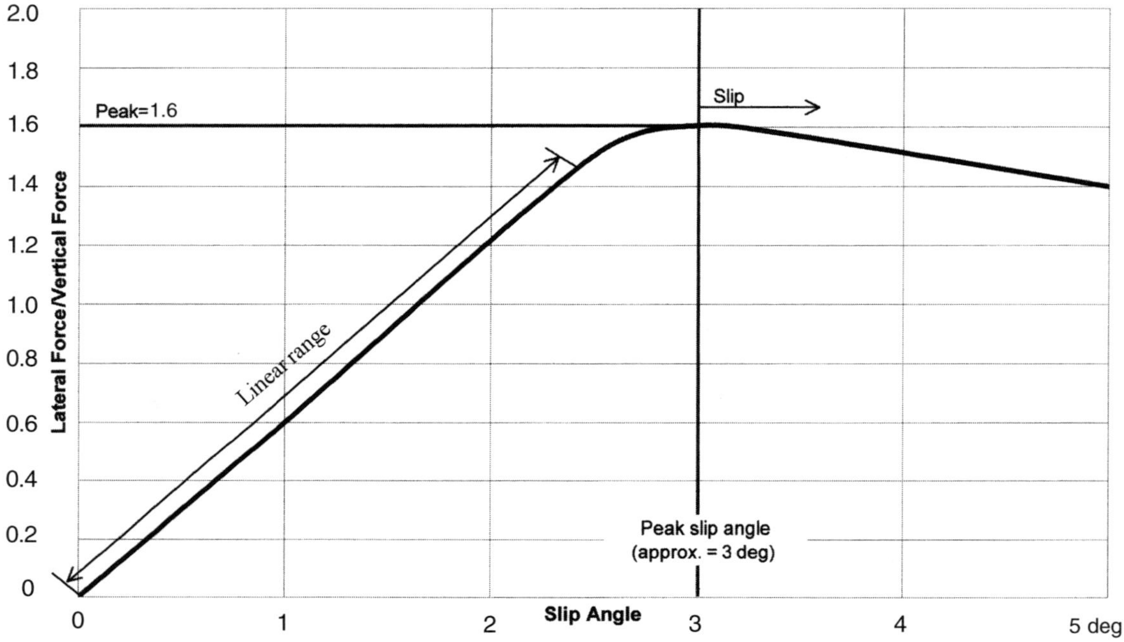

Fig. 7.3 Tire normalized lateral force versus slip angle characteristic. (Author's illustration)

Adding power or braking forces while the car is cornering means that the contact patch must do two jobs, and the available maximum force must be shared between them. The effect is to reduce the slope and peak of the lateral curve, roughly proportional to the magnitude of the longitudinal force. In reality, the relationship is more complex. For a more complete analysis, see Ref. 7.1.

The other tire characteristic of interest to the chassis engineer is the sensitivity of the tires to vertical load. Conventional friction theory would predict that the peak longitudinal or lateral load would increase proportionally with vertical load. However, in practice the peak load increases from very low loads and then falls again as a ratio of vertical load. There is a limit to the loads that the tire will carry. This characteristic is used by the chassis engineer to tune the cornering capability of each end of the car, by using the anti-roll bar to load up the outside tire on that axle and relieve the load on the outside tire of the other axle.

Construction

Tires are made of composites—very complex composites. Reinforcing fiber of carbon, Kevlar, glass, steel, manmade textile fibers such as nylon, or natural fibers such as cotton are laid up in the appropriate direction and bonded together in a matrix of rubber. The fiber materials used, the way they are bundled, their orientation, and the specification of the rubber are varied throughout the tire to provide exactly the desired properties for each part of it. To understand how in principle a racing tire is constructed, we can consider how the racing tire differs from the most simple tire.

If we took an inner tube and inflated it, the diameter of the ring it formed would increase as the pressure increased, and the diameter of the tube also would increase. It would neither stay on the wheel nor sustain a high pressure before bursting. The tire cover that is put over the tube must have two essential properties:

1. It must have bead rings that will support high circumferential loads to prevent the diameter of the tire from increasing with pressure. Steel wire is used in the beads.

2. It must have cords, stiff in tension but flexible in bending, running from bead to bead, to prevent the diameter of the tube from increasing. These are made from a variety of fibers—Kevlar or carbon for low weight in a racing tire.

If the wheel and beads seal properly, the rim of the wheel completes the sealed volume, and the tube can be eliminated (Fig. 7.4).

We now have a nicely rounded tire, capable of being inflated to high enough pressure to support the loads on it without excessive deflection. However, it is the wrong shape for a wide, racing tire. To provide a wide and flat tread, a cylinder of circumferential fibers embedded in rubber is placed around the tire (Fig. 7.5). This is known as the "belt," and again stiff, lightweight fibers are used. The belt must be capable of surviving continuous flexing, and Kevlar makes an ideal fiber for its construction. The belt not only supports the internal pressure loads but also the tread centripetal loads caused by the rotation of the tire.

The fibers running from bead to bead can run either perfectly radially (radial tire) or at an angle (the bias angle) zigzagging around the tire (cross-ply tire), as shown in Fig. 7.6. These fibers, usually twisted into cords, transmit the loads from the tread to the wheel. For the fibers to transmit torque and hence longitudinal load in the tire, they must have a bias angle. In a true radial tire, the tire twists on the wheel to generate a bias angle. Radial racing tires must

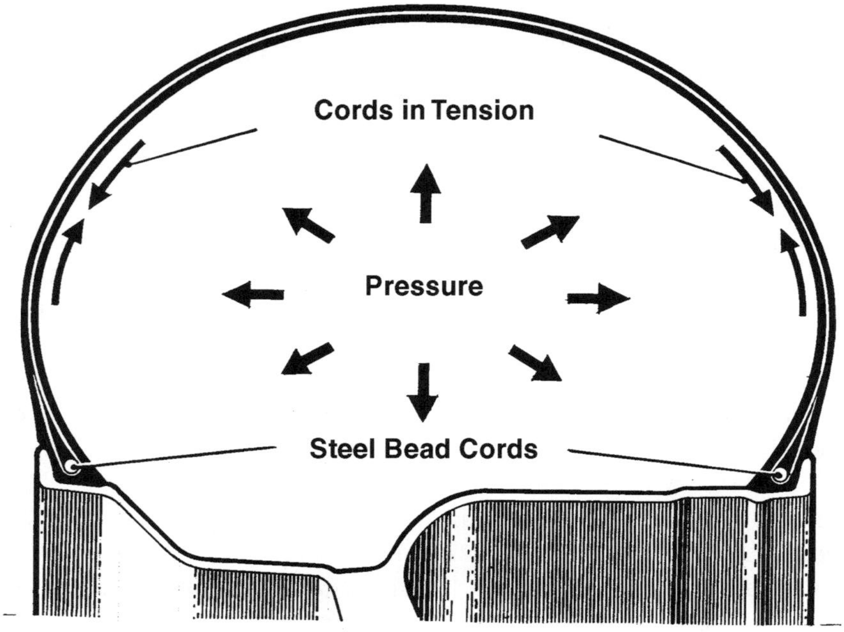

Fig. 7.4 Belt-less radial tire.

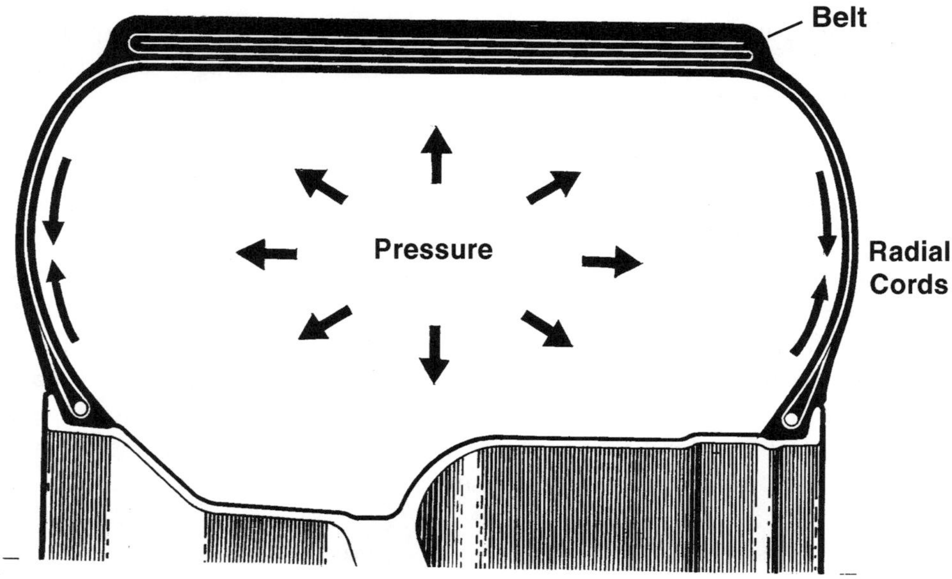

Fig. 7.5 Belted radial tire.

transmit large torques, and excessive twisting is undesirable because it can set up torsional oscillations as well as cause the peak slip ratio to be too high. Thus, even radial tires have an initial bias to the cords. Alternate layers of cords run at positive and negative bias angles in order to transmit both braking and accelerating torques.

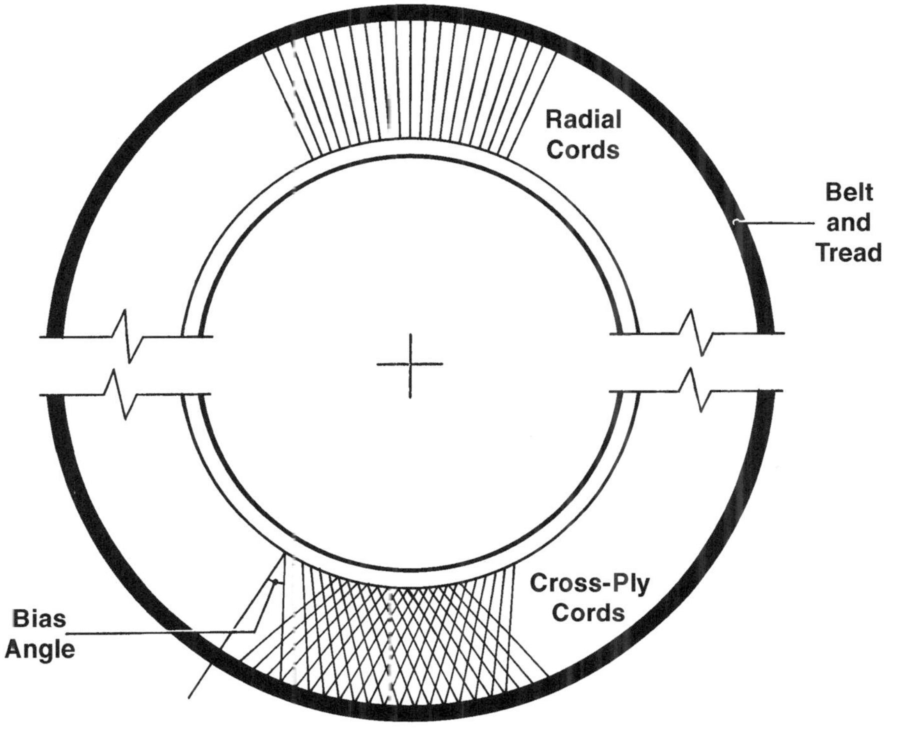

Fig. 7.6 Cord arrangement on radial and cross-ply tires.

The sidewalls of the tire, made up of the bead-to-bead cords, are not very good at transmitting lateral loads from the tread to the wheel, needing to deflect sideways a long way to provide a tension load path for the fibers. To stiffen the sidewalls as much as possible (and hence reduce the peak slip angle), they are kept as short as possible and additional layers of fiber-reinforced rubber are laid up on them (Fig. 7.7). The sidewalls now transmit lateral loads by bending as a cantilever beam. The graduated stiffness, from bead to shoulder, is carefully fine-tuned to give the tire the desired characteristics.

The tread rubber is wrapped round the belt.

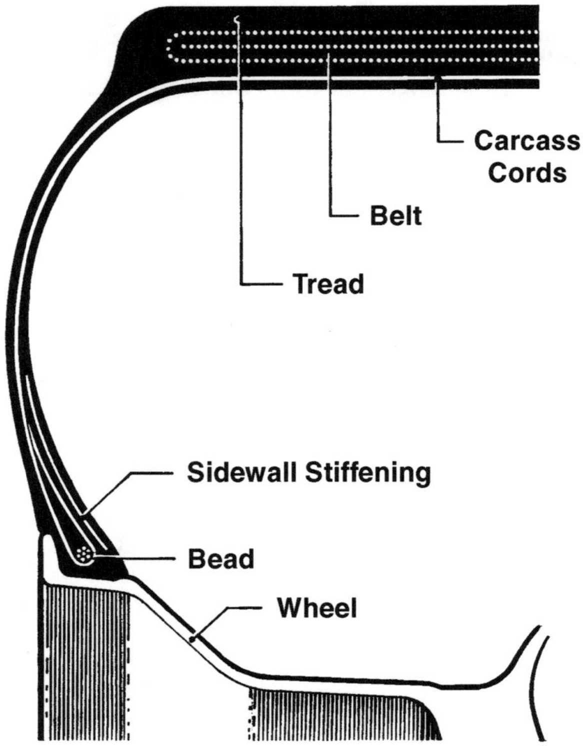

Fig. 7.7 Sidewall stiffening.

When the whole lay-up is fused together in a mold and cured to give the desired final properties, we have a structure that, after inflation to put all fibers in tension, has the stiffness characteristics and load-carrying capabilities necessary to transmit accelerating, braking, cornering, and vertical loads (weight and aerodynamic download) from the four contact patches to the wheels of the car, while rolling at surface speeds of up to 350 kph (218 mph) on a Formula 1 car.

The rubber compositions used in the construction of the tires are well-kept secrets. Each part requires a different formula, depending on whether it is tread rubber (friction properties dominate) or sidewall (stiffness and damping are most important). Not only does the power absorbed by the tire in transmitting loads heat the tire, but the flexing due to supporting vertical and horizontal loads while rolling also generates heat. It is desirable that the rubber in the carcass of the tire has some damping properties to damp out vibrations in the tire, but too much will cause overheating of the sidewalls—a disastrous and dangerous condition that often results in a sudden high-speed blowout of the tire. A delicate balance is required, and the rubber compounding chemists are regarded as magicians of tire engineering.

The tread rubber is formulated to run at a particular optimum temperature of approximately 110–130°C (230–266°F). If the temperature is lower, grip is degraded; if it is higher, the rubber will blister. The highest temperature is not generated at the surface unless the tire is spun or slides too much, in which case the surface is burned, but between the tread and the carcass, where the maximum distortion occurs. This is why tire engineers must measure tire temperatures with a needle probe. Persuading the tires to run at the optimum temperature is a major part of the work that tire and chassis engineers carry out together. Juggling factors such as load distribution, camber, toe-in, and damping to compensate for car weight changes and track and air temperature variations never ceases.

Wet Tires

To work effectively, racing tires depend on two things: (1) their tread must be pressed hard against the surface of the track, and (2) they use a tread rubber compound that generates a high coefficient of friction when running at its optimum temperature of 110–130°C (230–266°F). Water is a hydraulic fluid, a lubricant, and a coolant. It does everything in its power to stop a race car tire generating grip. The challenge for racing tire designers is to overcome the negative effects of rainwater where possible and to harness the properties of water to assist the working of the tire.

When a slick tire rolls along a track that has a layer of water on it, the tire must first push the water out of the way before the tread rubber can make contact with the track surface. The dynamic force needed to shift the water generates a pressure in the water layer which, if it is greater than the load per unit area in the contact patch, lifts the tread off the track. Because water is virtually incompressible, it must flow out of the way (Fig. 7.8). If we use the example of a 380-mm (15 in.) wide Formula 1 rear tire with a normal load of 5000N (1124 lbf) on it and an inflation pressure of 1.5 bar, the contact patch will be approximately 350 mm (13.8 in.) wide by 95 mm (3.74 in.) fore and aft. At the leading edge of the contact patch, the tread meets the layer of water, which must be persuaded to flow out sideways before the tread can make contact with the track. If this takes 2 milliseconds to clear, for example, then only the rear 40% of the contact patch will actually be in contact at 100 kph (62 mph). At 170 kph (106 mph), the water will not clear before the rear edge of the contact patch, and the tread will never actually touch the track—the tire aquaplanes. The greater the depth of water and the farther it has to move sideways (i.e., the greater the width of the tire), the lower the speed at which aquaplaning occurs.

If the tread of the tire has grooves around the circumference and transversely across the tire, the water on the track has a much shorter distance to move before it can flow from the area between the tread and the track into the groove (Fig. 7.9). If the grooves are large and numerous enough, they act as reservoirs that store water before discharging it behind the contact patch. If the grooves are insufficiently large to absorb all the water in the path of the tire, the water must flow along the grooves to exit from the contact patch, either behind it or at the side (Fig. 7.10). However, there comes a point where making the grooves larger reduces the remaining tread to such an extent that insufficient rubber remains to make useful contact with the track, and the remaining small tread blocks become overloaded.

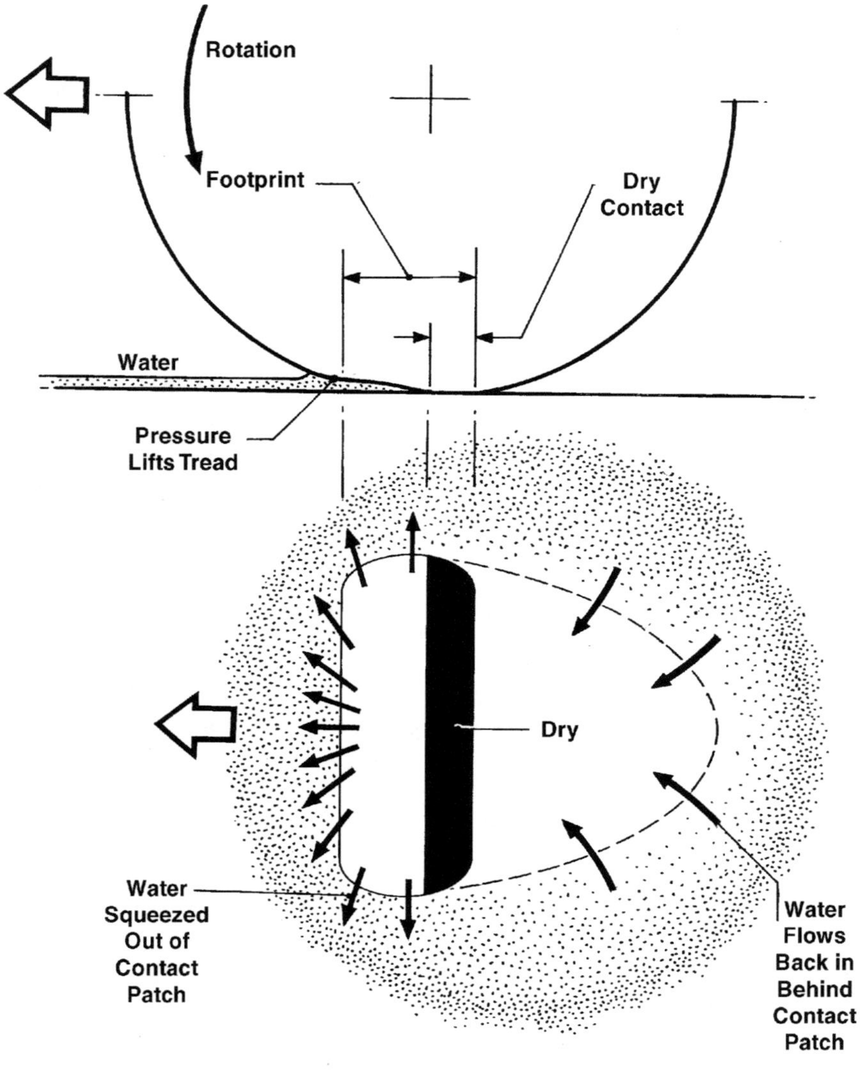

Fig. 7.8 Slick tire on a wet surface.

The size, position, and direction of the grooves is critical. Deciding how much of the water is to flow longitudinally to the rear of the contact patch and how much of it is to flow out sideways is the subject of much research. Traditionally, the movement of the water as the tire rolled along the track was visualized by fitting a glass window into the track, with an observation room below it. Automatic, high-speed photography of the tire contact patch as it passed over the window enabled the water flow conditions to be analyzed. The power of computers now enables the flow of water to be modeled, but it is an extremely complex dynamic system, and a good deal of black art is still involved. The solution will be optimum for only one water

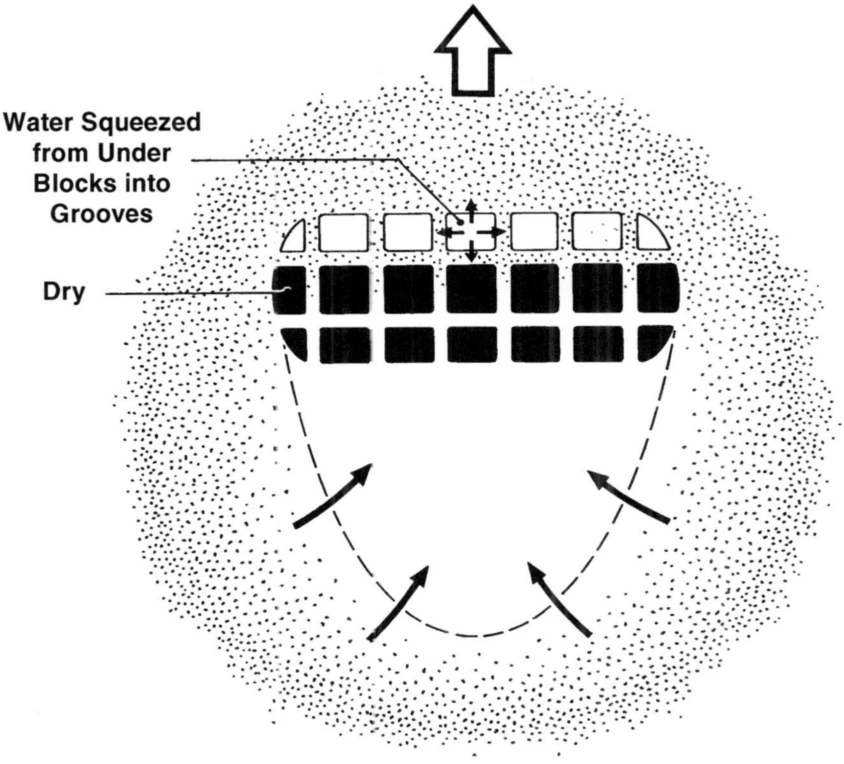

Fig. 7.9 Water absorbed by grooves in a wet tire.

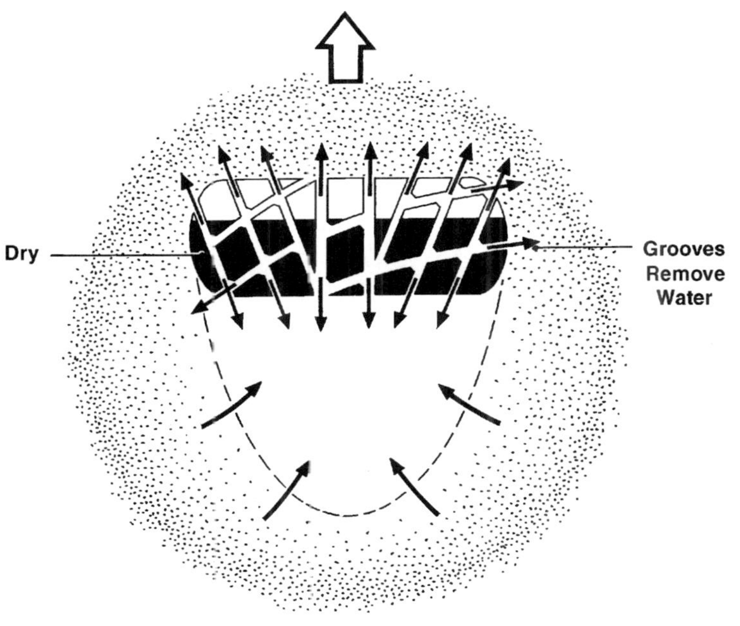

Fig. 7.10 Water ducted away from the contact patch in a wet tire.

depth and speed. Because of the inertia and viscosity of the water, moving the water out of the way quickly creates a pressure in the water film under the contact patch. This prevents the tread from contacting the track in the same way as with a slick tire. However, the speed at which the tire aquaplanes will now be much higher because the water is cleared from the contact patch in a much shorter time. Groove patterns may be asymmetric to compensate for the fact that the contact patch itself is asymmetric due to wheel camber. An element of styling is also present in the design of wet tire tread patterns, often related to the requirement for racing tires to provide image for the tire manufacturer's road car products.

Having removed the layer of water so that at least part of the contact patch is touching the track, the tread rubber is left with a thin film of lubricant (water) between it and the track surface. To develop the maximum possible grip, this lubricant must be removed. The mechanism employed is to wipe the track dry with the sharp edges of the tread blocks (Fig. 7.11).

As the tire slides across the track, the tread blocks are distorted and the pressure under the leading edge of the block is increased. If this edge is sharp, it wipes the track in a manner similar to the way a windshield wiper clears water from a windshield. The leading edge leaves the track dry for the remainder of the tread block to grip, and the following tread blocks benefit from the dry patch left by the preceding blocks. Cuts (sipes) may be made in the tread blocks to provide additional wiping edges.

Wet tire construction is softer than that of dry tires, and the inflation pressure is usually lower. Because the tire is not able to generate as much force in the wet, it does not need to be as laterally stiff. Soft construction and lower pressure also increase the size of the contact patch for a given vertical load, providing more time for the water to be removed and more tread rubber on the road.

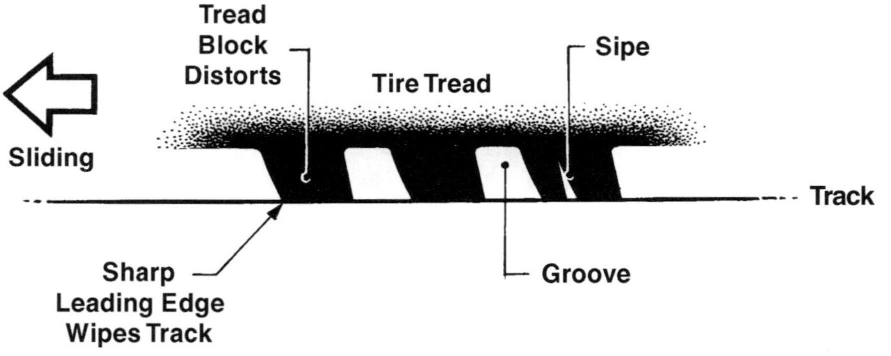

Fig. 7.11 The edges of a wet tire's tread blocks and the sipes wipe the contact patch.

The rubber compound used in the tread must work well at low temperatures because it is continuously cooled by the water. There is little chance of molecular bonding (sometimes known as "chemical grip") due to the water film. Thus, the rubber must be soft enough at low temperatures to lock onto the granular surface of the track ("mechanical grip"). Soft compounds such as these distort significantly under load and thus generate heat due to the hysteresis of the rubber. If the tread blocks are too small or the grip is too great (e.g., as the track dries), the blocks will distort, the sharp edges will wear quickly, and the rubber compound will overheat. Drivers can often be seen seeking the wet parts of the track on the straights to cool their soft-compound wet tires as the track dries.

The design of wet tires is a compromise. A wet track can be anything from damp to several millimeters of standing water, and even a combination of conditions due to "rivers" running across the track while elsewhere it is drying out. It is impossible to optimize the tread design and tread compound for all conditions and speeds. If the grooves are numerous and large enough to deal with standing water, the blocks will be too small to be stable in damp conditions. If the rubber compound is soft enough for a fully wet surface, it will distort and overheat if it is only slightly damp.

Tire manufacturers have developed a range of wet tires to deal with conditions ranging from heavy rain to a damp track. When it rains during a race, the driver and team must judge whether the conditions will worsen, remain constant, or improve, in order to choose the appropriate wet tire to fit to the car when they call it into the pits. To prevent the driver from having to change tires too often, the choice is made on the best estimate of how the conditions will change, and sometimes a wrong decision is made. A driver fitting an intermediate wet tire for light drizzle will have to make another stop if the rain becomes harder. A driver who remained on slicks until he sees how the rain develops before deciding to change to a full rain tire will make only one stop. A quick shower tests the mettle of both driver and team and is one of the wildest cards to be dealt in any race.

The track surface itself has a big effect on the performance of rain tires. An open or porous surface will drain away water, particularly when the pressure is increased in the contact patch. These special, porous surfaces are being developed for roads because they reduce both spray and noise. On the racetrack, they not only reduce spray, aiding visibility and safety, but also reduce the tendency to aquaplane. On any racetrack surface, the racing line builds up a layer of rubber, increasing the chemical grip of dry tires. In the rain, the wet rubber-to-rubber contact does not have a good coefficient of friction and there is little chemical grip; other, clean lines may give more grip. The great drivers in the rain—Senna and Schumacher stand out among drivers in recent years—always explore various lines as soon as it rains, seeking where the grip lies. They usually settle on completely different lines from the ones they use in the dry, and these are often very different to the lines taken by other drivers in the wet.

Retaining the handling balance of a car in the wet is difficult. The rear tires run partially in the tracks of the front tires, and under many wet conditions will experience a partially dried track, the water having been temporarily removed by the front tires. As the car corners, the tracks diverge due to the car taking a slip angle. As a result, the relative grip levels available, front and rear, may be very different. The changing balance of the car and continuously varying track conditions test the adaptability and skills of the drivers, providing an ideal playing field on which car-control skills can be displayed.

A Formula 1 car, travelling at 300 kph (186 mph) on a track that has 3 mm (0.12 in.) of water on it, must remove approximately 100 liters (26.4 gal) of water per second from the track surface. Much of this water is ejected sideways out of the contact patch; the remainder is entrained in the tread grooves to be flung tangentially from the rear of the tires. All this water is picked up by the turbulent, upward-flowing aerodynamic wake of the car and is a serious visibility hazard for following cars. The work done on the water, in moving it quickly out of the way of the contact patch, also increases the rolling resistance of the tires—by a factor of 10 under some conditions.

Grooved Tires

All the hard-won data and experience of the race performance of tires on individual tracks was rendered obsolete in 1998, when tires had to have circumferential grooves in the tread—initially three but now four on the front, and four on the rear. The purpose of this regulation was to provide a "simple" means by which the FIA could adjust the performance of the cars to maintain speeds at 1996 levels. The need for a much harder compound was recognized immediately, as the newly created "mini-shoulders" on the edges of the grooves wore quickly. This is exactly what happens to the edges of the tread blocks of a wet tire, as soon as the track dries and the grip increases. With a width limit that affected only the rear tire at that time, there was a trend toward moving weight forward and widening the front tire to compensate. It was the only way to obtain greater tire width and hence more rubber on the road. In response to this development, the FIA regulated the width of the front tires and added the fourth groove to them.

Designing and Developing Tires

Finite element analysis (FEA), as applied to the analysis of tire design, is one of the most sophisticated applications of this technique. The engineering properties of a tire vary in each direction, as with all composites. They also vary within the thickness of the material. Added to the large deflections due to the low modulus of rubber, these non-isotropic properties make FEA one of the technologies that have raised tire design from a black art to a science.

Armed with a computer, the tire designer can determine not only the stresses and strains but also the shape that the tire takes when loaded. He can then explore the effect of small changes in construction and material properties. Dynamic analysis is used to determine the ways in which the tire will vibrate (mode shapes) and how best to control them.

All the characteristics of a new design are tested on rigs, the most complex of which is the tire dynamometer. In its most advanced form, it consists of a high-speed rolling road onto which the tire is placed at a variety of vertical loads and cambers. The slip angle and slip ratio are varied, and the loads are measured. Graphs of cornering and traction characteristics are plotted, as well as the effect of speed and load on rolling radius—parameters of consummate interest to the aerodynamicist because tire rolling radius is a major factor in the calculation of ride heights.

Tire dynamometers that can run at Formula 1 speeds and loads are not common. Goodyear uses the TIRF facility at Calspan, while Bridgestone has one of its own manufacture. Running a racing tire at or near its design limit for any significant time causes the tire to overheat quickly. Even if the hardest compound is used for testing, the tread tends to wear very rapidly. It is extremely difficult to maintain a tire under constant conditions (particularly temperature) while a full range of tests is conducted. The only alternative is track testing, but most chassis designers require measured tire characteristics for vehicle dynamics modeling purposes.

Much of the work conducted on tire dynamometers is durability testing. It is not acceptable for a tire to blow in track testing; the consequences are too dire. Many hours of high-speed running at the maximum or overload conditions of the tire are carried out. The results—superb reliability—speak for themselves.

Track Testing

The role of the chassis, as far as the tires are concerned, is to generate and distribute vertical loads among the four tires, to present the tires to the road in the desired manner, and to feed the resulting loads back into the chassis and to the driver. The aerodynamics, inertia, suspension, and steering kinematics, and differential and brake characteristics are all involved in an interactive way. Modeling is beginning to unravel the complexities of these interactions, but it will be a long time before it replaces track testing.

Matching the chassis characteristics to those of the tires is the essence of chassis tuning. To do without it is to rely on luck. To understand the importance, consider the equivalent process for an aircraft.

Tires are the equivalent of both the wings and the propellers (also effectively wings) of an airplane. They generate all the support, propulsion, and directional control forces. They even have similar characteristics: a linear range of lift against incidence, followed by the stall. This is almost an exact analogy to the linear cornering force versus slip angle of a tire, followed by skidding. The performance of tires installed on a car is influenced by the way they flex under load, again similar to the wings of an aircraft. Imagine designing an airplane without wings and then looking for these components among suitable suppliers, bolting on the components and then trying to fly, without knowing anything about their characteristics. No pilot would test fly that airplane! This is exactly what happens with most race cars and even many Formula 1 cars: the car is often designed without knowledge of the characteristics or even the make of the tires. It is then sometimes raced without any testing on the designated tires because these may be delivered to the track only shortly before the race.

The advantage of working closely with a tire manufacturer at both the design and testing stages is obvious. The ability to adapt the chassis design to suit the tires and, when appropriate, to adapt the tires to the chassis, is really the only way to optimize the combination. For this reason, it is important for a Formula 1 team to tire test with the supplier to gain prior knowledge of the development direction and to be able to influence it. Where two or more teams are supplied, the manufacturer must tread a middle path to develop tires to suit all their teams. However, the manufacturer will inevitably be influenced by its leading team's design philosophy and chassis requirements. In a monopoly tire supply situation, the middle path is easier to define. However, when a competitive tire supply situation exists, the need to win dominates. Purely by way of example, a tire supplier may strive to provide the best tire for Ferrari, the performance of whose car may demand a very different tire than Williams would need, for instance. If development follows the Ferrari requirements, the rest of the runners on that make of tire will have to make do with a tire construction that is not necessarily optimized for their possibly lower-performance chassis, or they must attempt to change the chassis of their car to suit.

Chapter 8

Aerodynamics

We tried it with tail in front, behind, and every other way.
When we got through, Will was so mixed up
he couldn't even theorize.
—Orville Wright, writing about his brother Wilbur

The potential of aerodynamics to increase the performance of a race car is now well established. The means by which downforce and drag have been traded off during the past 30 years to expand the "g-g-V" performance envelope of a Formula 1 car was described in Chapter 2. However, there is more to race car aerodynamics than generating the maximum downforce for the least possible drag. The development of ground effect heralded the arrival of pitch sensitivity, and the problem has never fully receded. It manifested itself initially as a pitch instability—a coupling of the heave and pitch modes of the car on its suspension and tires, excited by the change of aerodynamic forces generated between the underside of the car and the ground, similar to structural flutter in aircraft. The problem that remains is to control the fore-and-aft distribution of downforce under all conditions of speed and acceleration.

When setting aerodynamic objectives for a Formula 1 car, it is impossible to do so without regard for vehicle dynamics. The balance among downforce, drag, and the movement of the center of pressure (i.e., the change in pitching moment or the distribution of downforce), and matching it to inertia, suspension, and tire characteristics, differentiates the achieved performance of a car as much as differences in power. It is only recently, since computers have found common usage, that advances in vehicle dynamic modeling have enabled engineers to really understand the requirements. However, arriving at the aerodynamic solution is still based very much on experimental techniques rather than theory. The aerodynamic data and knowledge bases of the teams, gleaned from thousands of hours of wind tunnel testing and tens of thousands of kilometers of track testing, are among their biggest technical assets.

Aerodynamics, as applied to open-wheeled, open-cockpit race cars, is an experimental science and will remain so for some time. The day when aerodynamicists will be able to calculate theoretically the shapes they require to generate the forces and flows needed for the dynamic and cooling performance lies far ahead in the future. Some components, such as the front and rear wings, do respond to theory, much of which was developed for aeronautical wing sections. However, as computational fluid dynamics (CFD) is further developed, that will change. In the meantime, Formula 1 aerodynamics is mostly about realistic objectives, experimental techniques and their application, and clear, objective interpretation of the results. The prime tool for experiments is the wind tunnel.

In 1903, the Wright brothers made the first powered flight. One of the reasons they beat all the other would-be pioneer aviators into the air was that in 1901 they built a wind tunnel. They used this, fitted with a purpose-designed wind tunnel balance, to solve an error in wing design that had long frustrated their efforts to fly. They also used the wind tunnel to systematically calibrate and compare more than 40 different airfoil sections.

The Wrights' wind tunnel was not the first; that is credited to Francis Wenham and John Browning in 1871. However, the careful work the Wrights performed in their tunnel enabled them to design their aircraft based on measured facts, taking them ahead of their competitors to achieve controlled flight.

With the development of the airplane from the earliest days so firmly wedded to the use of wind tunnels as a fundamental part of the design and development processes, it is surprising that it took another 60 years before most designers of Grand Prix cars appreciated what a wind tunnel could do for them. Part of the reason is that until the 1960s, the quest was for low drag, with little thought given to the generation of aerodynamic downforce.

Grand Prix racing from the 1930s through the 1960s often occurred on tracks where the maximum speed of the cars could be attained over a significant distance. A development aimed at drag reduction could easily be tested by measuring the effect on top speed. Mercedes and Auto Union focused designers on the need to streamline bodywork, and the chief tools of the trade were a practiced eye and wool tufts. In the 1950s and early 1960s, attempts were made to involve academic and aeronautical aerodynamicists, along with their wind tunnels, in the reduction of drag. Except for Frank Costin of Vanwall and Lotus fame, these aerodynamicists tended to produce body designs that generated lift and upset the handling balance of the car. The phenomenon was not well understood, and wind tunnels gained a bad reputation in motor racing.

Chevrolet R&D, working with Chaparral in the early 1960s, began to realize the enormous potential of downforce and applied it to the Chaparral Can-Am cars—first with body shape, then wings, and finally with fan-induced suction. Chapparal's success, much of which was attributed to the increased tire grip achieved by the aerodynamically induced downforce, startled Formula 1 designers and awakened them to the potential. Since that day, Formula 1 has led the way in innovative methods of generating downforce within ever more restrictive regulations.

Downforce is so fundamental to the braking, accelerating, and cornering performance of a race car that regulating bodies worldwide have had to address the issues and impose ever increasing limits on the magnitude of the downforce that can be generated, according to the class of racing involved. Downforce has been used to quadruple the size of the braking and cornering sectors or the "g-g-V" diagram, while leaving the V axis almost unchanged. The gains in horsepower described in Chapter 5 have been utilized to overcome the associated drag of the downforce-generating parts of the car rather than yield greater top speed. While technically fascinating and one of the most obvious features that sets Formula 1 cars apart from road cars, downforce and its associated g-levels are now coming under close scrutiny as the culprit in preventing overtaking and spoiling the racing spectacle.

The mathematics that describes what downforce accomplishes for the performance of a car is among the simplest of all, which is why it is so surprising that downforce took so long to be recognized and exploited. In the absence of downforce (or lift), the maximum horizontal force that can be generated by a tire is

$$F_{horizontal} = F_{vertical} \times \mu_{max}$$

where μ_{max} is the peak coefficient of friction of the tire and

$$F_{vertical} = M \times g$$

where M is the mass associated with that tire, and g is the acceleration due to gravity.

The acceleration that this force will generate on the mass M is

$$\text{Acceleration} = \frac{F_{horizontal}}{M} = \frac{M \times g \times \mu_{max}}{M} = g \times \mu_{max}$$

Thus, the maximum acceleration attainable is a function solely of the peak coefficient of friction and is independent of mass. This is completely true in the braking and cornering senses, but only true in acceleration when the tire is the limit, not the available torque.

If downforce is added, then the equations become

$$F_{vertical} = [M \times g] + \text{downforce}$$

and

$$\text{Acceleration} = \frac{([M \times g] + \text{downforce}) \times \mu_{max}}{M}$$

$$= g \times \mu_{max} + \frac{\text{downforce} \times \mu_{max}}{M}$$

where $\frac{\text{downforce} \times \mu_{max}}{M}$ is the incremental acceleration due to the downforce. Thus, mass becomes significant in the presence of downforce, for all conditions in which peak tire grip is the limiting factor.

The reasons the potential to increase acceleration levels through aerodynamic downforce was ignored for so long must be that the possible magnitude of the downforce that could be generated was not appreciated, despite all the data being available within the aeronautical industry. It was probably also considered that the tires would not be able to sustain the additional work and that the drag penalty associated with generating the downforce would result in a net loss of average speed. On the high-speed circuits that predominated prior to the 1970s, this may have been true. It may also have been simply an attitude among designers, most of whom came from production car backgrounds—the cars did not have wings. It took the free spirits of Frank Winchel and Jim Hall, and their measurement and simulation groups at Chevrolet R&D and Chaparral, to cast aside convention and explore the possibilities of this new and now obvious way to increase performance.

The aerodynamic performance of a car can be expressed as the force coefficient of lift, drag, and sideforce, and the moment coefficients of pitch, roll, and yaw. Lift, drag, and pitching moment are the most important to a race car designer, but also of great interest are the derivatives of these coefficients as the car moves along and about the three axes. These derivatives, the change of lift, and pitching moment as the car rises and falls along the vertical axis (heave) and pitches about the transverse axis are the most significant. The presence of the ground and its influence on the downforce affect the magnitude and distribution of the vertical force component on the tires and hence affect the statics and dynamics of the car.

Conventionally, the aerodynamic force coefficients on a car are defined with reference to the frontal area. On a road car, the drag coefficient is of greatest interest, with the lift coefficient having a small value, either positive or negative. On a race car, the lift coefficient dominates performance, with the drag coefficient generally referred to as a lift-to-drag ratio—a measure of the efficiency by which the downforce is generated. For this reason, it is more relevant to reference both lift and drag to the plan area of the car, treating it in the convention by which the characteristics of wings are measured.

Both the frontal and plan areas of a Formula 1 car are limited by the regulations. To gain the maximum downforce from the car and use the maximum permitted track width, all of the available frontal area is used, there being seldom any attempt to minimize it. The maximum plan area is defined in every respect except for the wheelbase. However, not all of that available is used, as full-length sidepods tend to cause undesirable pitching moments. Defining the force coefficients according to both of these reference areas, for a current Formula 1 car, gives the following typical figures:

	Maximum Frontal Area	Maximum Plan Area
Area	1.5 M^2	5.8 M^2
Lift Coefficient (C_L)	2.7	0.7
Drag Coefficient (C_D)	0.9	0.3
Lift/Drag Ratio (L/D)	2.5	2.5

A Formula 1 car can be described aerodynamically as a very low aspect ratio (0.38) bluff body in close proximity to the ground (gap/chord = 0.005). It is surrounded by low aspect ratio rotating cylinders in contact with the ground, and it is equipped with leading and trailing edge airfoils, the leading edge device also being in close proximity to the ground. There are major internal flows, with heat addition (water, oil, and brake cooling), an air intake, hot gas injection, and a large open cavity (the cockpit). The flow is three-dimensional and has extensive areas of separated flow and vorticity. The various components interact with each other, both beneficially and detrimentally, and the whole defies conventional aerodynamic analysis. For this reason, it is almost impossible to analyze how well the downforce and efficiency potentials of a Formula 1 car are being exploited. It also explains why it is currently mainly an experimental science.

When Chevrolet R&D first started working on the Chaparral Can-Am cars, it did so without the benefit of electronic sensors, data storage, and plotters. It had to develop its own tools to measure the forces generated. Using rods attached to the suspension and poking through the bodywork, the driver was asked to watch a scale on the rod and note the amount of suspension deflection at a given speed. For further information, pressure tapings were placed on the surface of the body and connected to a water-filled manometer. At the required speed, the tubes were clamped by the driver and the readings used to calculate the pressure distribution. These were crude test techniques that were difficult to perform, and it was not until the mid- to late 1960s that the quest for downforce made designers turn to wind tunnels to seek techniques that would yield reliable results.

At the historic First Symposium on Road Vehicle Aerodynamics (Ref. 8.1), held at the City University in London in November 1969, a number of papers were presented relating to wind tunnel testing of race cars. They represented the first successful steps in adapting wind tunnels to the needs of race car design and cover most of the activities relating to race car aerodynamics worldwide. These papers included:

- Work at City University on wings and spoilers on sports cars.
- Testing of the Piper GTR Le Mans car at Kingston Polytechnic in Surrey, England.
- The work on Carl Deutsch's little CD Le Mans cars, carried out in the tunnel at the Laboratoire Eiffel in Paris.
- Testing of rotating wheels on a Ferrari Formula 1 car at the Politechnico di Turin wind tunnel.
- Comparison work by the Motor Industry Research Association (MIRA) of the effect of a moving ground plane in the wind tunnels of the Royal Aircraft Establishment (RAE) and Imperial College in London. The models used included one of the Rover-BRM gas turbine Le Mans cars.
- Some very fundamental work by Stollery and Burns, using the moving ground plane wind tunnel at the Imperial College, streamlined bodies near the ground. They noted that as the body approached the ground, it generated negative lift (i.e., ground effect). The paper also reviewed tests on Donald Campbell's Bluebird Land Speed Record car, a GT car, and a Formula 1 car with rotating wheels.

A couple of years prior to this conference, BRM had started to become involved in wind tunnel testing. Using the same 5 ft × 4 ft one-quarter-scale wind tunnel as Stollery and Burns, but without the moving ground plane or rotating wheels, the shape of the BRM T139 V-12 car was developed. Subsequently, following a request from the chief engineer Tony Rudd to find a way to avoid fitting wings to the car, a monocoque shaped like an inverted airfoil was investigated in the tunnel. BRM continued to use the Imperial College tunnel, and I believe that BRM and Ferrari were the only Formula 1 teams performing regular wind tunnel testing at that time.

Also in the late 1960s, Fackrell and Harvey (Ref. 8.2) conducted experimental work at the Imperial College, which defined the flow around a rotating wheel in contact with the ground, putting in place another fundamental piece of the foundation of the aerodynamics of Formula 1 cars and indeed any open-wheeled car.

Having joined Lotus, Tony Rudd initiated an aerodynamics program for the "clean-sheet-of-paper" Formula 1 car that Colin Chapman had tasked him with producing in 1975. This car was to be the ground-effect Lotus T78 for the 1977 season.

A fundamental program of wind tunnel tests was planned to investigate the interaction of the nose and bodywork of an open-wheeled car with the ground. To ensure that the tests were realistic, the one-quarter-scale tunnel and moving ground plane at the Imperial College were used for all tests. This basic decision was essential in discovering ground effect and how to generate large amounts of downforce through the use of side-skirts. The wind tunnel program started in the middle of 1975, and from that date Team Lotus never developed a Formula 1 car without the use of a wind tunnel.

Surprisingly, it was not until the T79 revealed its dominance in 1978 that other designers started looking for wind tunnel facilities with moving ground planes. Today, one-quarter-scale (25%) models are too small and 40% is the minimum, with nearly all Formula 1 teams testing at a scale of 50%. Many tunnels now exist worldwide in universities, research establishments, and racing teams, which are dedicated to the development of race cars. These tunnels often run day and night in the quest for more downforce and less drag.

Much to the chagrin of many designers and engineers, wind tunnels do not design race cars nor do they tell the aerodynamicist what to do. They are simply measuring instruments, or, to be more precise, they provide a controlled environment in which a number of different aerodynamic measurements can be made on a stationary object. They permit the pressures and forces generated by the flow of air past the object to be quantified and compared with other objects or the same object at a different attitude. They also permit the velocity, total energy, static pressure, rotation (vorticity), and turbulence level of the air flowing past to be measured and studied. All these measurements are difficult to perform on a moving object, but the wind tunnel solves this problem by moving the air past a stationary one.

Ideally, the aerodynamicist would like to test a full-size model of the car at air speeds up to the maximum speed of the car. However, the cost and power requirements of a full-size tunnel are enormous, models are almost as expensive to make as the prototype car, and those models must be strong enough to withstand full-size aerodynamic forces. Although many road cars are tested in full size, the norm for race cars is to use scale models from 40–50%. This allows many variations of body shape and wings to be built and tested at a reasonable cost and within a reasonable time frame.

It is possible to test at a smaller scale and lower speed than the full-size car, provided great care is taken to ensure that the flow characteristics are the same. At low speed and small scale, laminar flow will extend farther back along the surface of the model than on the real car. This will have a very small effect, provided that the laminar boundary layer does not separate from the surface of the model, if it does not do so in full size. If the Reynolds number (i.e., the ratio of the inertia forces to the viscous forces)

$$\text{Reynolds number} = \frac{(\text{airspeed} \times \text{length} \times \text{air density})}{\text{viscosity of air}}$$

is too low, it is possible for the wings and underside of the car to stall. This is particularly likely on a Formula 1 car, where the aerodynamicist is trying to push these surfaces to generate as much downforce as possible. A stalled model in the wind tunnel will produce totally erroneous results and will seriously mislead the aerodynamicist. Some of the aerodynamic instability experienced by past and current Formula 1 cars is a consequence of testing at too low a Reynolds number. The aerodynamicist wants to test at as representative a Reynolds number as possible, and this has resulted in the increase in scale and the development of higher wind tunnel speeds. In 1998, Formula Benetton commissioned a new 50% tunnel at its facility in Enstone, Oxford. This $20 million tunnel, illustrated in Fig. 8.1, sets new performance standards for the Reynolds number at which models can be tested.

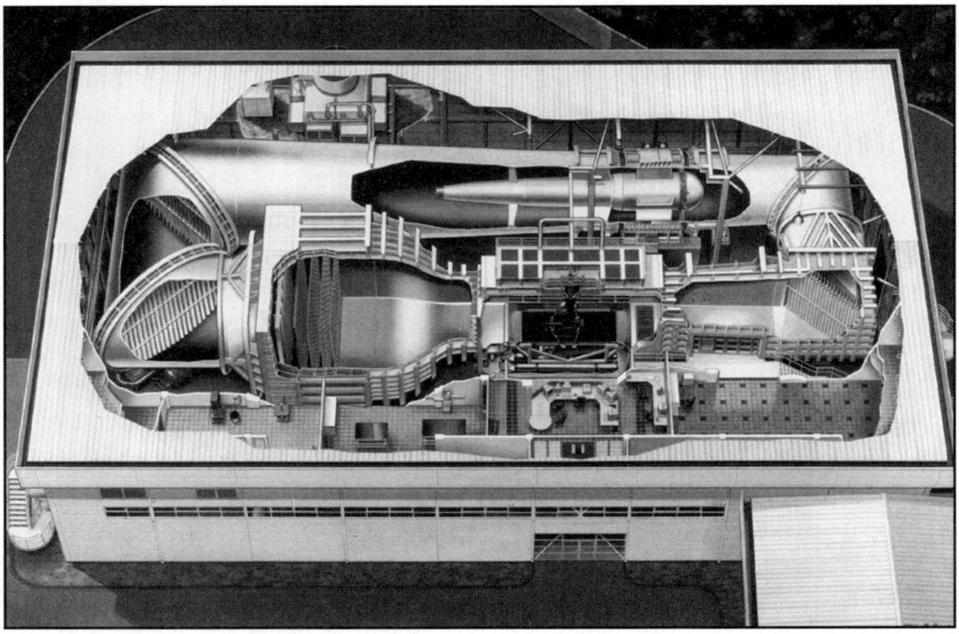

Fig. 8.1 Formula Benetton's pressurized, half-scale wind tunnel.

With a 3.5 × 3.5 meter working section, and a maximum air speed of 215 kph (134 mph) (the moving belt and fan are designed for 250 kph ([155 mph], but available electrical power currently limits it to the lower figure), it can achieve a Reynolds number equal to a third of that of the full-size car, at its top speed of 350 kph (218 mph). However, the tunnel has been designed such that it can be pressurized to 1 bar, doubling the density of the air and doubling the Reynolds number to 0.6 of that of the full-size car. The steel structure of the tunnel forms a massive pressure vessel weighing 700 tonne (689 tons), and all access apertures and doors are sealed. Pressurization takes 15 minutes. The power required to drive the tunnel at its rated speed is 10,000 hp (7460 kW) and a large refrigeration plant removes the heat that this puts into the air, maintaining the constant temperature that precise aerodynamic testing demands.

Early car and race car wind tunnel tests tried to establish to what degree it was necessary to replicate the airflow over the ground surface. Tests were performed on long and short ground planes, ground planes with boundary layer suction, and moving ground planes, each giving a different boundary layer characteristic on the tunnel floor/simulated ground. The real car, while travelling through the still air and over the ground, experiences a ground plane with no boundary layer. (A strong wind is a wholly different and much more complicated case to analyze.) Tests with cars that generated no underbody downforce and ran at ground clearances of 100 mm (3.9 in.) or more showed little sensitivity to ground plane boundary layer conditions. The erroneous conclusion that a moving ground was not necessary meant that it was unlikely that ground effect would be discovered in these wind tunnels!

Ground-effect-generated downforce depends largely on the amount of air flowing under the car and whether that air remains attached to the underside surface of the bodywork. If ground clearances are small and the ground plane has a significant boundary layer on it, very little air will flow under the car. The air that does go under the car is subjected to conditions that are conducive to boundary layer flow separation (stalling), and the boundary layer on the ground may even separate—a condition that cannot occur with real cars. Thus, a good moving ground plane became an essential part of a race car wind tunnel.

The moving ground consists of a flat belt running over rollers, with the top surface flush with the floor of the wind tunnel and travelling in the same direction as the airflow. The floor boundary layer is removed by suction, ahead of the belt, and additional suction removes the boundary layer from the belt itself. Today, the speed of the belt is electronically synchronized to the air speed. The best tunnels achieve full stream air speed down to within 1 or 2 mm (0.04 to 0.08 in.) of the ground plane. Because the underside of the model generates low pressure, this tries to suck the belt upward, changing the shape of the region under the car disastrously. To counteract this effect, the belt runs on computer-controlled suction boxes that distribute an equal and opposite low pressure to hold the belt flat. As speed increases, the heat generated by the belt being sucked downward onto the running surface becomes a problem and requires a liquid-cooled platten. It also becomes difficult to prevent the belt from running off the rollers sideways, particularly when the wheels are running at a slip angle. A servo-controlled, third tensioning roller is steered to counteract the sideways movement (Fig. 8.2). Although there are full-size moving ground wind tunnels for aircraft, the belt does not require suction because the pressure under the wings holds the belt flat.

The cost and complexity of solving belt lift and the demand on these tunnels by aerospace companies has precluded their use for race cars, although it is reported that the 18.3 × 9.1 meter (60 × 30 ft) tunnel at NASA Langley may be fitted with a moving ground plane suitable for car testing.

The provision of a moving ground makes it relatively simple to fit the model with rotating wheels—the final component necessary to faithfully represent the actual car. The wheels must be mounted such that the forces imparted to them by being supported and driven by the belt are not measured by the wind tunnel balance mechanism. This virtually precludes mounting the wheels on the

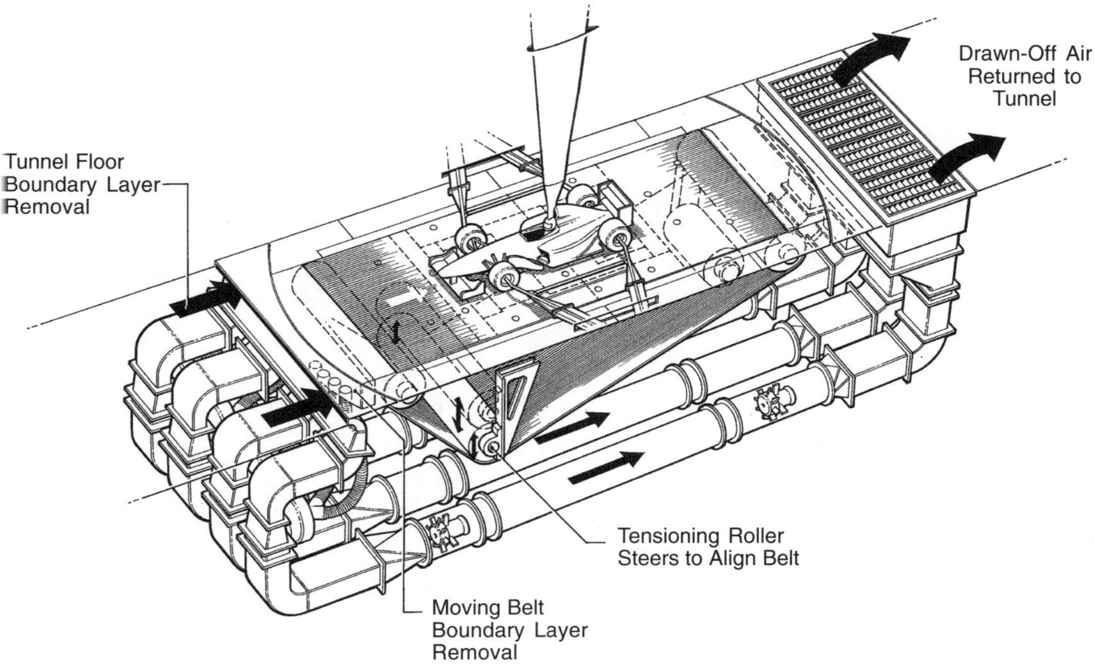

Fig. 8.2 The moving ground plane fitted to the Benetton wind tunnel.

model, and they are supported on their own suspension system, fixed to the tunnel structure. Measuring the forces on the wheels must now be conducted separately from the measurement of forces on the model.

Race car wind tunnels fall into the class of tunnel known as "low speed." Two types exist: (1) open circuit, and (2) closed circuit. Currently most are closed circuit. The open-circuit tunnel is fundamentally a straight duct, open at each end. Air is drawn in through a bell mouth intake, speeded up, and passed through the working section before expanding again to be expelled by the fan. Open-circuit tunnels suffer from disturbed airflow at the intake and from dust and cannot be pressurized. The closed-circuit tunnel passes the air from the fan around a return duct and back to the intake. The whole duct is closed, and the flow conditions are carefully controlled by turning vanes, flow straighteners, and screens to reduce the turbulence level. The structure is much larger and more expensive to build, but the improved quality of airflow in the working section is worth the cost in critical applications.

Not only is it beneficial to be able to test a 50% model in these large tunnels, but the tunnels will also accommodate 100% models of race car wings, permitting precise research into wing profiles and, for cooling flow measurements at reduced accuracy, a full-size car.

The 50% wind tunnel models of Formula 1 cars are constructed to the same or greater precision than the full-size cars, and they cost approximately the same (Figs. 8.3 and 8.4). They are

Fig. 8.3 Half-scale wind tunnel model mounted in wind tunnel. (Courtesy of Sutton Motorsport Images)

Fig. 8.4 Half-scale model without bodywork or wings fitted.

built around a central structural spine, and the entire shape of the car is modular, such than any part that is licked by the airstream can be changed. Between the spine and the mounting strut is a six-component, strain-gauge balance, mounted within the spine. The balance measures the three forces and three moments. Also mounted within the spine are multi-channel electronic pressure sensors or scani-valves, for sequentially sampling pressures on the model. The position and attitude of the model car and the incidence of the wings and flaps can be remotely changed via stepper-motor actuators in the model.

Suspension is fully detailed and articulates as the model changes attitude. Brake cooling ducts and the wheel uprights are aerodynamically correct and embody airflow measuring sensors. Radiators are represented by matrices with the same pressure drop characteristics as the actual radiators and are also fitted with airflow sensors. High-pressure air is fed to an ejector within the model, driving an airflow that draws air from the engine's ram intake and exhausting the enhanced flow through fully representative, five-into-one exhaust pipes. Any surface can be pressure tapped, but it is the front and rear wings and undertray that are of particular interest. CNC machined from aluminum pressure ports are laser-milled into them, such that it sometimes appears that an aluminum-eating cousin of the woodworm has taken up residence inside the components. The attachment points incorporate a manifold system, and these key components can be changed without the need to replumb the multiple pressure tapings. Other bodywork parts are often made in CFRP, laid up in CNC-machined tools. Rapid prototyping techniques are being employed widely to make model parts such as exhaust systems.

Wheels are mounted on their own force measuring systems, external to and not touching the model. They run on the moving belt and are supported by rollers beneath the belt to prevent a local heat buildup due to friction. Their drag forces are measured by their mounting system, but lift forces can be taken only via their support rollers which do not provide precision measurement due to the presence of the belt.

Data from the sensors is multiplexed, the model being one node of a network that forms the data and control systems for the whole tunnel complex. The streamlined support strut, locating the model to the roof of the tunnel, houses the power and data wiring loom and the duct for the ejector air. A typical test run will involve the 30–40 test points needed to compute an aeromap of ride heights and will take approximately 20 minutes. The whole control of the tunnel and test is pre-programmed into the central computer. Fan startup, airspeed, moving-ground startup, belt speed, boundary layer control, belt suction, air temperature, model position, force and pressure data sampling, tunnel shutdown, data analysis, and plotting are all computer controlled via the network. Test results are available seconds after a run, enabling aerodynamicists to make judgments about results prior to subsequent tests.

Given a model in the wind tunnel, with a fully representative airflow moving over it, in what is the aerodynamicist interested? Primarily, he wants to know what forces the air imparts to the car. (Actually, it is the vehicle dynamicist who wants to know this.) To understand how these forces are generated, he will also want to know the pressures on the surface of the model and the flow conditions that have caused those pressures. In addition, he will want to investigate details of the flow passing through the internals of the car, particularly inside cooling ducts.

For most of the tests, on single-seaters at least, side force and roll and yawing moments are not measured. They are measured only when the model is yawed. A moving ground that can be rotated to be at an angle to the flow direction is a very large increase in complexity, particularly in a pressurized tunnel. However, it is necessary to investigate the effect of yaw angle on downforce and pitching moment. Two conditions exist: (1) the car and road both yawed relative to the airstream, simulating a cross wind, or (2) only the car yawed, simulating the car cornering at a slip angle. The basic aerodynamic yawing stability of single-seaters is fairly good due to the large rear-wing end plates.

To visualize the flow of invisible air, the aerodynamicist has a number of simple, traditional techniques available. Filaments of smoke, introduced into the airstream, mark streamlines as they bend round the model. As soon as the flow becomes turbulent, which is around the front wheels of a Formula 1 car, the smoke filaments break up and disperse.

Wool tufts, attached in rows to the surface, indicate flow direction, areas of vorticity, and flow breakaway. Care must be taken not to make them too long, or the "tail wags the dog" and the results can be misleading. A single tuft, on the tip of a wand, is a useful tool for finding the separation region and flow direction away from the surface. The wand must be very fine, less than 1 mm (0.04 in.) in diameter, or it will disturb the flow and cause separation.

Fine powder, dispersed in oil and painted on the surface, reveals surface flow details. However, they are exactly that—they represent the flow right on the surface, within the boundary layer. If a vortex lies along the surface, the picture may be confusing and require a great deal of skill to interpret. The technique also makes a real mess of the model and the tunnel, and is likely to block fine pressure tapings.

Measurements of flow direction, speed, energy, vorticity, and turbulence can be made, using a variety of tiny probes introduced into the flow under investigation. A five-axis, computer-controlled gantry places and points the probe precisely, anywhere around the model or in the wake. Surveys can be made, sampling the flow on a pre-programmed grid and then using analysis software to plot the data as contour maps or using arrows representing flow velocity to visualize the flow in detail.

The aerospace industry is developing many new test measurement techniques, particularly for fine-mesh surface pressure mapping. Computational fluid dynamics produces colored pressure contour maps, and there is a need to match them with tunnel measurements. Boeing has developed a technique employing a pressure-sensitive paint that changes color under certain light according to the surface pressure. Laser measurement of air density is another technique that enables the surface to be scanned and the local pressure determined. These methods are sure to find use in Formula 1 wind tunnels as they become more widely disseminated, and they will be invaluable for correlating CFD and wind tunnel results.

Computational Fluid Dynamics

Although much is claimed by the proponents of CFD, its contribution to the development of Formula 1 car aerodynamics remains limited (Ref. 8.3):

> Due to the necessarily low aspect ratios of the lifting surfaces and the dominance of tip flows, the flow structures around a race car are characterized by large areas of vortical, unstable and separated flows. Understanding the interaction of tire jet vortices produced at varying angles of slip, with a three-dimensional high lift wing in very close ground proximity is just one example of the enormous challenges faced by the F1 aerodynamicist. Even components on the car such as the high lift multi-element wings or radiator ducts, on which there is much available literature, interact so greatly with other areas of the car that it renders these studies of limited value. It should also be borne in mind that often the most performant aerodynamic "package" may sacrifice the inherent performance of one component for the benefit of another in its wake. This consideration often renders accepted aerodynamic logic invalid.
>
> The wealth of experimental knowledge, an emphasis on relative performance improvement and the complexities of the flows around the generic Fl car, have meant that the pursuit of theoretical understanding of the flow problems faced has been compromised. However, as experimental methodology is honed by all the teams, prospective competitive advantage is minimized and more teams are looking toward numerical methods in order to enhance understanding of the more complex flow structures present around the car. Of all the viable solvers, it is the Navier-Stokes solvers which are now most in favor with the Fl teams, primarily due to the fact that their increasing commercial availability has led to increasingly available expertise in their use.
>
> As suggested above, of increasing importance to the Fl aerodynamicist is the understanding of vortex interaction and thus methods of control and enhancement. A good example of this is the realization that the control of the front wing tip vortex/front wheel jet vortex interaction can lead to very large gains in downforce at the rear of the car, to the detriment of front wing performance. With this in mind, it is clear that the ideal numerical solution would capture flow details in all the fluid volume around the car, not only close to the generating surfaces. This requirement precludes most potential flow methods due to their inability to model vortex interaction, and indeed at extremes even the latest Navier-Stokes solvers with current turbulence models, since all still fail to adequately predict vortex burst. However, on the acceptance that all currently available codes will not fully capture the flow structures present, "validated" and "calibrated" solvers are becoming of great use as parametric tools complementing the wind tunnel testing program, and offering valuable insight into the nature of the flow around parts of the race car. Current limitations in accurate mesh production and detailed geometry capture, together with other factors restricting speed to solution,

mean that the improvements in aerodynamic performance are still a consequence of experimental wind tunnel testing. However, computational techniques are being used more and more in harmony with experimental development to help the aerodynamicist improve understanding of the discipline and ultimately the performance of the car.

Full-Size Testing

In spite of the growth in the use of wind tunnels to derive aerodynamic force maps, and the increasing fidelity of the models and test conditions under which they are measured, full-size aerodynamic testing has continued. With the development of suspension load and position sensors, laser ride height sensors, accurate air speed and pressure measurement, acquisition and storage by electronic data systems, and computational analysis of the results, the process of validating aeromaps has become almost routine. During the 1980s and early 1990s, when active suspension was under development, accurate data concerning the variation of forces with ride height was essential. Full-size aeromapping and the proving of resultant suspension control laws were an integral part of active suspension development. The suspension system itself could be used to maintain precise ride height during testing. By testing on a suitably flat, smooth surface, multiple data points could be measured during a single run by automatic changing of the front and rear ride heights. Since the banning of active suspension, the suspension is fixed in a given position and a single test point measured on a run.

Total drag force figures are not only difficult to measure in the wind tunnel due to the problems associated with measuring aerodynamic forces on the rotating wheels while in contact with the moving belt, but are also difficult to measure full size on the track. Rolling resistance of the tires, losses in their bearings, pads rubbing on brake disks, and transmission losses all contribute to the overall resisting force. Laboratory tests to isolate and quantify individual components of the resistance, torque measurements in the drive train, and the measurement of the deceleration during coast down tests (analyzed using parameter identification techniques) must be combined to derive the aerodynamic drag figure (Ref. 8.4).

The following is an excerpt from Ref. 8.3.

Aerodynamic Components of a Formula 1 Racing Car

To attempt to describe in detail the different components that make up an Fl car, and more importantly the flow around them and interaction between them, is beyond the scope of this document. Instead, the remainder of this report tries to focus on areas that can be isolated, and that present problems of direct interest to the Fl aerodynamicist.

As a basic design philosophy, the aerodynamic designer is trying to maximize the downforce and minimize the drag produced by the car, while maintaining a constant aerodynamic balance over the whole working range of the car. These rather simple ideals prove to be rather less simple to obtain in reality.

Front Wing

The front wing of an Fl car is typically a two-element high-lift wing with an adjustable flap to vary the downforce obtained from the wing, and hence balance the car with different rear-wing configurations. The front wing runs in very strong ground effect, with typical ground clearances ranging from 70 to 100 mm [2.75 to 3.94 in.]. Due to the fact that the rest of the car operates in the wake generated by this wing, and that this wake severely affects in particular the undertray and diffuser of the car, it is important to tune the wake profile of the front wing. Experience has shown that the front wing should have a constant distribution of lift across its chord, and due to tip effects, greater incidence and chord are required outboard to achieve this. Inevitable tip effects are minimized and harnessed by the front-wing endplate, which can influence large gains at the rear of the car. The lower to the ground, the better the sealing effect of the front-wing endplates and the more efficient the front wing. However, the height of the endplates is governed in the regulations at 50 mm [1.97 in.] above the flat bottom of the car.

Due to the presence of the ground, the front wing is very efficient (in race car terms), with a lift/drag ratio in the range 7–9. The performance of the front wing is also strongly dependent on the presence of the front wheel. A rotating wheel produces strong crosswise flow areas close to the ground in front of the wheel due to a squeezing or jetting effect. These jet vortices are highly influential in understanding the form of the front wing wake, and their effect changes with the steering angle of the front wheel. This is still a little understood, transient aerodynamic effect, which is difficult to reproduce accurately in a wind tunnel test. For this reason, such a problem may avail itself to slightly more ambitious numerical investigation.

The front wing produces 25–30% of the total downforce of the car. The amount of downforce produced is very dependent on the front ground clearance, and it is this fact that produces most of the aerodynamic stability problems. As the front of the car moves lower, the front wing produces more downforce due to increased "ground effect" and more effective sealing of the endplate. This in itself produces a forward shift in balance, but also the increased upwash from the front wing reduces the rear downforce, increasing the balance shift. Designing a front wing that is both efficient and is not sensitive to ground clearance does greatly improve the driveability and efficiency of an F1 car.

Undertray and Bodywork

From the FIA regulations, the floor of an F1 car must occupy two clearly defined planes between the rear edge of the front wheels and the front edge of the rear wheels. The surfaces of the floor in these planes must be flat, rigid, and impervious. Downstream of the front edge of the rear wheel, a diffuser section is used to lower the pressure under the car and thus generate downforce. About 40% of the total downforce of the car is produced by the undertray and bodywork. In general, the body of the car can be thought of as a bluff body close to the ground, with a large wake and associated form drag. In general, improving the design of the diffuser and producing lower pressures under the car does not lead to an increase in drag, and so the production of downforce by the body of the car is normally the key to an aerodynamically efficient car.

Again, the aerodynamic stability characteristics that the diffuser implies on the car are of critical importance in vehicle driveability. Thus, stall phenomena should be predictable and tuned to specific requirements. The lower elements of the rear wing can be used as a "spoiler" to set a base pressure condition and thus control stall characteristics. The diffuser angle can also be used to tune stall characteristics, where a higher angle generally gives more downforce over the pre-stalled operating range, but causes the diffuser to stall at higher rear ride heights, and vice versa. The particular diffuser angle chosen is normally a compromise between aerodynamic efficiency and stability.

Due to the low aspect ratio of the undertray, and also due to the presence of the rear wheels, the flow under the diffuser is characterized by large areas of highly vortical three-dimensional flow. [Fig. 8.5] In general, the full flow in the underbody is little understood and is very difficult to investigate experimentally. Surface pressure measurements, surface flow visualization, and numerical modeling are possible methods in attempting to understand some of the aspects of this flow.

Rear Wing

A typical rear-wing configuration on a current F1 car consists of a two- or three-plane wing, with the upper element set varying between a low-chord single-element wing at comparatively low incidence, to a large-chord, highly cambered three-element wing. The range of available rear wings is so as to allow tuning of downforce levels to particular circuits.

The rear wing produces approximately 30–35% of the total downforce of the car, and approximately 25–30% of the total drag of the car. Optimization of the individual elements and the interaction of the lower elements with the diffuser are of critical importance to the aerodynamic design of the car. The addition of the lower elements actually reduces the downforce produced by the total rear wing layout itself, but it increases the efficiency of the car by increasing the downforce produced by the undertray and bodywork. [Fig. 8.6] Optimization of rear wings is one of the few areas where the use of isolated numerical studies is applicable, especially for the highly cambered multi-element upper wings which run in relatively "clean" upstream flow conditions.

Wheels

As suggested previously, the open wheels of an Fl car cause much of the complexity in the flow around the car. They produce about 40% of the total drag of the car and also produce lift which is very difficult to measure experimentally. This can often be a source of confusion in experimental assessment. They further affect the car's aerodynamics by producing strong cross flows in critical areas of the car. Ultimately,

Fig. 8.5 A Ferrari undertray. Circumstances such as this provide the best opportunities to study competitors' aerodynamics. (Courtesy of Sutton Motorsport Images)

a better understanding of the interaction of the flow field around the wheels with the rest of the car, including the effect of steering angles on the flow field, could indicate ways of harnessing these flow characteristics.

Without radical regulation changes to reduce drastically the possibility of generating downforce, aerodynamics will rank as one of the four key parameters in determining the performance of a Formula 1 car, alongside power, weight, and tires. The allocation of resources by the teams to aerodynamic development (e.g., aerodynamic departments that are staffed by a dozen or more highly qualified engineers, \$15–20 million wind tunnels, and \$0.5 million models) remains necessary to maintain the yearly gains in downforce and lift/drag. Greater accuracy and more sophisticated numerical tools are needed to find the 6–10% improvement necessary each year to remain competitive (Fig. 8.7).

Fig. 8.6 Tip vortices shed by a rear wing on a humid day provide a vivid illustration of the downforce being generated by it. (Courtesy of Sutton Motorsport Images)

Fig. 8.7 Adrian Newey's McLaren-Mercedes—the epitome of Formula 1 aerodynamic design. (Courtesy of Sutton Motorsport Images)

Chapter 9

Suspension

All real systems have an infinite number of degrees of freedom and this renders impossible a complete vibration analysis.
—The Mechanics of Vibration
by R.E.D. Bishop and D.C. Johnson

The earliest wheeled vehicles had no suspension but survived without it because they were both slow and heavy. In man's instinctive quest for speed, his carriages became both lighter and faster. Soon it became apparent that a lack of suspension not only caused discomfort to the passengers but also caused the wheels to leave the surface of the road and thus become unable to provide traction control and stability. It did not take long to determine that if the wheels were allowed to move up and down in response to road irregularities, while the main carriage was supported above the wheels on soft springs, great improvements in comfort and control would result.

If roads were truly flat, suspension would not be needed. However, roads are not flat, and even if they are created flat initially, they do not remain so for long. To begin with, only springs were used to attenuate the vertical forces fed into the carriage by the vertical motion imparted to the wheels by their horizontal motion over bumps. However, because the forces could not be reduced to zero, due to limits on available wheel motion, it was soon found necessary to extract energy from the oscillatory motion of the excited masses, and dampers were added to the springs.

Once the wheels are no longer firmly attached to the chassis or a steerable axle, not only must the vertical motion be controlled by springs and dampers, but the other five degrees of freedom must also be controlled. The number of ways this can be accomplished is very large, and most of the possibilities have been tried at one time or another, not always successfully. For a long time, pairs of wheels were connected together on an axle, thereby simply constraining four of the six degrees of freedom, leaving only vertical motion and steering. However, the advantages of independently suspending the wheels soon became apparent, especially in motor racing where divorcing the motion of one wheel from its axle pair yields benefits in tire contact patch control. The kinematic control of the motion of the wheels, combined with the contact patch load control via springs and dampers, forms the two pillars of suspension design and development.

On a race car, the requirements for the linkage system that attaches the wheel to the chassis are:

1. Light weight
2. High stiffness to minimize compliance effects
3. High strength
4. Provision of the desired kinematic characteristics
5. The ability to configure the links to feed the wheel loads into the chassis at appropriate points
6. Minimum aerodynamic influence (open-wheelers only)

The optimum way to meet these requirements is with six pin-jointed links. Not surprisingly, this is exactly how it has been done on Formula 1 cars, at both the front and rear, since the late 1950s.

The use of pin-jointed links ensures that they are subjected only to pure tension and compression loads without bending, which enables the first three of the above objectives to be met and, by using airfoil section links, the sixth objective as well. The pin-jointed pivots are either ball joints or, since John Barnard pioneered the concept on the Ferrari in 1994, flexure pivots to reduce stiction and increase joint stiffness in the direction of the load axis.

The double wishbone and push-rod suspension of a Formula 1 car is a triumph of simplicity and elegant design. Each link is optimized for strength, stiffness, weight, and aerodynamic influence (low drag and turning effect on the airflow, within regulatory limits), infinitely possible thanks to the use of CFRP for their construction. The arrangement provides for all the geometry subtleties and adjustment needed to endow the wheel movement characteristics that the designer desires, in all planes and about all axes. Using ADAMS and other linkage-analysis software, it is possible to optimize the layout and kinematics. It is unlikely that any significantly different suspension arrangement will appear on a Formula 1 car in the foreseeable future, nor indeed on any competition car where the regulations provide the freedom.

Analysis of the vertical load variation at the contact patch as a consequence of bumps is well established, using two-degrees-of-freedom dynamic analysis for one corner of the car, and seven-degrees-of-freedom analysis for the whole car. Given the stiffness and damping of the tires, it is possible to optimize the spring and damper characteristics for a given input frequency and to arrive at a suitable compromise for the range of frequencies to which the car will be subjected on a circuit. Although ride qualities are almost irrelevant, it is the load variation in the contact patch and how it affects the ability of the tires to grip the track surface that are most important. Almost as important is the control of the heave and pitch attitude of the car relative to the track surface and the resultant effect on ground-effect aerodynamic characteristics.

Two basic and fundamental features of any four-wheeled vehicle couple the tire contact patch loads to the horizontal forces on the car, and add to the complexity of the dynamic system, making it much more difficult to analyze its characteristics:

- The primary forces—tire contact patch forces—do not pass through the center of gravity, resulting in the vertical loads on the tires being dependent on the horizontal loads.

- The distribution of the vertical component of those forces on the four wheels is indeterminate; there is an infinite number of possible solutions.

Bicycles do not have either of these problems during cornering, and three-wheelers have only the first of these features. The analysis of these vehicles is thus simpler than that of the car, and only the advent of low-cost computing has enabled the car to catch up during the last 30–40 years. Even so, suspension development on both road and race cars remains somewhat a black art rather than a science. At the core of the problem is lateral load transfer distribution (LLTD) and the stiffnesses that determines it. The stiffness and damping requirements conflict with those for optimum contact patch force control, which may also conflict with their requirements for optimum control of the vertical motion of the body induced by road irregularities.

To understand the influence of the horizontal forces on wheel load control, the analysis of vehicles without suspension—karts—reveals a great deal. The effect of adding suspension to the underlying principles can then be examined.

If a completely rigid chassis mounted on four solid wheels and rolling over a perfectly flat road is subjected to horizontal loads through the CG, it is possible to determine the vertical loads on each wheel from the geometry. However, if the road is not flat, one wheel will always leave the surface, and its vertical load will be zero. The need for compliance in tires, in order for them to be able to generate forces and to provide them with an ability to absorb small road irregularities, means that they enable the rigid-chassied vehicle to avoid wheel lift on a non-flat surface. A kart also has a non-rigid chassis, and these two compliance systems acting in series mean that it is no longer possible to determine the vertical forces on the wheels from geometry alone. It also means that, by adjusting the compliance of both tires and chassis, the kart setup engineer is able to influence the distribution of vertical loads on the tires as a function of the horizontal force vector and thereby influence the control and stability of the kart. The effect of chassis stiffness is complicated by the fact that the torsion and bending deflections depend on the distribution of the loads along its length, and these will vary according to the position of the main mass, the driver, the position of which can be varied while the vehicle is moving. Setting up the handling of a kart is so complex, even if the actual analysis is more straightforward than for a car, that it will probably be one of the last of the black arts of motorsport to become a science.

Kart regulations forbid the fitting of suspension systems, but all other forms of motorsport do not. Rather than rely on chassis flexibility for absorbing road irregularities, a quasi-rigid chassis and four-wheel suspension is universally employed. Given the compliance required to deal with bumps, it makes no sense to add the confusion of a third compliance in series with the tires and springs. Thus, the stiffest structure possible within weight constraints is used. Provided the stiffness is an order greater than the equivalent tire or wheel rate, the chassis torsional stiffness ceases to have a significant effect. Typical chassis torsional stiffness of a Formula 1 car is in excess of 40,000 Nm/deg, ensuring that it neither influences load distribution nor that its natural frequency coincides with any ride frequency.

The four functions of the compliances in a four-wheel independent suspension have conflicting requirements:

- Isolation: Soft, approximately 70% critical damping, long stroke.
- Load distribution control: The greater the required variation from the rigid-body, rigid-wheel, no-suspension distribution, the stiffer the requirement.
- Transient response: High stiffness, around critical damping.
- Aerodynamic ride height control: Stiffness depends on required characteristics. Generally stiff front, soft rear. Negative stiffness (spring extends as load increases) could be desirable at the front.

The designer must provide the setup engineer with the means to combine the available compliances in such a way that he achieves the best possible (i.e., fastest) compromise among these conflicting requirements. Table 9.1 shows typical characteristics of the compliances available in the system.

TABLE 9.1
TYPICAL CHARACTERISTICS OF CHASSIS COMPLIANCES

Item	Component	Stiffness	Damping
1.	Tire Carcass	~50 N/mm	~5% Critical
2.	Tire Pressure	~200 N/mm	~0
3.	Spring/Damper	~175 N/mm	~70% critical
4.	Bump Stop	Rising rate to 500 N/mm	~0
5.	Anti-Roll Bar (ARB)	0 to 1000+ N/mm	0 to ~70% critical
6.	Chassis Torsion (at Wheel)	1500+ N/mm	~0

The two compliances of the tire (Items 1 and 2 in Table 9.1) work in parallel; the springs and bump stop (Items 3 and 4 in Table 9.1) work in parallel; and the ARB (Item 5 in Table 9.1) is added in parallel for roll and single-wheel bumps. The three resulting groups (Items 1 and 2; Items 3, 4, and 5; and Item 6) are combined in series. The tire, spring, and chassis characteristics are designed in and can be changed only by replacing them with a component having different characteristics, whereas the other characteristics (tire pressure, damping, bump stop, and the ARB) can, within certain limitations, be adjusted on the car while it is in the pits. Prior to the present Formula 1 regulations, the driver could also make adjustments while running.

With such a complex set of tools and such a conflicting set of requirements, it is only since the analytical power of portable computers became available to engineers that they have been able to apply science to the art of setting up a race car chassis. As soon as on-car computers became available, it was inevitable that they should be applied to a control system that set out to remove the compromise inherent in any passive suspension setup—active suspension. Active and semi-active suspensions, which are best defined as the computer-controlled addition of energy to the system, permit the various functional requirements to be optimized independently. Banned by the regulations in 1994, active suspension is no longer relevant to Formula 1. However, its importance as a development and the insight it provides into suspension requirements are such that a full description is included in this book as Appendix C.

Active suspensions enabled the components of suspension stiffness and damping to be separated and optimized individually. Individual wheel rates, axle vertical rate, and roll stiffness could all be set to suit the individual requirements of the tires and aerodynamics, providing the race engineer with a whole new set of tools. When active suspension was banned, engineers were reluctant to give up their newfound tools and thus set out to design mechanical analogs. The result is the addition of a third spring and/or damper to the traditional two-spring/damper and anti-roll bar arrangement.

On a road car, the cornering characteristics of each axle are governed by the suspension kinematics and compliance characteristics (e.g., roll steer, lateral force compliance steer). Tuning of these characteristics allows the suspension engineer to determine the tire slip angles under any set of conditions and thereby balance the side forces of the front and rear axles, endowing the car with the "handling" so often referred to in road tests. Apart from ride requirements, the spring and ARB stiffness determine wheel and body deflections and their associated kinematic positions. This works well while the tires are in the linear range. Unfortunately, the race car cornering characteristics of most interest to the driver are when the tires are well into the nonlinear range and therefore insensitive to small changes in slip angle. Tuning of kinematic characteristics ceases to be an effective means of controlling axle cornering forces.

The problem is made worse by the need to control axle ride heights very precisely, using stiff springs for aerodynamic reasons. Thus, only small kinematic movements are available.

To compensate, racing tires are designed to have relatively high sensitivity to vertical load—the peak cornering "μ" falls off with load—enabling the axle cornering force to be adjusted according to the magnitude of the load transfer across it. By adjusting the LLTD, front to rear, the cornering balance of the front and rear axles can be set to give the desired overall understeer/oversteer characteristics to the race car. This is by far the most powerful means of tuning the mechanical balance of the car, but it remains a second-order effect compared to aerodynamic force balance, fore-and-aft load transfer, and the braking or accelerating torque applied to an axle.

The vertical stiffness of a conventional, two-spring, and ARB suspension is determined by the sum of the two wheel rates; the roll stiffness is the sum of the contribution of the ARB added to the roll stiffness generated by differential deflection of the two wheels as the car rolls. Because of the need to control ride height with speed, the vertical stiffness is nonlinear, rising with wheel deflection to enable large aerodynamic loads to be supported without excessive loss of ride height. As the vertical stiffness increases, so does the wheel rate contribution to the roll stiffness and thus also the total roll stiffness. This may be desirable, but it equally well may not be.

Rising wheel rates can be achieved in a number of ways: rising-rate springs, bump rubbers, pneumatic springs (which are naturally rising-rate), or through the geometry of the linkages connecting the wheel to the spring (i.e., the velocity ratio). The roll rate due to the ARB can also be varied by linkage design, either rising or falling rate; however, the scope to counteract a typical rising vertical rate is limited.

The purpose of adding a third spring to the system is to decouple as much as possible the roll rate from the vertical rate, so that each can be optimized individually throughout the speed range. Before examining what is possible, it is worth reviewing what the objectives might be.

The braking performance of the current breed of Formula 1 cars is such that the front tire is sized primarily to provide sufficient braking grip. As a result, it is oversized for all but the highest speed corners, to such an extent that in the slow, second-gear corners, only one front tire is needed! As much roll stiffness as possible is applied to the front axle, relieving the rear from carrying too much roll moment and hurting traction out of the corner. Because the tires have a radial stiffness of approximately 250 N/mm, there is a limit to how much front roll stiffness is possible. If the whole axle was solid, the roll stiffness would be determined by the tire stiffness alone. In practice, the maximum percentage roll stiffness that the front of a Formula 1 car can carry is approximately 63–65%.

With an active suspension, where the spring units do not have to follow normal spring laws (i.e., the deflection does not have to be proportional to load and can even be of the opposite sign), it is possible to achieve a higher percentage front roll stiffness by twisting the axles in opposite directions. At high speeds, when the axle load is high due to aerodynamic downforce, it was possible to have the front axle carry 100% of the roll moment, without the inner wheel leaving the ground, and thus the rear wheel loads remain equal under cornering loads. These are interesting possibilities but are denied with a passive suspension.

A high front roll stiffness is thus desirable for low-speed corners. However, an even higher stiffness—the natural consequence of rising-rate springs—may not be so good in high-speed corners.

The rear has different problems. The 380 mm (15 in.) wide rear tires are too small for the torque of 800+ bhp engines at low speeds. The desire is to keep the rear tire loads as equal as possible under hard acceleration. Unfortunately, hard acceleration is accompanied by squat; therefore, the suspension and also the roll stiffness tend to become stiffer. Under braking, the rear rises and stiffness reduces, which is generally desirable to increase stability, compensating for the forward movement of the aerodynamic center of pressure. High-speed cornering requirements will depend on the mass and aerodynamic balances.

Of course, each car has its own requirements, which may differ from those mentioned here Certainly, they will change with the type of circuit. However, it is evident that some of the roll stiffness characteristics are the opposite of what is desired. A potential solution lies with a third spring.

Most anti-roll systems employ the differential deflection of the two sides of the suspension to twist a torsion bar. Somewhere in the system, there exists a point of zero torsional deflection due to roll but which moves with vertical axle motion. Adding a third spring at this point serves only to add vertical stiffness. In practice, this point is either halfway along a conventional transverse ARB or in the middle of the transverse lever of a T-configuration vertical ARB.

Setting up such a system, with its infinite possible characteristics, is fraught with wrong paths to follow. Some fundamental principles are needed as guidance and to provide a starting point. The two primary springs should be as soft as is required to generate the minimum roll stiffness. This will also determine the single-wheel bump rate, which affects the ability of the car to ride over curbs. The range of the ARB (or selection of ARB's) adds to the spring-induced roll stiffness to give the total range of roll stiffness desired. The third spring adds to the sum of the other two to raise the axle vertical stiffness and to endow it with the rising-rate characteristics required. The third spring may consist only of a bump rubber (Fig. 9.1).

Additional nonlinearity can be built in via the geometry of the bell-cranks and linkages, but the stiffness of the system will only be as good as the installation stiffness. The loads in the linkage of a push-rod suspension are high, any deflection is likewise multiplied upward and detracts from the total installed stiffness.

The range of available characteristics is literally infinite, and the task of tuning the suspension for a circuit is commensurately large. Knowledge of the characteristics of the system, which are derived from rig testing, combined with a systematic approach is absolutely necessary.

As if the three-spring suspension setup is not complicated enough, all of this reasoning can be applied to dampers, complete with their own nonlinear and asymmetrical characteristics. In a

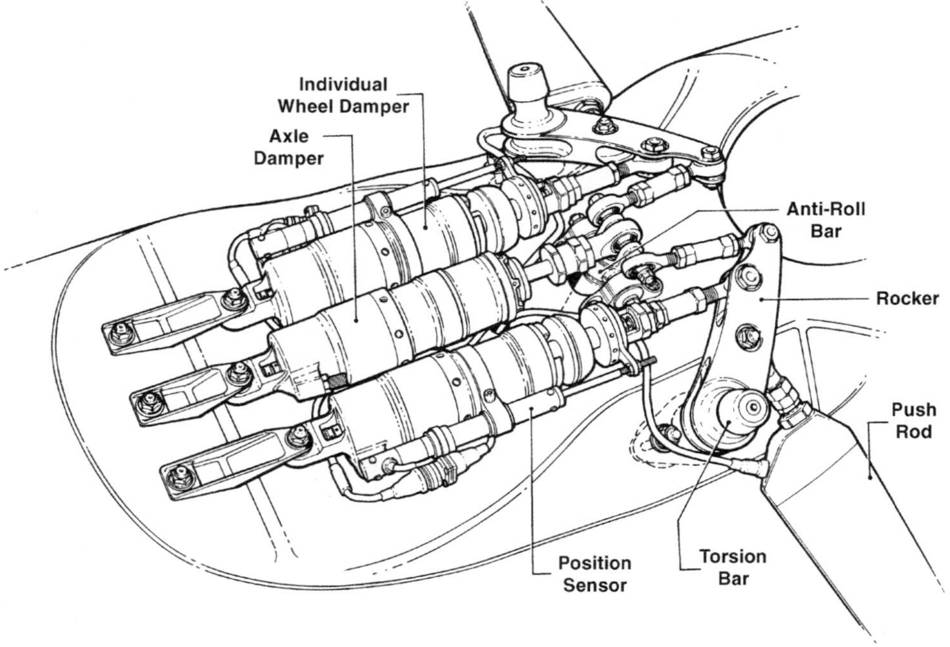

Fig. 9.1 Williams FW-19 front suspension springs and dampers.

two-spring/damper arrangement, the roll mode is usually over-damped if vertical motion is correctly damped. Many road cars exhibit roll-rock (felt at the driver's head) as a result of this inherent characteristic. Reducing the damping of the individual wheels to optimize roll and adding vertical motion damping via a third damper, coaxial with the third spring, is attractive if sufficient track time is available to sort it all out.

Involving the dampers opens new possibilities. All gas-pressurized dampers have an inherent spring rate, generated by the piston rod displacing fluid into a gas spring. If a single gas spring is connected to the two dampers instead of individual ones, then only vertical motion compresses the gas, roll motion resulting in a transfer of fluid from one damper to the other and no net transfer to the gas spring. By adding a coil spring inside the gas spring, it is possible to make the rate more highly rising-rate than the characteristics of the gas spring and thus endow the suspension vertical rate with all the desired nonlinearity. Now we have a rising-rate vertical stiffness and zero contribution to the roll stiffness.

The downside to this arrangement is that the fluid pressure rises, with associated increased seal friction. Likewise, damping characteristics become more complex with interaction between the two sides of the car.

The potential benefits of a well-set-up three-spring suspension will make themselves felt in the handling balance of the car, under all conditions of speed, accelerating, and braking. They will not significantly affect the overall grip but rather how much of it the driver can use. The task of optimizing the setup is much more difficult than with a two-spring suspension and contributes to the non-universal adoption of this arrangement. However, having explored the possibilities of separating the vertical and roll characteristics, it is perhaps surprising that the same has not been tried for pitch motion. Alec Moulton showed what is possible, using hydraulics on the Austin/Morris 1100. However, I know of no even moderately successful application to a ground-effect car, with its premium on pitch control. Maybe it will come.

Although the six-link independent suspension arrangement has been used at both ends of the car in Formula 1 since the 1960s, the way in which the spring, damper, and ARB loads are transmitted to the wheel has evolved steadily and continues to do so. Perhaps because every designer or engineer brought up on Meccano loves to design a linkage system, a variety of arrangements have been seen. However, a number of layouts have achieved state-of-the-art status for a period, until the requirements have changed and a new layout has been pioneered by the most forward-thinking designer of that time.

The adoption of concentric coil springs and telescopic hydraulic dampers as standard on Formula 1 cars in the late 1950s led quickly to the most structurally efficient way of applying the forces they generated. Mounted as the diagonal within the trapezoid formed by the suspension links, the installation was inherently stiff and the spring force helped unload the lower wishbone of the outside wheel when highly loaded during cornering (Fig. 9.2). The spring stiffness was sufficiently low that buckling moments in the damper did not unduly load the damper rod bearing, causing friction. The limitations of this light and efficient arrangement are the aerodynamic drag of the exposed spring/damper and the limited range of motion ratios between the wheel and the spring/damper that are possible.

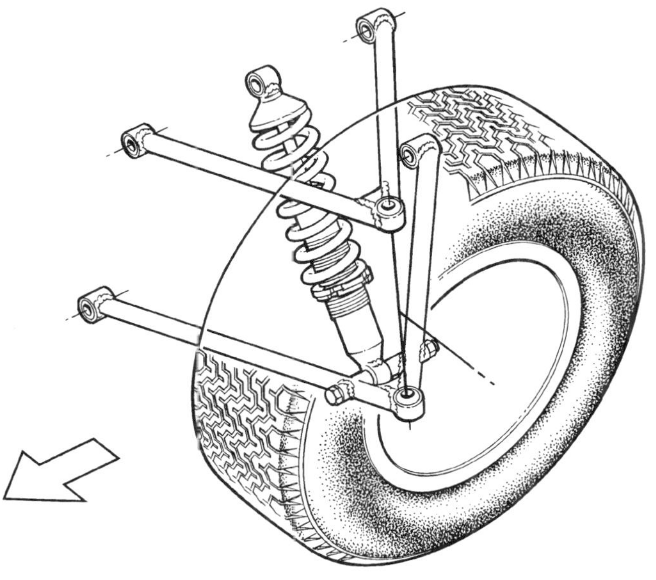

Fig. 9.2 Double wishbone and outboard spring/damper suspension.

These limitations were overcome in the top rocker arrangement that first appeared on the front suspension of the Lotus T21 in 1961, but did so at the expense of installation stiffness and structural efficiency (Fig. 9.3). By turning the top wishbone into a rocker, which operated the spring/damper unit mounted inside the chassis, Colin Chapman not only moved the unit out of the airstream but was also able to design the geometry of the rocker to provide rising-rate characteristics. However, the top wishbone was now subjected to bending loads, while the spring load no longer relieved the peak compression loads of the lower wishbone. Because it was considered that airflow was already disturbed by the time it reached the rear of the car, and because of space restrictions, the rear suspension arrangement retained outboard springs/dampers. This combination of front and rear spring/damper arrangements held sway until 1970, the period during which aerodynamic downforce was introduced, when Colin Chapman redefined Formula 1 cars with the Lotus T72.

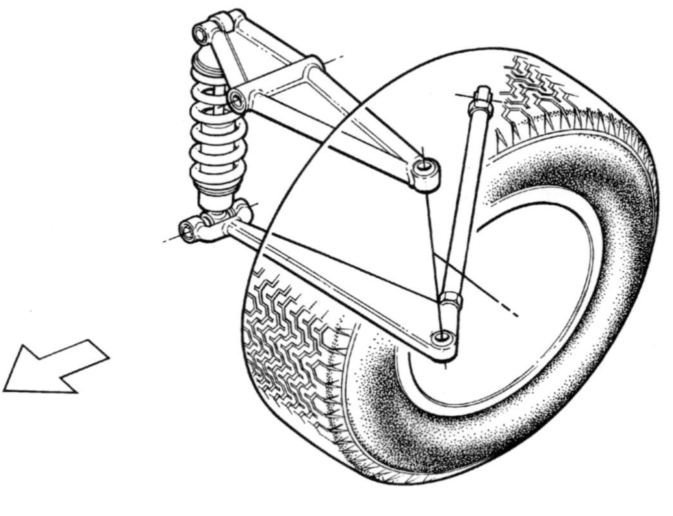

Fig. 9.3 Top rocker and inboard spring/damper suspension.

Chapman wished to use softer-than-usual springs on the T72, without excessive body movement due to aerodynamic and fuel loads, necessitating a greater degree of rising-rate spring characteristics than had been used until then. The inboard front brakes of the car also presented a packaging problem for a combined inboard coil spring and damper. For these reasons, Chapman chose to use compound torsion bars, front and rear, operated by short rockers. Pull rods transferred the loads to the top wishbone (Fig. 9.4). At the front, the rocker also operated the damper, while at the rear a conventional outboard damper installation was used. The arrangement is aerodynamically very clean, and a variety of motion ratios are achievable.

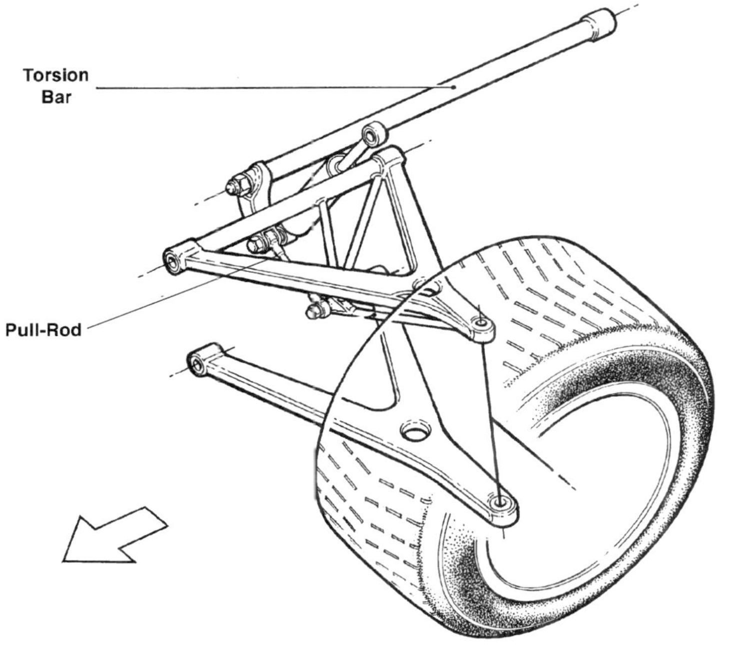

Fig. 9.4 Lotus T72 suspension: pull-rod operating an inboard torsion bar and damper.

Gordon Murray, probably the most innovative Formula 1 designer to succeed Chapman, picked up the pull-rod arrangement for the Brabham BT44 of 1974 but eschewed the torsion bars in favor of coil springs (Fig. 9.5). He was also able to achieve the desired rising-rate characteristics with a long pull-rod that connected the rocker directly to the outboard end of the top wishbone, cleaning up the whole layout. Although the pull-rod is structurally very efficient, its tension loads increase the peak compression load in the top wishbone, which occurs during braking. For the BT46 four years later, Murray changed to push-rods, operating the spring/damper via small rockers (Fig. 9.6). Functionally very similar to the pull-rod arrangement, the compression load of the push-rods relieved the bottom wishbone in a manner similar to the outboard spring/dampers of the 1950s. The need to relieve the lower wishbone had increased dramatically with the high cornering g-forces that ground effect had brought, and the high spring rates and forces that had accompanied them. Thus, the disadvantages of moving the suspension units inboard were finally overcome. This functionally is the arrangement used today, both front and rear. However, pull-rods continued to be used by some designers into the 1990s and reappeared on the front of the 2000 Arrows to lower the mass of the torsion bars.

Refinement to increase installation stiffness without increasing weight, and to improve packaging for a slimmer monocoque and better access, has created further evolution. The rocker and spring/damper was turned through 90° to lay it on the top of the monocoque. This arrangement permitted a narrower monocoque between the wheels, deemed important for aerodynamic

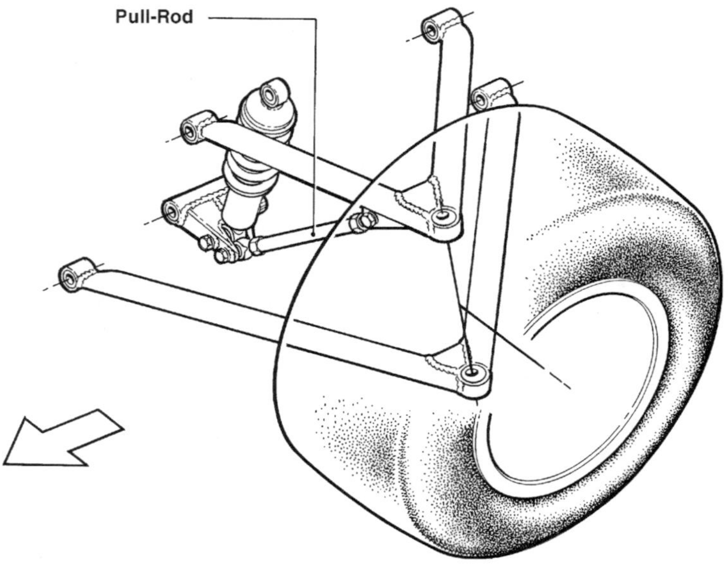

Fig. 9.5 Pull-rod and inboard coil spring and damper.

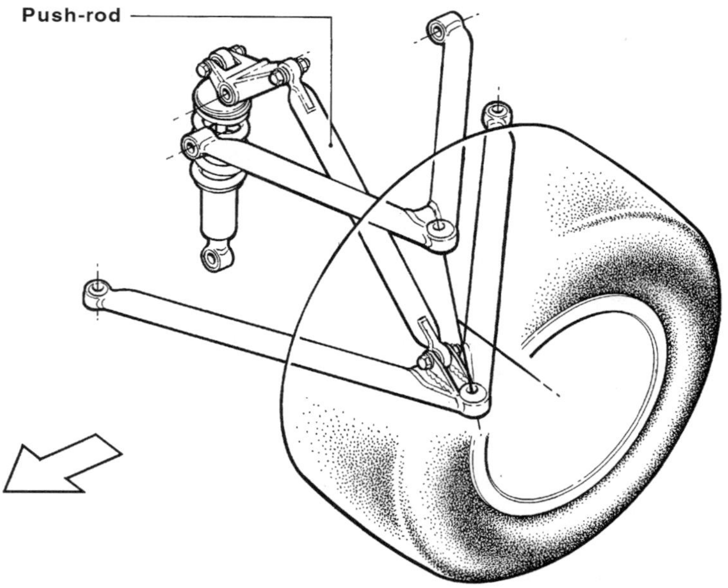

Fig. 9.6 Push-rod and vertical coil spring and damper.

reasons, and put the suspension units in dead space above the driver's lower legs where they were easily accessible for adjustment. The disadvantages were that the mass of the suspension units was raised, and the rocker pivot shaft was in single shear, making it inherently less stiff (Fig. 9.7). Coil springs were replaced by torsion bars operated directly by the rockers, which continued to work the dampers (Fig. 9.8). An additional advantage was that it removed the stiff springs from the dampers which, if the ends of the springs did not remain absolutely square to the damper axis, side-loaded the damper shaft and caused friction. The final evolution to date, which Rory Byrne introduced on the 1998 Ferrari F300, was to turn the rocker through 90° again and lay the torsion bar horizontal, while the damper reverted to its previous location (Fig. 9.9). A regulation minimum monocoque width in this area meant that there was no aerodynamic penalty. The rocker gained a second bearing, greatly increasing its installation stiffness. Likewise, the third damper and bump rubbers, mounted transversally connecting the two rockers, were almost aligned with the push-rod, minimizing its reaction loads in the rocker pivot.

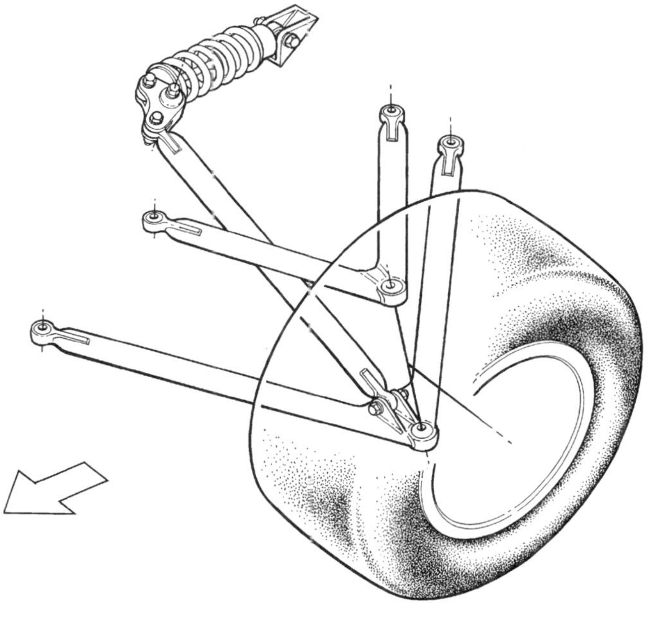

Fig. 9.7 Horizontal coil spring and damper, mounted on top of the monocoque, operated by push rod.

A number of tools are now available to the suspension designer and development engineer. The first is linkage system analysis software, the best known of which is probably ADAMS. Equipped with such tools, the desired kinematic properties can be determined quickly within the packaging constraints. The same software also will calculate the loads in all the components and, when combined with FEA on those components, will yield the compliance characteristics. Thus, the suspension engineer can know the attitude of the wheel and tire, relative to the chassis, under all loading conditions. At least, he can know this in theory.

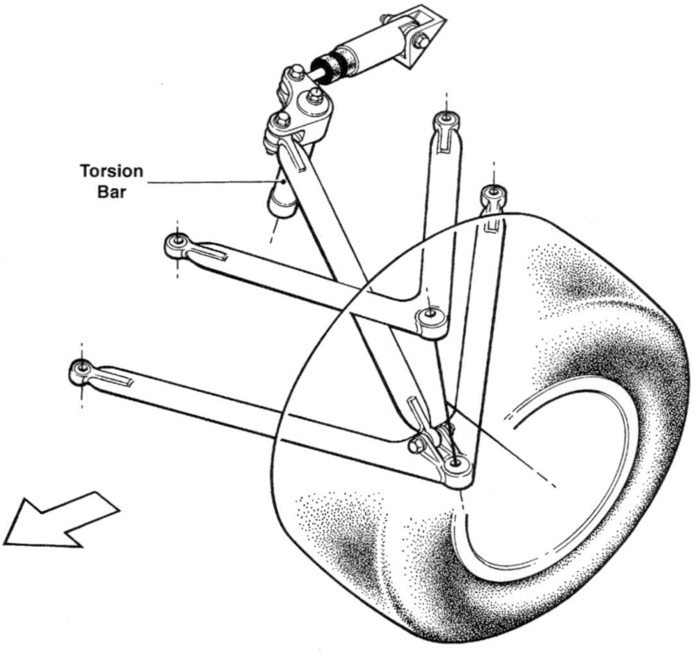

Fig. 9.8 Vertical torsion bar and horizontal damper, operated by push rod.

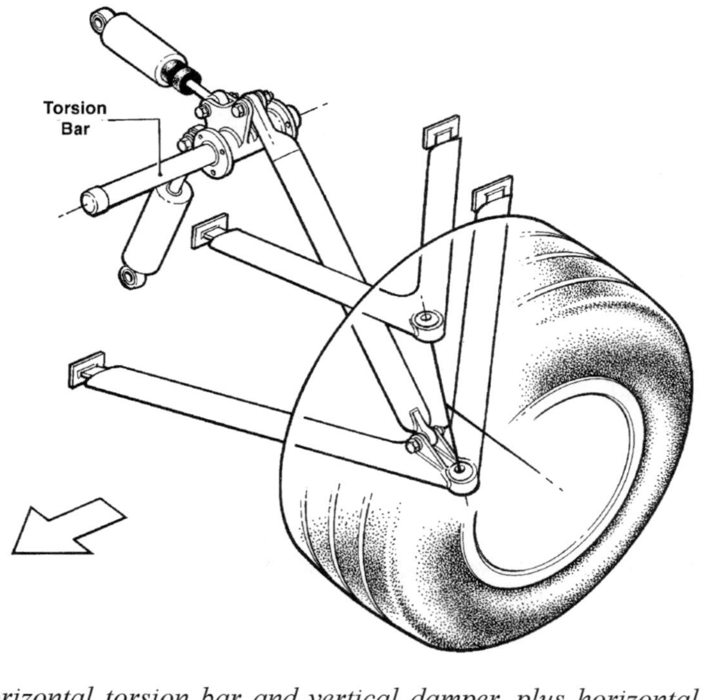

Fig. 9.9 Horizontal torsion bar and vertical damper, plus horizontal third damper (connected to the opposite side's rocker), operated by push rod.

Fig. 9.10 CFRP and flexure pivot front suspension. (Courtesy of Sutton Motorsport Images)

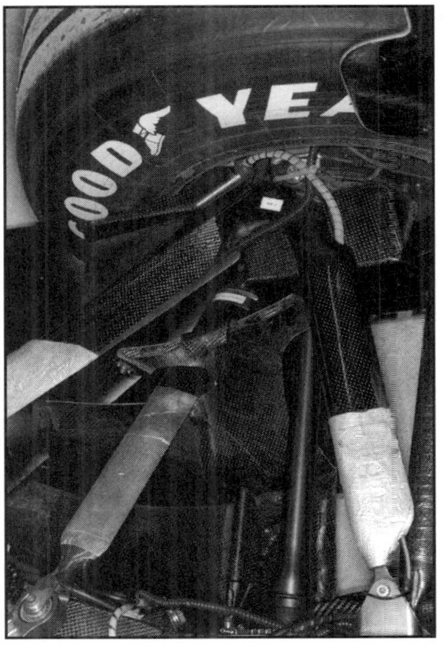

Fig. 9.11 CFRP rear suspension. Note the insulation to protect CFRP linkages from exhaust heat.

Fig. 9.12 Novel anti-roll "bar" system fitted to the Ferrari F300.

To validate the output of the linkage program, a suspension kinematics and compliance test machine (K&C machine) is required. Such machines have been in common use in road car laboratories for many years but are quite new on the racing scene. The chassis of the car to be tested is rigidly mounted to the K&C machine, and three-axis loads are applied to the contact patches of the tires via friction pads. The motion along and around the three axes is measured as the loads are applied, the whole process being computer controlled and the data stored digitally. Plots of kinematics and stiffness characteristics are output, including free play and hysteresis. Few design faults escape the scrutiny of the K&C machine.

The K&C machine is a quasi-static test rig. To test and analyze the dynamic properties of the suspension, a seven-post servo-test rig is needed. The car is supported by four electro-hydraulic servo-actuators under the tires, and three more servo-actuators apply vertical loads to represent the aerodynamic downforce. The downforce actuators have closed-loop load control and are programmed to apply the desired aerodynamic distribution in heave, pitch, and roll. Load cells and low friction pads are placed between the tires and the four actuators, which can be either displacement or load controlled. The car is excited via these actuators in the four modes—heave, pitch, roll, and warp—with either a discrete frequency, white noise, or a schedule derived from tests on a given track. The transfer function of the tire, suspension linkage system, and spring/damper, as well as the modal characteristics, can all be measured and computed. Endurance testing of the suspension can also be conducted on this equipment.

Because the wheels do not rotate, and hence their dynamic characteristics may not be simulated, care must be taken in interpreting the results from the seven-post rig. The greatest value of this rig is in deriving the dynamic characteristics of the suspension and the structure of the car for dynamic modeling, as well as for validating the models. Many teams also use the rigs for setting up the suspension, particularly the damping, using road profiles of the tracks on which they will race.

Although these tools have banished much of the black art that surrounded race car suspensions, the interaction between the damper and the tire, and the influence of the former on the ability of the latter to generate grip on a given surface, are not yet fully understood by either suspension or tire engineers.

We have seen that the suspension springs influence several potentially conflicting characteristics of the car. The same is true for the dampers. Not only do they affect the oscillatory motion of the chassis in heave, pitch, and roll, but they also influence the single-wheel motion or wheel-hop. On top of this, the dampers determine the rate at which the car pitches when braking or acceleration is varied, and the transient roll motion when cornering. At the same time, they determine the instantaneous distribution of wheel loads during these maneuvers, affecting transient understeer or oversteer. The chances of incorporating the ideal characteristics for each and all of these functions in one mechanical device are small.

Using the viscous properties of a fluid to extract energy from the suspension motion has been the preferred damping method since friction dampers lost favor in the 1930s. The basic characteristics of a piston driving fluid through an orifice is that the load will increase as the square of the velocity, giving a progressive load-velocity characteristic, whereas it has been long established that a digressive characteristic is usually more desirable. To generate this characteristic and the desired curve shape, multiple orifices or valves are placed in parallel—some fixed, some adjustable, and some pressure relieving, with separate sets for bump and rebound motion (Figs. 9.13 and 9.14). A great deal of work has been done in modeling and accurate calibration of these orifices and valves because their importance was highlighted by active suspension. However, the critical features of the load-velocity curves of a damper are also affected by static and running friction, fluid compressibility and temperature characteristics, fluid cavitation, and damper structural stiffness. This makes determination of the characteristics extremely complex, even if the desired characteristics have been established.

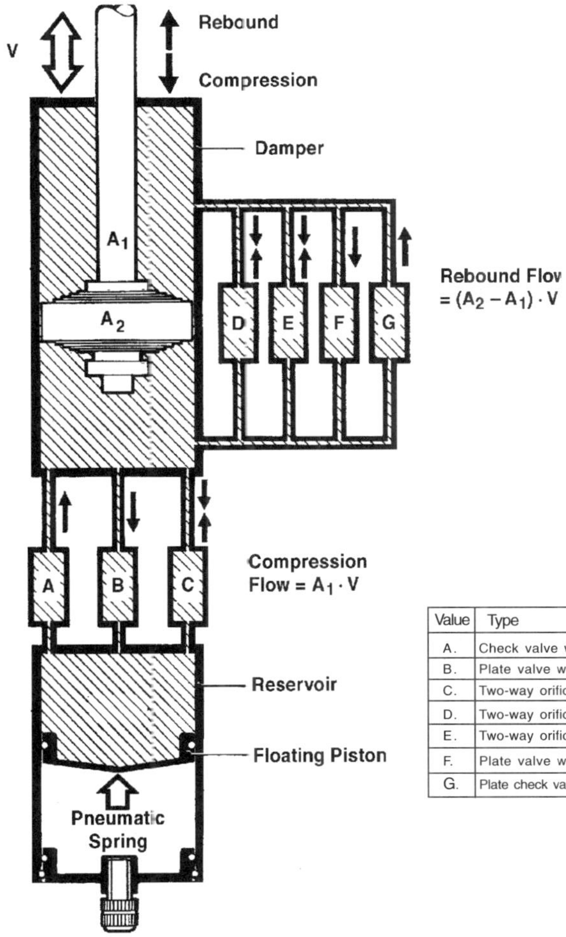

Value	Type	Adjustability
A.	Check valve with preload	No
B.	Plate valve with preload	Yes
C.	Two-way orifice	Yes
D.	Two-way orifice	Yes
E.	Two-way orifice	No
F.	Plate valve with preload	Yes
G.	Plate check valve with preload	No

Fig. 9.13 Hydraulic valve schematic of a racing damper.

Fig. 9.14 Penske four-way adjustable damper. (Courtesy of Penske Shocks, Inc.)

Computer-controlled valves to actively modify the load-velocity characteristics are commonly used in road cars but were banned in Formula 1, along with active suspension. However, the properties of electro-rheological and magneto-rheological fluids are being investigated, in which the alignment relative to the flow direction of micron-sized particles suspended in the fluid is changed by high-voltage electrical or magnetic fields to change the viscosity. Closing a loop around velocity would eliminate many of the variations inherent in a conventional damper. However, the temptation to use such a device to provide active damping will have to be resisted.

Major performance gains will not be achieved via suspension design now that active suspension is banned. However, the suspension is the key to yielding the potential of both the tires and the aerodynamics, and to providing the driver with a stable and controllable car, which he can take to the edges of the performance envelope as much of the time as possible. Suspension seldom adds performance. However, it often detracts from it.

Chapter 10

Brakes

Praise without end the go-ahead zeal
Of whoever it was invented the wheel;
But never a word for the poor soul's sake
That thought ahead and invented the brake.
—Howard Nemerov

When a Formula 1 driver lifts his foot off the throttle pedal at 300 kph (186 mph), which is the speed achieved on most short straights, the car will slow at more than 1g due solely to the aerodynamic drag—a rate equivalent to the maximum braking on a typical road car. By the time the driver has put his foot on the brake pedal and the brakes have heated to their working temperature, the car will have slowed by 20 kph (12 mph) and the brakes will add approximately 3.25g to the effort to slow the car, instantaneously absorbing more than 2200 hp. Slightly more than two seconds later, the car will be travelling at 100 kph (62 mph), and the driver will be ready to ease off the brake pedal as he approaches the apex to the second-gear corner ahead.

During these frantic couple of seconds, apart from changing down four or five gears, the driver must modulate the load on the brake pedal to reduce the braking as the aerodynamic downforce comes off the car and to make sure that the balance does not change too much and destabilize it. He might possibly overtake another car while all this is occurring. These two or three seconds, as drivers slow their cars and set them up to enter a corner, are probably the drivers' busiest periods in a complete lap. These moments distinguish the great drivers from the merely good ones as the deceleration forces work against the natural stability of the vehicle, and make it want to swap ends, just as the driver turns in. Drivers inevitably complain most about the performance and characteristics of the car under braking, and most incidents when control is finally lost occur in the braking phase. Thus, it is not surprising that designers seek to provide the most sophisticated controls and control systems that they can devise, within the regulations, to assist the driver in getting the most from his car under braking.

The numbers quoted here make the brakes, rather than the engine, the most powerful system on a Formula 1 car and one of the most influential in respect to the stability and control of the car. The engineering disciplines involved in the design and development of braking systems include mechanical design, material sciences, aerodynamics, heat transfer, and vehicle dynamics. Ideally, the driver will have under his control a system powerful enough to reach the braking limit of the tires at all speeds, and yet be able to modulate it such that the limit is never exceeded, in wet or dry conditions. The energy absorbed must be managed and dissipated without unduly changing the performance of the system. The distribution of braking torque among the four tires should be adjustable, within the limits set by the regulations, such that the effect on stability and control can be optimized.

The maximum longitudinal force that the four tires together can generate ultimately limits the magnitude of the braking force that can be generated. This is a function of their local and instantaneous coefficients of friction and the individual vertical loads on them. Aerodynamic downforce, longitudinal load transfer, and suspension characteristics all play significant roles in determining the vertical loads and how they vary during the braking cycle. The development of aerodynamic downforce and high coefficient of friction tires during the 1970s led to the grip of the tires being able to overpower the cast-iron brake discs and their copper/asbestos pads at high speeds. The maximum rate of energy dissipation (i.e., the time taken to brake down to a given speed) is a function of the thermal inertia of the disc and pads and the rate of heat dissipation, which is primarily a function of the temperature of the disc and pads. The simple solution to both the power and energy absorption limitations is bigger brakes—larger effective radius, greater pad area, greater mass, and greater cooling surface area. The penalty is mass. Unless mounted inboard, brakes are part of the unsprung mass, any increase of which has a negative effect on tire grip. In the same way that the first Dunlop disc brakes to be used on race cars were developed from aircraft brakes, the technology for improved braking has continued to rely heavily on the aerospace industries. Gordon Murray, the designer of Brabhams during the late 1970s and early 1980s, was the first to see the potential of the carbon-fiber reinforced carbon matrix materials being developed for aircraft brakes (Fig. 10.1).

Carbon-carbon maintains its friction properties up to much higher temperatures than copper-asbestos/cast iron and is almost a quarter of the density. For racing series that prohibit carbon-carbon brakes, carbon-metallic pads have been developed for use with cast iron discs, and these maintain the friction to similar temperatures, as all carbon-carbon systems. Their performance is similar to that of carbon-carbon, but the mass penalty of the disc remains. Designing brakes to run optimally at these high temperatures allows better use of the cooling air for heat dissipation but presents its own problems in prevention of heat reaching the wheel bearings, calipers, and brake fluid. The original carbon-carbon material that Brabham used in 1976 was fitted into a disc design similar to the aircraft application, but the different duty cycle caused temperature, cooling, and mounting/drive problems that were not completely solved until the mid-1980s.

Fig. 10.1 Carbon-carbon disc brakes first appeared on Brabhams in the late 1970s. (Courtesy of Phipps/Sutton)

Aircraft brakes are designed for one high-energy stop. They rely on multi-discs and use the mass of the cold wheels and tires as heat sinks, with almost no air cooling. They more closely resemble multi-plate racing clutches, the design case for which is the start—a single, high-energy application. Racing brakes must absorb a continuous series of high-energy stops with short recuperation periods between them. Material development and discs machined in one piece from carbon-carbon have resulted in lightweight brakes with consistent performance. The uprights, calipers, wheels, and tires stabilize at a temperature that is limited by the characteristics of individual components, and their ability to absorb energy at a greater rate is small. The disc and pads have a limited thermal inertia and rise in temperature quickly when the brakes are applied. Heat must be conducted away from the friction surfaces and transferred by forced convection into the airstream. Although the friction properties of carbon-carbon are maintained at high temperature, the wear rate rises sharply when the oxidation temperature of 650–700°C (1200–1300°F) is reached. Wheel uprights are designed to duct air from the brake cooling ducts on the inside of the wheels, through the vaned internal structure of the upright (cooling the wheel bearings in the process), and to pre-swirl the cooling air into the eye of the disc. Radial holes in the disc centrifuge the air through its core (Fig. 10.2). Air is bled from the ducts to cool the pads and critical areas of the caliper. Ducting air directly onto the friction surface is not desirable because it increases the speed of the oxidation process and hence increases wear. The whole process is monitored by sensors measuring critical temperatures, including disc face temperature, disc thickness (disc wear), and caliper piston extension (pad + disc wear).

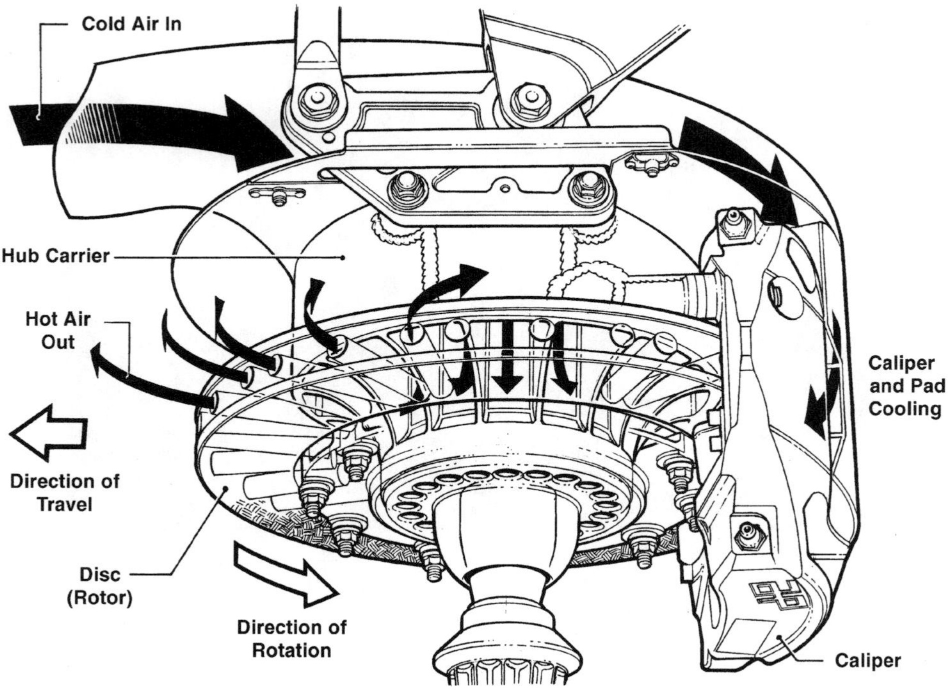

Fig. 10.2 Front upright and brake, showing brake cooling airflow.

The development of carbon-carbon and the technologies that enable the material to be applied to both the brakes and clutch are now fairly mature (Fig. 10.3). The detail friction-temperature-wear properties can be tailored to suit individual applications, and Formula 1 teams have a choice of materials for brakes, to suit the variety of braking cycles on different circuits. Specialist suppliers of carbon-carbon (such as SEP and Hitco) and brake systems (such as AP and Brembo) take most of the load of developing brakes off the teams, leaving the latter to concentrate on providing solutions to the problems that the driver experiences in applying and controlling this powerful system during the two to three seconds in which he slows the car and sets it up for a corner.

The force with which the driver is able to clamp the pads to the disc is limited by the strength in his leg and foot and the stiffness of the system linking his foot to the brake pads. If the system was infinitely stiff, the force ratio between his foot and the pads could be such that he could generate whatever force was required to apply the maximum braking torque that the tires were capable of transmitting to the road, at all speeds. The hydraulic system that applies pad clamping force through a caliper bridging the disc is not infinitely stiff. As a result, the force ratio is limited by the available stroke of the brake pedal, which itself is limited by the articulation of the driver's foot. High downforce and high-grip tires in the early 1990s led to the development of servo-brakes, thereby sidestepping the stiffness limitation by the addition of energy. With a system that was now powerful enough to lock the wheels at any speed, ABS was included to control locking and hence brake distribution as well.

Fig. 10.3 Carbon-carbon rear brake.

Power brakes and ABS were banned at the end of 1993, leaving the drivers with a brake system for which they had inadequate force × distance at the right foot to gain the maximum braking available from the tires at high speeds. This led to a focus on all aspects of the stiffness of the system, particularly the caliper. Aluminum caliper material gave way to metal-matrix aluminum, and finally beryllium-aluminum alloy in the search for stiffness and low weight. As cost escalated and all of the competitors followed this route, thus eliminating any competitive advantage, the FIA stepped in and limited caliper material to a modulus of elasticity less than 80 GPa. The resulting expensive search for the lightest material with this modulus has caused the regulations to further limit all metallic materials used in the construction of the chassis to a specific modulus less than 40 GPa/(gm/cm^3) which was introduced for 2000. The engine was included in this regulation the following year.

This leaves the distribution of braking among the four wheels as the dominant braking issue confronting the system designer. Figure 10.4 shows how the percentage of vertical load on the front axle varies with speed, as well as the braking g-level. The g-level must fall with speed as the aerodynamic downforce reduces with the square of the speed. If the aerodynamic distribution is almost the same as the weight distribution and does not change much with car attitude or ride height, then the main contribution to the changing front axle percentage is due to weight transfer. Approximately 10% of the retardation force is transferred from the rear axle to the front while decelerating. As speed reduces, the retardation force falls and the balance changes toward the rear. In reality, most Formula 1 cars have aerodynamic configurations that move the aerodynamic balance toward the front as the ride height reduces and the car pitches nose downward. The shift in load distribution toward the rear as the speed falls will be more pronounced than that shown in Fig. 10.4. Formula 1 cars, and indeed all race cars, have two independent brake circuits (front and rear) that come together at the brake pedal. A pivot on the pedal apportions pedal load to the twin master cylinders according to the position of the pivot on the cross bar that links the pedal to the master cylinder pistons.

Because the setup of the brake balance between front and rear is so important to the stability of the car under braking and turn in, it is left up to the driver to adjust the balance while driving. Mechanisms to move the pivot point, via a cockpit-mounted control lever or knob, permit fine adjustments to be made to the balance, compensating for changes in track condition, fuel load, tire conditions, and driver preference. However, it is beyond the capabilities of even the finest drivers to make such adjustments while actually braking to compensate for the changes in vertical load distribution shown in Fig. 10.4.

During the 1970s, a new breed of young and talented drivers were learning their trade in karting, often before reaching the age that permitted them to drive a car. With only two pedals, distributed one to each foot, only the left one could be used for braking. With no preconceived

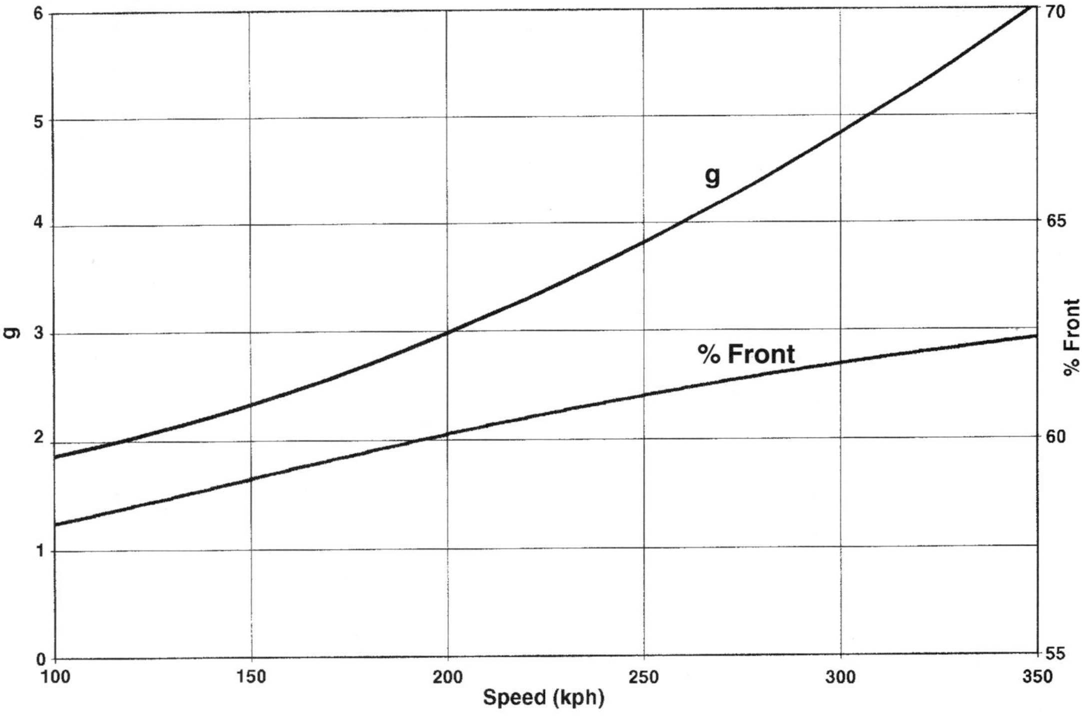

Fig. 10.4 Braking deceleration and optimum brake balance versus speed. (Author's illustration)

ideas on how to drive fast, these young Ronnie Petersons *et al.* discovered that by using the throttle and brakes at the same time, it was possible to tune the brake balance and thereby alter the balance of the kart as they braked and turned into a corner. By applying power to the rear axle rather than the front, they made the front tires do more of the braking than the rear. This stabilized the kart under braking and prevented it from tending to oversteer during the critical turn in.

Peterson rose quickly through the lower racing ranks and joined Team Lotus in 1973. He then found Colin Chapman possessed a receptive ear for his ideas about using left-foot braking in Formula 1. Lotus had had some experience with two-pedal race cars in the 1971 T56B gas turbine Formula 1 car. In this application, power was not applied while the left foot was used to brake to adjust brake balance because this would not have worked in the same way, being a four-wheel drive car. Instead, the driver had to keep the throttle open during braking to maintain rpm in the gas generator section of the two-shaft turbine, so that power was available when he needed it in a corner. Without doing so, a lag of several seconds occurred before the turbine spooled up to generate the torque required.

Chapman's answer to Peterson's requests was to design the 1974 T76 Lotus with a Y-shaped brake pedal, the branches of which lay on either side of the steering column, making the brakes available to either foot. The clutch pedal was retained but was only used for starting. Thereafter, gear-changes were performed using a button on the gear lever to activate the clutch via an electro-hydraulic actuation system. Chapman had Peterson and Ickx's road Lotus Elan+2's fitted with identical systems for practice. Although Peterson was particularly enthusiastic about the potential of the system, both it and the car proved problematic.

Chapman did not give up easily. By the time Peterson rejoined Lotus in 1977 after a brief sojourn at March, Chapman had revived the Lotus "queerbox"—a tiny gearbox concept that, in theory at least, did not require use of the clutch to change gears. It appeared briefly on the T78 but gave the drivers such a severe case of tennis elbow, resulting from vibrations fed back through the lever during attempted changes, that it was abandoned so as not to compromise the promising new car.

At the end of 1993, when the FIA gave the job of controlling the car back to the drivers, and brake systems reverted from power/ABS systems to traditional cockpit-adjustable passive systems, one area that was not banned was the use of a control system to change gears, provided it responded only to a signal from the driver. The clutch could be operated automatically during downchanges (it is not needed at all during upchanges), but it had to be operated by the driver during starts. It did not take long for designers to question the need for a clutch pedal simply to enable the driver to get away from rest, but which effectively prevented drivers from left-foot braking. Experiments with a steering-wheel-mounted paddle, similar to the gear-change paddles, showed that a driver could make as good a start (sometimes) controlling the clutch with his hand, as he could with his foot. Instead of operating a switch as per the gear-change, the paddle controlled the clutch position proportionately; thus, a new skill had to be developed because the driver received no force feedback to tell him when the clutch started to take up the drive. With the brake pedal placed ideally for the left foot, Formula 1 cars were similar to karts—20 years after Colin Chapman had tried to achieve this for Ronnie Peterson! With nearly all of the current breed of Formula 1 drivers brought up in karts, it did not take long for the concept to catch on (Fig. 10.5).

Exactly how each driver uses his left foot on the brake, and his right on the throttle, depends on the characteristics of his car, his setup, and his driving style. No driver tells everyone how he obtains the most from the car, not even his team "mate." One thing that is sure is that it is different from the technique used in karts, because aerodynamic downforce changes everything. Without downforce, the g-level and load distribution on each axle remains fairly constant during braking. For maximum retardation, the front-to-rear brake balance will be set up to brake the front and rear tires, so that they reach the limit together (i.e., near or on the point of locking). Under these conditions, the car is marginally stable; any disturbance that puts more load momentarily onto the front axle and takes it off the rear (e.g., a bump in the track) will give the front axle more potential side grip and may instantaneously lock the rear, so that it is unable to generate any side force. The car will now be unstable and want to spin. In a straight line, under these maximum braking conditions, this may not be a problem and should be well within the driver's capabilities to control. On turn in, while continuing to brake hard, the driver may not be able to prevent the rear from overtaking the front.

However, if the driver applies power to the rear axle, the braking effort on the rear tires is reduced and the car becomes stable. Adjusting the power gives the driver a means of controlling

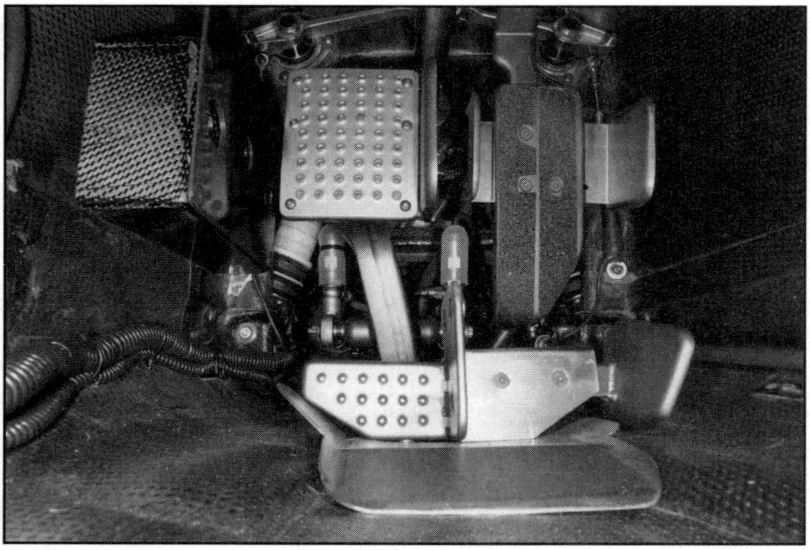

Fig. 10.5 The two-pedal arrangement for left-foot braking in the Ferrari F300.

the stability, and hence his ability to control the car during the entry phase to the corner. This is something to which the driver is accustomed, steering the car on the throttle, even if the sense is reversed: more power causes understeer, and less power leads to oversteer. Of course, it is the other way around with a front-wheel drive car—that is, the same response as when normal steering with the throttle.

Add downforce to all this, and the technique becomes more complex. The g-level and axle loading do not remain constant during maximum braking. At high speed, the front axle load is higher and will accept a larger proportion of the braking effort, but it must be reduced as the speed falls off, along with the overall braking deceleration. In theory, if the driver could apply power at high speed to reduce rear braking, he could maintain the correct balance. That is really too difficult to adjust during the brief braking period, along with gear changing, and it does not make sense to add power at a point when it is impossible to apply enough pedal load to achieve the full braking potential of the tires. Until the correct gear for the corner has been selected, and the driver starts to turn in to the corner, it is not feasible to use left-foot braking to adjust the brake balance. The driver needs help from the engineer.

Control systems as such are not banned in Formula 1; it is what can be done with them that is limited. With the means to inspect hardware and software, it is possible to control how such systems are applied. In the case of brake systems, power braking is not permitted, nor is any form of anti-lock nor one that feeds back a dynamic parameter (e.g., speed, wheel rpm, lateral or longitudinal acceleration, or yaw rate). Thus, none of the ABS or stability augmentation systems appearing on road cars are permitted. If the engineer has any doubts about whether his proposed system is legal, he is encouraged to check with the FIA before spending too much time and money on it.

What is permitted is to provide an open-loop system that emulates a passive, mechanical system. The only advantage that employing computer-controlled electro-hydraulics can bring to the performance of the system is the ability for either the engineer (in the pits) or the driver (in the cockpit) to adjust the built-in open-loop characteristics. Both parties can press buttons or turn knobs to select a different setting—basically no different from the traditional, mechanical, cockpit brake balance adjuster, but more variable and precise.

The interesting question is: What constitutes a passive, mechanical system?

In an unrestricted world, it would be desirable to know how much of the available capacity of each tire was being used for braking and how much was left for controlling and stabilizing the car. To measure this, it is necessary only to measure the slip ratio of each tire and compare it with the known tire characteristics. This is what ABS does (and traction control in the accelerating case), adjusting individual slip ratios according to the instantaneous conditions and demands. However, even a purely mechanical ABS system is expressly forbidden. Lucas once marketed such a system, but it was not popular in competition with the falling cost and greater precision of electronic ABS.

During 1997, McLaren developed a lateral-biasing rear brake system, designed to enable the driver to steer the car using one or the other rear brake, in much the same way that yaw stability/anti-skid systems on some road cars use the ABS system to apply a rear brake and control a skid. On the McLaren, the clutch pedal was replaced by a second pedal (the clutch being operated by the paddle behind the steering wheel). The driver selected the inner rear brake for the approaching corner and when the car began to understeer into the corner, as most Formula 1 cars do, he applied braking effort to this second pedal to brake the inner rear wheel. The resulting yawing moment thus generated turned the car into the corner. The driver could apply power at the same time and use both feet—on the throttle and the brake pedal—to assist with the steering control of the car. On exit from the corner, this same inner rear wheel will tend to spin as power is applied. Thus, a further application of the pedal will brake it, helping the differential transfer torque to the loaded outer wheel and preventing the inner wheel from spinning. This interesting development, at the heart of stability and control issues, was protested and withdrawn before its eligibility was fully assessed, unfortunately denying an age-old method of controlling vehicles from being applied to race cars.

The Technical Regulations now state that the pressure in the calipers of either the front or rear brakes must be equal across the car. Thus, distribution control is limited to front-to-rear passive changes.

Chapter 11

Measurement, Control, and Data Systems

Data, data everywhere, but is there time to think?
—Professor Harold Woolhouse,
with apologies to Samuel Taylor Coleridge

If a single technology is responsible for the shift from black art to science in motor racing over the last two decades, it is measurement and data acquisition. The development of the high-speed digital computer, both vehicle-based for acquisition and storage and desk-based for analysis, has brought this technology to center stage. In the late 1960s and 1970s, the data engineers and their cumbersome, unreliable analog systems were treated with tolerance at best, as they struggled to keep their boxes and wires on the car. Results were viewed with suspicion and seldom relied on, experience proving more reliable than the theories produced by the "boffins" as they strove to interpret their noisy data.

The need to control systems on the car, initially the fuel injection and ignition on the turbocharged 1.5-liter engines of the early 1980s, and active suspension heralded the digital onboard computer with its solid-state memory. It brought a discipline to the engineering of these systems and integration with the rest of the car.

Measurement is an essential engineering technology—the first wheel would not have fitted the first axle without it—but only recently has it been possible to measure dynamic parameters that change faster than it is feasible for a human to record the readings. The development of dynamic sensors, chart recorders, and analog tape recorders revolutionized engineers' ability to determine what was actually happening to the mechanisms they had designed. These systems were first run on Formula 1 cars in the 1960s to obtain pressure and temperature data and to explore the effect of the emerging development of aerodynamic downforce (Fig. 11.1).

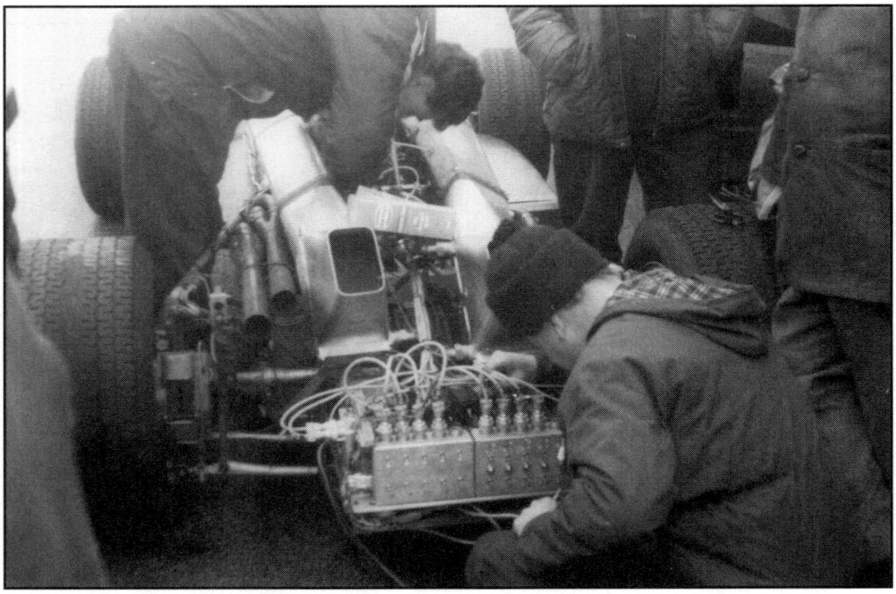

Fig. 11.1 Analog data system mounted on a BRM in 1969. The technician is adjusting the eight-channel chart recorder. (Author's collection)

Today, engineers are able to measure parameters every few microseconds during events that may last only a millisecond or so, for instance the pressure during the combustion process, within the cylinder of an engine turning at 18,000 rpm. Routine onboard measurements every millisecond are common (e.g., the acceleration pulse during a crash), and a car with all its systems may have 100 or more sensors monitoring every function. Data can be stored in memory (400 mb or more, at 120,000 × 16 bits per second) or transmitted to the pits by either continuous or burst telemetry for almost real-time display to engineers seated in front of PC screens. Almost every rig and piece of test equipment is fitted with data acquisition equipment, the most complex being those fitted to wind tunnels and engine dynamometers for routine testing (Fig. 11.2).

Measurements are made for three main purposes: (1) display, (2) storage for subsequent analysis, and (3) input to control systems.

Driver displays take two forms: (1) displays to enable the driver to monitor the health of the key systems, particularly those that may require him to modify his driving, for either safety or reliability reasons; and (2) those that assist him in the driving task. After digital data systems were fitted, it became possible to replace analog instruments with an LCD screen, on which selectable parameters are displayed. It also became possible to monitor certain parameters and display warnings only if they went outside limits. The main instrument for assisting with driving is the rev-counter, and this too was displayed on the LCD screen, albeit in analog

Fig. 11.2 A 100+ channel, digital data system in 1999. (Courtesy of Sutton Motorsport Images)

format. As soon as stored data became available, the engineers no longer relied on the driver to report readings of pressures and temperatures, or peak rpm on the straight. Less and less did the driver need to look down in the cockpit at his display, and thus the display soon evolved into a systems warning panel, checked by the driver only when alerted to do so. Absolute rpm is also no longer required by the driver, and thus the rev-counter has been replaced by an LED display that informs the driver when to change gear.

Telemetry has provided the engineers, back in the pits, with a much higher level of systems monitoring than any driver could manage while driving. Almost real-time plots of key system health parameters unfold across PC screens in front of systems engineers. These experts know what patterns are normal as the car circulates around a given circuit. The engineers will notice any deviation from the normal, and the driver can be alerted to destructive trends by radio. Numerous engine blowups have been saved by timely warnings of rising temperature or falling pressure but not all, because the life expectancy of an 18,000 rpm engine when something goes wrong is measured in seconds.

The trend today is toward monitoring, rather than the slavish gathering of megabytes of data. Fault-finding algorithms in the car computer can identify the cause of many problems and communicate the findings to the pits by telemetry. There, a pop-up window alerts the engineer, and, if necessary, the local system computer is instructed to modify or shut down that system. This approach bypasses the decision-making and relatively slow actions of human intervention, but it is open to false diagnosis and has lost races.

The development of sophisticated software to perform real-time monitoring, diagnosis, and corrective or precautionary action may save on the amount of data collected during a competitive event, but it is unlikely to have an impact on the data acquired during testing. The move to limit the amount of Formula 1 testing, on the basis that it is the same for everyone and the cost keeps escalating without real benefit, is putting more emphasis on getting as much from a day of testing as possible. Each department—engine, transmission, control software, vehicle dynamics, structures, design, and aerodynamics—jostles with the others for test time. The more tests that can be run simultaneously, the better. This results in test cars being covered in sensors, as engineers from each department try to maximize the test time they are allocated during a day. Data is transferred to computers in the pits and is either initially analyzed there, influencing the next test point, or is electronically transferred to specialists at the team's base.

Until recently, reliable telemetry of data back to the pits could be achieved only by using line-of-sight transmission. A 10-mb burst is transmitted as the car passes the receiving aerial mounted by the track. Semi-continuous telemetry provides almost real-time transmission; however, on some parts of most circuits, reception fails. Intelligent strategies enable lost data to be retransmitted, but the process is inhibited by the regulation that prevents information from being transmitted from pits to car. The latest digital broadcasting techniques, using multiple digital signal processors (DSP's) to unscramble signals, are able to use pieces of signals that have been reflected off buildings, etc., to provide complete coverage. Voice and data transmissions are combined in a single unit, and video will follow.

The concept to track test process for any part or system of a Formula 1 car tends to follow the steps illustrated in Fig. 11.3.

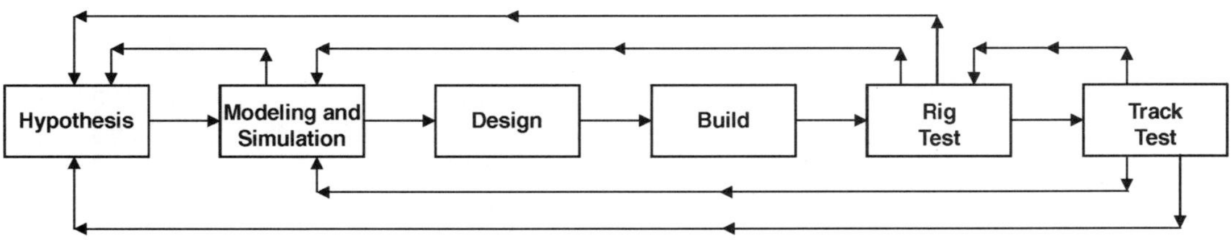

Fig. 11.3 Concept-to-track test process.

The modeling or simulation stage has been introduced since the computer entered design and engineering offices, and it has increased in importance as the software for modeling more and more of the car and its systems has become available. Simulations require real data as inputs, and rig and track testing must be used to validate their outputs. The feedback loops are the test data files that enable the concept and design to be refined by more faithful simulation, and they enhance engineers' understanding of the performance of the system.

However, the control systems for the engine and chassis drove the development of small and reliable sensors, wiring systems, and computers and electronics able to withstand the harsh environment of a race car. A high-quality measurement system is the heart of any control system. Formula 1 engines have large out-of-balance forces, and ±100g at 300 Hz can be measured at typical electronics mounting locations, with higher levels at sensor locations on the engine itself. High-energy ignition systems generate significant EMI, which can be picked up by high-gain sensors and their wiring. A great deal of development work went into anti-vibration mountings, EMI protection, and signal-conditioning filters, to enable measurements to be fed back in closed-loop control systems. This work has raised the integrity of all measurements on the cars.

Many of the sensors used on Formula 1 cars and throughout racing are derived from either aerospace or road car sensor technology, e.g., temperature, pressure, acceleration, linear and rotary position, rpm, engine airflow rate, and gas analysis. With the proliferation of control systems on road cars, many of these sensors have dropped in price from thousands of dollars to only a few dollars, gaining reliability at the same time. Micromachined silicon sensing elements mean that sensors can take their place alongside other electronic components, mounted to circuit boards if required.

However, motor racing also has different measurement requirements from other industries, and Formula 1 has developed specialist technologies to suit these needs, some of which have pushed sensor development in directions that are now becoming relevant to road vehicles. Examples are:

- Temperature, strain, and acceleration of reciprocating and rotating engine components.
- Measurements of the combustion process (temperatures, pressures, and optical gas analysis).
- Torque measurement of both gearbox input shaft and axle half-shafts. Strain measurement techniques are giving way to a variety of systems that detect variations in the magnetic field generated by a torsion-loaded, magnetized shaft, which do not require electronics to be mounted on the rotating shaft.
- Moving surface temperature: tire tread and brake disc surface.
- Aerodynamic and suspension loads: strain-gauged suspension elements.
- Ride height: laser distance sensors for moving surfaces.
- Yaw rate: Mechanical gyros are unreliable in high-vibration environments. Vibrating element and micromachined silicon devices are replacing gyros. Greater precision will come with fiber-optic laser gyros.

- Car position and attitude: Inertial sensing plus Differential Global Positioning Systems (DGPS) are capable of providing car position on the track to within centimeters, as well as vehicle heading and track, enabling vehicle slip angle to be calculated.

- Tire pressure: This requires telemetry from each rotating wheel.

- Video: High-speed video of rig and car dynamic systems requires high-bandwidth telemetry for on-car use.

The closeness of the relationship between the engine supplier and the team will have an influence in determining the level of integration of the engine and chassis computers. Much engine control software is based on road car manufacturers' algorithms, and, unless a significant long-term involvement with the team exists, the engine manufacturers may be unwilling to incorporate them into a single computer. Whatever the level, all modules on the car share data and receive commands on the CAN network which, for control purposes, requires precise timing.

A further influence on the hardware design is the demand of the Technical Regulations for hardware and software inspection. The facilities for any module to be reprogrammed, and its ability to self-overwrite code must be approved, along with the control laws themselves.*

Computer hardware, for data acquisition and control, is more than up to the task of performing these jobs on a Formula 1 car. With little competitive advantage to be gained by faster, higher-precision, larger-memory devices, the focus is on integration, weight, and reliability, plus ease of programming and software development. Modular boards for data logging, control, and I/O can be used to build a computer for most applications, complete with operating system, in much the same way as a PC. High-level setup programming, using facilities such as MATLAB, enables engineers to develop code without having software-writing skills.

The vast pools of data that are acquired must be organized such that engineers within the team are able to access the data easily. To be of any use, they must be interfaced with engine and chassis setup information, wind tunnel, dynamometer and rig data, simulations, MATLAB, and other design and engineering software, via an information manager. Any designer or engineer with the appropriate level of access can then validate models, input data to rigs, derive outputs for driver information, perform design analysis, and other such actions (Fig. 11.4).

The third function that, allied with measurement and computation, makes up a control system is actuation. Actuators are devices that turn the milliwatts of electrical signals into kilowatts of motion. They depend on their force × velocity (i.e., power) requirements to determine the specifications of their power supplies. Electric, pneumatic, and hydraulic power have all been used, but hydraulics predominate. Electrics are used for valves (e.g., fuel injectors) and

---

* See author's note on page 290.

Fig. 11.4 Data analysis is performed at computer terminals in the garage at a racetrack. (Courtesy of Sutton Motorsport Images)

low-power applications, or as the first stage of two- or more stage hydraulic amplifiers. Pneumatics are used in place of hydraulics, where the compliance of the system is an advantage that outweighs the inevitable lower bandwidth that results. They have been used in gear change systems, where it was felt that the compliance would mean the system was gentler on the engagement dogs in the case of a balked gear-change, endowing the system with some of the sensitivity of a manual gear-change.

However, the hydraulics, with its combination of high power density and frequency response, provides the actuation power of most systems currently permitted. These include:

- Gear change
- Clutch
- Differential
- Drive-by-wire throttles
- Power steering
- Brake balance adjustment
- Variable length intake trumpets

Added to these in the past, but now banned, are:

- Active and semi-active suspension
- Power brakes and ABS
- Variable aerodynamics
- Rear steer
- Active brake balance
- Traction control (via drive-by-wire throttles)

A single hydraulic power supply, using an engine-driven pump, supplies all the systems. The pump is either variable displacement or fixed displacement, plus off-loading valve. The sizing of the pump and hydraulic accumulator is determined by the circumstances as the car pulls away from a standstill, at which point pump rpm is low and the clutch, gear change, throttles, trumpets, and steering may all be working hard at the same time, demanding hydraulic fluid from the accumulator. At speed on the track when the engine is at high rpm, the pump is more than able to meet the intermittent demand. Thus, minimizing power losses becomes the critical issue.

The key interface between the computer output and the hydraulic actuator is the electro-hydraulic servo-valve. Developed in the late 1940s for aircraft flight control systems, it has not needed to change significantly. It provides conversion of an electrical signal into a pressure differential, which drives a spool-valve. The spool-valve controls the flow, and hence the pressure and flow velocity of the hydraulic actuator or cylinder. Capable of great precision and high response, it is the enabling technology of all the servo-hydraulic control systems fitted to Formula 1 cars. Figure 11.5 illustrates the operation of the Moog servo-valve universally used in Formula 1. Fig. 11.6 shows three Moog valves in a Williams FW14B.

The trend in aircraft toward all-electric or electric/hydraulic control systems (individual electric-powered hydraulic power supplies integrated into each actuator) is unlikely to apply to race cars, because the distances over which power must be transmitted are relatively small. An all-electric system has the potential of saving on the weight of a hydraulic system, but power density of electric actuators and their associated power electronics are not yet competitive with hydraulics. The main developments are toward smaller and lighter assemblies, using materials such as titanium and filament-wound Kevlar, for accumulators and cylinders.

(a)

Torque Motor

- Charged permanent magnets polarize polepieces
- DC current in coils causes increased force in diagonally opposite air gaps
- Magnetic charge level sets magnitude of decentering force gradient on armature

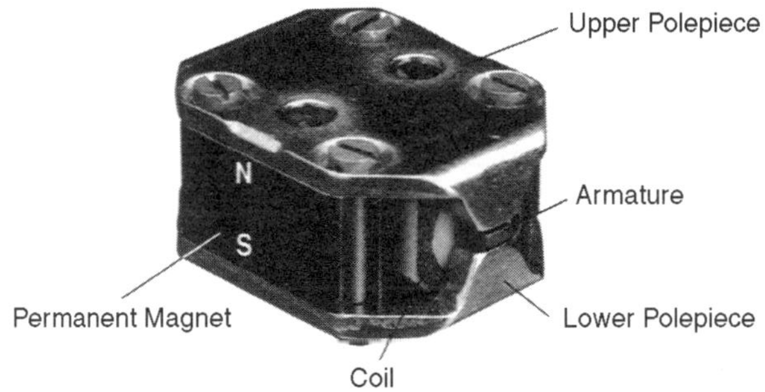

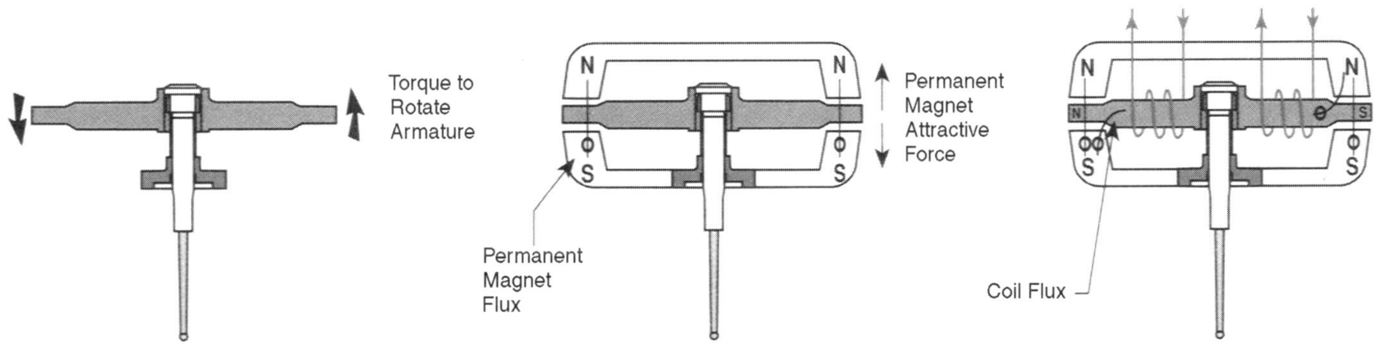

Fig. 11.5 Operation of Moog electro-hydraulic servo-valve. (a) Torque motor. (Courtesy of Moog Controls Ltd.)

(b) ***Hydraulic Amplifier***

- Armature and flapper rigidly joined and supported by thin-wall flexure sleeve
- Fluid continuously flows from pressure P_s, through both inlet orifices, past nozzles into flapper chamber, through drain orifice to return R
- Rocking motion of armature/flapper throttles flow through one nozzle or the other
- This diverts flow to A or B (or builds up pressure if A and B are blocked)

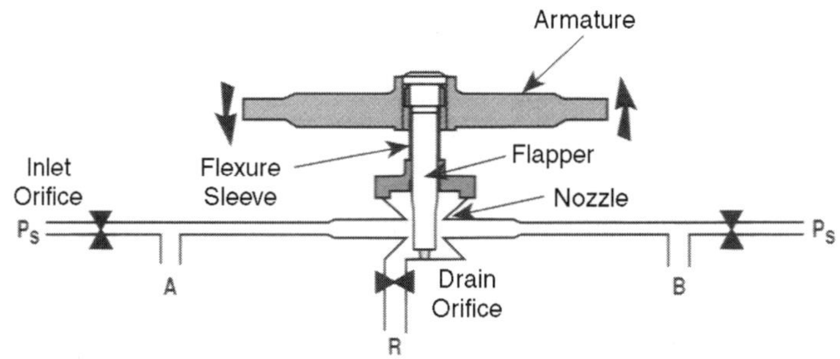

(c) ***Valve Spool***

- Spool slides in bushing (sleeve), or directly in body bore for Series 30
- Bushing contains rectangular holes (slots) or annular grooves that connect to supply pressure P_s and return R
- At "null" spool is centered in bushing; spool lobes (lands) just cover P_s and R openings
- Spool motion to either side of null allows fluid to flow from P_s to one control port, and from the other control port to R

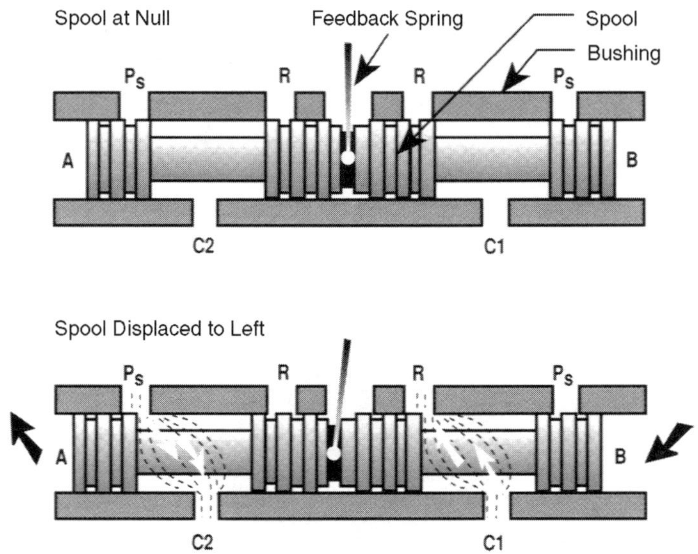

Fig. 11.5 (b) Hydraulic amplifier. (c) Valve spool. (Courtesy of Moog Controls Ltd.)

(d) ***Operation***

- Electrical current in torque motor coils creates magnetic forces on ends of armature
- Armature and flapper assembly rotates about flexure sleeve support
- Flapper closes off one nozzle and diverts flow to that end of spool
- Spool moves and opens P_S to one control port; opens the other control port to R

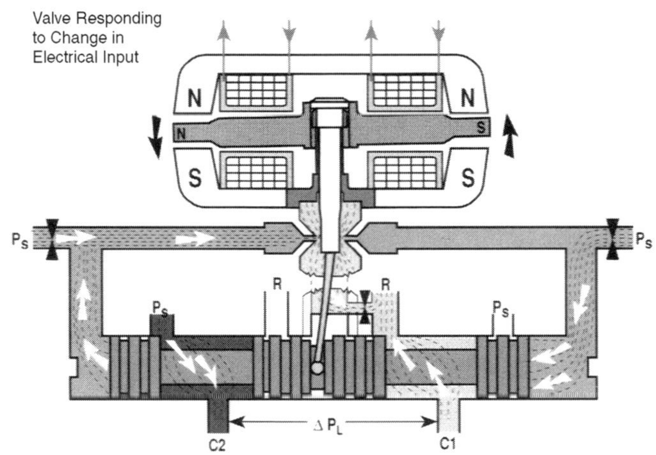

- Spool pushes ball end of feedback spring creating restoring torque on armature/flapper
- As feedback torque becomes equal to torque from magnetic forces, armature/flapper moves back to centered position
- Spool stops at a position where feedback spring torque equals torque due to input current
- Therefore spool position is proportional to input current
- With constant pressures, flow to load is proportional to spool position

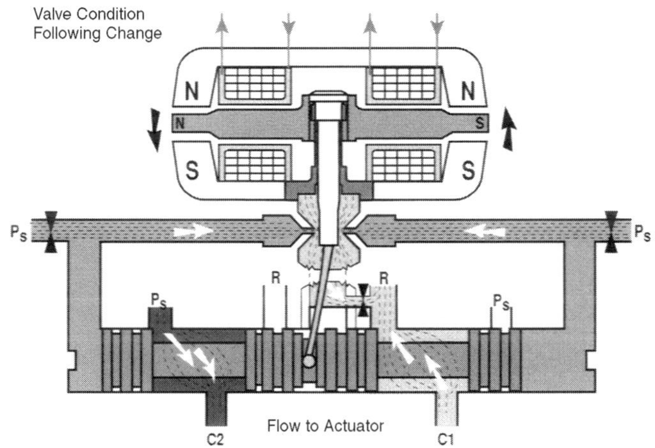

Fig. 11.5 (d) Operation of the system (Courtesy of Moog Controls Ltd.)

Fig. 11.6 Three Moog valves, with tempertaure stickers attached, fitted to the active suspension system of a Williams FW14B.

Fig. 11.7 The Ferrari F300 steering wheel—the driver interface.

Fig. 11.8 The rear of the Ferrari F300 steering wheel, showing gear-change and clutch control paddles.

Chapter 12

Simulation and Design Tools

However, undaunted by reality we pressed on.

—Louis Faix

"What if..." is an often repeated expression that takes moments to say, but in all spheres of engineering, costs so much and takes so long to answer. In motor racing, where the engineering objective is to do as much as possible in the time available and within the set budget, the question of "What if..." cannot always be answered by building a car or rig to test the hypothesis. As with so many things, the computer has come to the rescue of the engineer and enabled him to go "virtual racing" with his ideas, without leaving his office. If the physical laws that describe the function of all parts of a system are known, the power of computers today enables the system to be modeled. Most static and dynamic systems, as well as design and manufacturing processes, are now modeled. A vast library of software is available to perform routine tasks, however complex, with most of it having been created within the aerospace industry. Where bespoke models are required, engineers are able to write and develop their own software to perform particular simulations. Formula 1 has inevitably demanded specialist versions of industry-standard software. However, it has also had to develop one particular simulation, not required by other industries: lap simulation.

Calculating the performance of a race car based on weight, engine power, tires, and aerodynamics is not difficult, and engineers have done so since racing began. However, to do so, point by point around a circuit to calculate the predicted lap time, was almost impossible using hand-calculation methods and a slide rule. Such a task is exactly what a computer does so well. In the 1970s as the first Apples and PCs arrived on racing engineers' desks, it was among the first tasks the computers were set to perform.

These early lap simulation models were simple by today's standards, but they sufficed as tools to estimate the sensitivity of changes to the fundamental parameters (weight, weight distribution, power curve, tires, aerodynamic lift, drag, and center of pressure) on the lap time for different circuits. Sensitivity checks could be made (e.g., change in lap time per kilogram of weight) for each circuit. Circuits could be graded by factors such as the importance of weight, power, braking, and fuel consumption (time on full throttle). A full set of sensitivity runs for a given circuit required a whole night of number crunching on these early desktop computers.

Although simple models were used in the simulation, there was little point in making them more sophisticated, if the car characteristics on which they were based were not or could not be measured. Data systems on race cars were in their infancy, and extensive rig testing was rare. The lack of good tire characteristic data illustrates this point. The coefficient of friction between tire and track, and how it changes with load, temperature, traction, and variations in the track surface, dominate the performance of the car. The early lap simulations were not very good at predicting the exact lap time, but they adequately quantified the effect of changes. To "tune" the model, tire coefficient of friction was adjusted to give the correct lap time.

I recall the first Detroit Grand Prix in 1982 where no one knew what the lap time would be. Goodyear, Renault, and Lotus had a friendly competition to see whose lap simulation would be nearest. Although Lotus' estimate was closest, within one second for the first practice session, it was based on pure guesswork in choosing what value to use for the tire performance on a dirty street circuit. Several years later, when Lotus first ran an active suspension car there in 1987 and was able to measure tire loads and hence coefficient of friction, it became clear that two types of surfaces existed at Detroit: one with a coefficient of friction for the tires in use at that time, equal to 1.7, and the other around 1.9. Back in 1982, we simply guessed!

To understand the limitations of these early lap simulations and how they have developed over the past 20 years, it is necessary to describe how they work. To perform the simulation, the software consists of three basic models: (1) the circuit model, (2) the car model, and (3) the driver model.

Circuit Model

In its simplest form, the circuit model consists of a series of straights and radii, which describe the line the driver takes in driving around the circuit. It assumes a completely flat circuit, with no gradients, camber, or bumps. The driving line can be derived from a circuit map, with an estimated line drawn on it (actual race drivers are best at doing this). Alternately, using modern data systems, the line can be calculated from car speed, lateral acceleration, and, if available, yaw rate data. Some software engineers have attempted to develop programs that actually calculate the optimum driving line from the plan of the circuit by trying a number of lines and using intelligent algorithms to converge on the optimum line. However, they are only as good as the track data used—many drivers will use a less than optimum line to avoid a bump in the track that unsettles the car or will cut across a curb that does not.

The next level of sophistication includes gradients and road cambers—essential for banked ovals—but does not usually include the effect of crowning of the track surface.

The final level of complexity, now being applied, is to include surface bumps and the details of curbs. To gather this data by Differential Global Positioning System (DGPS) survey techniques and laser profile measurement is costly. Furthermore, as will be seen in the next section, the car model to use this information also is very complex. At circuits such as Monaco, the performance of the car is affected significantly by the road profile, and experienced drivers exploit it to alter loads on the four wheels and modify the handling of their cars.

Car Model

The car model is made up of a number of modules, each of which may be set to various levels of complexity:

- Engine—Torque against rpm; inertia; throttle and fuel consumption characteristics; throttle response (delay) and drive-by-wire characteristics; airbox pressure effects.
- Gearbox—Gear ratios; inertia and losses; gear-change times.
- Differential—Mechanical characteristics; control system characteristics.
- Tires—Full set of characteristics, with load, torque, camber, slip angle, pressure, and temperature; rolling resistance; rolling radius against speed and load.
- Brakes—Friction characteristics against pressure and temperature; thermal and cooling characteristics; control system (balance) characteristics.
- Suspension—Track and wheelbase; kinematics and compliance characteristics; stiffness and damping characteristics.
- Inertia—Weight, weight distribution, CG height, pitch, roll, and yaw inertia; unsprung masses.
- Structure—Chassis torsional stiffness.
- Aerodynamics—Lift, drag, and center of pressure against ride height front and rear, and roll and yaw angles; cooling characteristics; airbox ram pressure characteristics; exhaust flow characteristics (effect on lift and drag).
- Air density, wind speed, and direction also may be input because they will affect aerodynamic forces.

Driver Model

A number of driver characteristics can be modeled, which will modify the way the car model behaves, usually at less than the optimum. The most common is the choice of a non-optimum gear ratio for a corner, allowing the rpm to drop way below that for best torque. Drivers do this to reduce the gain of the throttle pedal, and hence the response of the engine when accelerating away from the apex with the car at the limit.

The way a driver brakes into a corner also may be different from the theoretical optimum. Most simulations do not include all these models in their full complexity. In the early days, lap simulations used "bicycle" models in which even lateral load transfer effects were not included. Brakes were assumed to be either off or on at the limit of the tires. Aerodynamic characteristics did not include ride height mapping. Transient effects were also ignored. The car was accelerating, braking, or cornering, and transitioned instantly from one state to the next.

Motor racing computer games use exactly these same car and circuit models, with various levels of sophistication to generate the dynamic responses of the car. In these games, the driver model is supplied by the player.

Calculation

To calculate the lap time for a given car setup on a particular circuit, the computer must start at the exit from any corner where the speed is limited by lateral acceleration, as opposed to how fast the car can arrive at that corner. Maximum acceleration is applied, limited either by traction or power, starting at the maximum cornering speed and all the way down the following straight, until it arrives at the next corner. The speed for that corner is calculated and the car is accelerated backwards from that speed, at the maximum possible braking value, until the speed matches the speed on the straight at a particular point. That point is the braking point. Thus, the speed at each point on every straight—whether accelerating, braking, or at maximum speed—and in every corner can be calculated, and the total time to proceed all around the circuit, step by step, is calculated. All this adds up to the lap time. The distance interval is chosen as the best compromise between accuracy and computing time; however, on the simpler models, it may be as high as 5 meters (16.4 ft). Modern Pentium® PCs will complete the calculations within a few seconds. This is the Stage 1 approach.

Progressing from a 2 degrees-of-freedom "bicycle" model to a four-wheel, 5 degrees-of-freedom model with lateral load transfer, aerodynamics influenced by pitch and heave but no roll, yaw, or transient effects nor suspension kinematics, requires much smaller steps and may increase calculation time by a factor of at least 10. Too large a step size may cause the mathematics to become unstable as the computer tries and fails to determine a quasi-steady-state condition for each iteration. This is Stage 2.

The next step is to add roll and yaw, both of which affect aerodynamics. Suspension kinematics also are added to model accurately the wheel camber and tire slip-angles. This is Stage 3.

Stage 4 is the modeling of the transients (i.e., the conditions as the car transitions from one state to the next) and adds at least another order of complexity and complication. The dynamic characteristics of the aerodynamics, tires, suspension, engine, and even the structure of the car must be known and accurately measured to make such a step worthwhile. However, the transient characteristics of a car (more correctly termed "stability and control") often determine how much of its performance potential can be utilized by the driver. At this stage, the driver model needs to be more sophisticated and include driver response characteristics. Knowledge of the effects, by simulation, would be invaluable.

The final step, Stage 5, is to include road surface bumps and to model the way in which they disturb and excite the car. Television vividly shows how much the cars are disturbed by some surfaces, but the modeling task is enormous. However, with the power of PCs increasing almost exponentially, and the top teams using seven-post servo-hydraulic rigs to simulate road surfaces in the laboratory to greatly increase their understanding of how the car responds to bumps, it is coming.

For a long time, the Stage 1 approach sufficed. It was a case of 10% effort yielding 90% of the result. However, Formula 1 is about gaining the last half of 1%, and now most teams use at least a Stage 3 approach. This will give a fairly accurate assessment of the performance potential of the car on a smooth track, if the tire performance is accurately known. What it will not do directly is to explain why a particular driver cannot extract that performance on certain tracks. However, these simulations provide enormous insight into how the car is working, if they are properly validated to check both the models and the data on which they are based.

If the models are sufficiently detailed, the lap simulation can output the values of parameters in the same way as data systems; indeed, many of these parameters will be the same in both systems. By running a simulated lap under the same conditions as the car, comparison among the outputs can be made. They should be the same, but what is interesting is when they are not. Use of the lap simulator, to determine why they are not, is a start in exploiting the real power of this technology. However, lap simulators can be and are being put to several simpler uses. As mentioned previously, the first use of lap simulators was to carry out multiple runs on a given circuit, to determine the theoretical optimum settings for parameters such as CG, position, wing settings, power curve characteristics, roll stiffness distribution, and brake balance. Lap simulators were also used to determine the sensitivity of the lap time, to small changes in these parameters. Differences between the theoretical optimum settings, and those being used on the car, require careful investigation with particular reference to the fidelity of the simulation. The inability to match the speed of the lap simulator at any point around the circuit could provide a good focus on particular aspects of the performance of the car that were not being achieved due to handling or aerodynamic imbalance.

Lap simulators are also used to investigate the reasons for the better performance of competitors. For example, they are used to model the power needed to achieve the top speed of the other car, or the downforce needed for cornering speeds.

Comparison among circuits gives an insight into relative requirements for power, aerodynamics, fuel consumption, cooling, and braking. New circuits or changes to an existing circuit can be evaluated to establish initial setups, gear ratios, etc. The FIA uses a lap simulator to model the speed of new and modified circuits and to establish the required circuit safety requirements (e.g., gravel beds and barriers).

Tire companies can gain insights into the type of compound needed for a given circuit, by computing the work done by each tire, based on tire loads and speeds. This application, more than any other, requires an accurate estimation of the coefficient of friction between tire and track.

Engine manufacturers tend to use actual data to run simulations on a dynamic test bed, as real drivers are on and off the throttle much more than a "model" driver. However, for unknown tracks, the lap simulator will give pre-testing information on rpm ranges, time in given rpm bands, fuel consumption, and cooling requirements.

Refueling strategies are based on predicted lap times, with given fuel loads and tires. The limited amount of data available during practice and qualifying is input into the lap simulator, and alternative refueling and pit stop strategies are tested and integrated for the whole race, to determine the optimum strategy based on available data.

The main use of the more sophisticated lap simulators today is to guide and evaluate the results of R&D programs. Formula 1 cars are so optimized that changes to almost any aspect of them can easily detract from their performance. Any proposed development must be quantified in terms that can be input into the car model, and its effect on lap times must be evaluated on a full range of circuits. It is particularly useful in assessing control system algorithms (e.g., brake balance, differentials, drive-by-wire, and any proposed new gizmo from R&D).

This ability tempts forth the announcements of how much quicker next year's car will be than this year's car. After six months of conceiving and designing a new car, and armed with the latest power curve for the new engine, downforce figures from the wind tunnel, and all of the clever bits and pieces dreamed up by R&D, it is irresistible to go virtual racing while waiting for it all to be put together and tested. There are drivers, backers, and sponsors whose confidences need boosting through the winter months of waiting. Good figures from the computer are much too tempting for some team bosses to keep them secret. The reason their cars do not always perform as predicted is that the higher the pinnacle of performance, the more peaky it is and the more difficult it is to get to the top of that peak and stay there. This brings us back to the integration of lap simulations with the car data system.

If we take a Stage 3 model and run it under the same conditions as a real car, the output should match the data system of the car, give or take a few bumps in the track. If it does not, it is possible to make changes to the car model setup until it does. The necessary changes will indicate the possible areas where the car settings do not correspond correctly to those input into the model, and will act as pointers to the settings on the car that are not performing as expected. What is perhaps more useful, is when the output of the lap simulator broadly matches that of the data system of the car, but the driver is critical of the handling of the car. Matching the point in the outputs from the two systems from which the car does not respond in the way the driver believes it should, enables the output of the model to be examined closely to determine car parameters that cannot practically be measured by the data system (e.g., tire slip-angle, rear-wheel torque distribution, and wheel cambers). The problem may now become clear, and potential solutions can be tested in the lap simulator to determine if the altered response meets the driver's requirements. This is not yet the sort of thing that can be done during a practice or test session, but it can be carried out in the evening between practice and qualifying or qualifying and the race.

Another application is to turn the lap simulator model into a virtual tire dynamometer. We have already seen that the most important input parameters are tire characteristics and that they are also the most uncertain. However, if all other parameters are well validated, it is possible to compare the output of the lap simulator and the car data systems (very good data for wheel loads, slip-angles, and such factors are needed) and to use parameter identification techniques to adjust the tire characteristics in the model, until all aspects of the outputs match. Thus, the actual performance characteristics of the tires on the track can be established.

A good Stage 3 simulation will make an accurate prediction of the performance potential of a car, but it will not assist the race car engineer in understanding what a highly skilled driver is complaining about when he describes what he has to do to extract that performance from the car, nor why a less-skilled driver cannot extract it. For that, the stability and control characteristics of the car must be modeled, including the transient response to any input or disturbance—Stages 4 and 5. For these, each subsystem that influences the transient responses must be modeled and, of course, validated by rig or track data. The development of accurate simulations of the dynamics of the car and its components is controlling the pace of development of Stages 4 and 5.

Stage 4 simulation models the response of the car to driver inputs. The lap simulation may use filters to model the effects, but the characteristics of these filters must be established. This can be done either by measuring the response characteristics of the car or by modeling them. The latter approach requires accurate dynamic characteristics for the aerodynamics, tires, dampers, and any control systems that may affect vehicle response, and the structural stiffness of the chassis, suspension, and steering. Models and validation rigs are now common for most of these systems, and a library of characteristics should be available, except for the aerodynamics and tires.

Predicting the non-steady-state flow characteristics of a body that incorporates so much vorticity and separated flow is extremely difficult. Dynamic model tunnel testing is almost impossible at the frequencies of interest, because the dynamic characteristics of the force balance and surface pressure sensor system preclude accurate measurements. Computational fluid dynamics is at its weakest when it comes to modeling vorticity and separation, and these may both vary as the car moves.

Stage 3 simulation requires only steady-state tire characteristics, and these are made available by tire manufacturers for a given construction. The characteristics are measured on a flatbed tire test rig and are correct only for the compound, temperature, and belt surface used in the tests. Stage 4 requires transient tire characteristics, which are difficult to measure; thus, the data will depend on the sophistication of the manufacturer's tire model. It also requires a thermal model of the tire, to calculate correctly the tire temperature, at all points on the circuit where the full capability of the tire is required. Heat input due to rolling (a function of speed, load, and pressure), longitudinal and lateral force generation, and brake heat, less the transfer of heat to the track surface, conduction to the wheel, and airflow over the surface of the tire, determine the temperature and hence compound characteristics. Validation is difficult because measurement of the temperature of the tread below its surface requires purpose-built tires and telemetry.

The development of Stage 4 simulation is invaluable in terms of understanding the characteristics of the car, but it may offer little improvement in the lap time estimation. Modeling the response of the car to control system inputs is somewhat meaningless, if a major part of the control system (i.e., the driver) is omitted. Because circuit racing is now conducted on tracks with only 10–15 corners, the drivers learn them within a few laps and soon drive them open-loop. Control inputs include lead to compensate for phase lags in the response, and the magnitude of the input will be adjusted according to the stability and damping of the system. Only when the conditions are changing, or if the driver is at the limit searching for the last 0.5% of performance, does he drive closed-loop. The effect of a skilled driver on the transient response of the car when calculating lap time is to virtually eliminate transient effects. Even worse, the driver may actually drive in a non-optimum manner to overcome a handling shortcoming in the car, achieving a lower lap time than the simulation.

Stage 5 simulates the effect of disturbances on the car, at two levels, and requires circuit topography (e.g., bumps, camber, and curbs). The effect of these on the horizontal motion of the car is an extension of Stage 4 but uses circuit feature inputs instead of driver inputs. It is sensitive to the driver's response, as discussed previously, and the fact that the driver may modify his line to avoid or exploit circuit features, particularly curbs.

Of greater importance is the effect of undulations in the track on the grip of the tires. While the horizontal response characteristics are generally less than 7 Hz, the vertical oscillations that affect the forces on the tire contact patch are in the range of 3–30 Hz and higher. Knowledge of the resonant modes in the tires, suspension, chassis structure, engine, and transmission is necessary. If doubt exists that a rubber tire is sensitive to these higher frequencies, remember that Honda discovered that altering the firing order on its 500-cc Grand Prix motorcycle to reduce the frequency and increase the amplitude of the 200+ Hz torque pulses dramatically increased traction.

A tire has more than 50 modes, some in the hundreds of hertz range. A suspension has two fundamental modes, one at the sprung mass natural frequency (3–7 Hz) and the other at the unsprung mass frequency (12–15 Hz). The structure also has a large number of modes, but first torsion and first bending, at 30+ Hz, are the most relevant. The dampers are the primary means of damping these modes but may cease to function above 25 Hz. Modeling and measuring these characteristics is a mammoth task, but it is being undertaken by most Formula 1 teams that have seven-post excitation rigs. The road input consists of white noise, and most of the modes will be excited. Exactly how the frequency and magnitude of the variation in vertical tire loads affects the mean coefficient of friction and the heat input to the tire is not yet fully understood. Tuning the chassis to a particular tire for a particular circuit is one of the last black arts of motor racing. When it becomes a science, Stage 5 can be completed.

The future of lap simulators lies in developing, refining, and validating the models used in them, and in providing user interfaces that facilitate and increase the speed of the range of applications described. One day, they will become the race engineer's primary tool, providing him with accurate information, almost in real time, about what the car is doing and, more importantly, what it is not doing. The engineer will be able to obtain car data for a section of the track where the driver complains of undesirable characteristics, and simulate various solutions from the point where the car exhibits the problem, in order to find the best setup to cure the problem.

For a lap simulator to run in a reasonable time, the subsystem models take the form of either look-up tables of characteristics or equations that describe their characteristics. Behind these models are the full models of the individual systems that generate the simpler versions. Most of these are based on industry-standard software such as ADAMS, MATLAB, MADYMO-3D, and various FEA and CFD programs. Others are in-house developments, particularly those used for modeling the engine, and are often based on software developed for road car or aerospace applications. In some areas, Formula 1 is at the forefront of this development—for example, CFD as applied to the car as a whole and for predicting turbulence in the ports and cylinder of the engine; FEA of composite structures; FEA of dynamic composite structures, such as the tires; and FEA of composite structures during dynamic failure (i.e., during impacts). Modeling techniques for the impact-absorbing structure, driver restraint system, driver's protective equipment, and the driver's body have been adapted from the extensive research on road vehicle crashes and applied to the high-speed, high-g impacts experienced by well-restrained and protected racing drivers.

Very little of a Formula 1 car is built today without first being modeled to confirm the performance of those aspects that differ from the established state of the art. Tradeoffs can be quantified, lap time reduction on a variety of circuits established, and the design refined without cutting metal. The designer has many tools to assist with this process, from the most sophisticated 3-D CAD, to virtual imaging. Prior to committing to manufacture, the casting and NC-machining process can be simulated, and rapid prototyping technologies can present the designer with a "part" within hours of completing a design. Indeed, rapid prototyping is a technology that so suits motor racing, that it is used for the low-quantity production of casting patterns, wind tunnel model parts, and rig components.

Formula 1 cars have become extremely complex systems, combining high-speed mechanical, hydraulic, and electronic subsystems. It is well known that the uncertainty about a system increases exponentially with the complexity. Computer simulation provides a lifeline to the designers and engineers grappling with the whole and its parts, but pitfalls exist. It has been said that scientists and engineers, in the same way as artists, often become infatuated with their models. Too much faith in software models, without full understanding of the limitations, can easily lead to disaster. Engineers who write their own software for modeling a system gain invaluable insight into the workings of the system itself. However, even then, the model must be validated, either by rig testing or on the track.

Chapter 13

Test Rigs

Good. Now go and show them to the guy with the checkered flag.
—Colin Chapman, on being presented with wind tunnel test results for the Lotus T80

Designers and engineers work first and foremost with numbers: dimensions, mass, forces, stresses, strains, pressures, flow rates, temperatures, acceleration, volts, amps, and the ultimate arbiter of race car performance, time. When calculation took over from trial and error, the need to measure the performance of a component, sub-assembly, or the whole system became essential to the validation of the theory. The building of test rigs, to reproduce the conditions under which parts must perform, enables those parts and systems incorporating them to be observed as well as measured, as they are put through a series of operating cycles. The ability to watch a dynamic system working under laboratory conditions is one of an engineer's greatest assets.

Although computer modeling has taken over much of the role of test rigs, providing engineers with complete control over test conditions and the means to quantify and visualize any parameter, whether physically measurable or not, it is still necessary to validate the models by rig testing. As models have become more complex, so have test rigs, and the investment in these facilities by a front-running Formula 1 team may extend to tens of millions of dollars. If the full-size car could be run under representative aerodynamic and inertia loads in the laboratory, it would be run, particularly as track testing time is becoming limited by regulation. The technologies to subject the chassis and power train to all the loads experienced on the track are available, except steady-state acceleration; however, these technologies have not yet been brought together into a single test rig.

Thirty years ago, few rigs could be found in the chassis workshops of Formula 1 teams. Engine manufacturers had dynamometers to test and tune the engines, and development was backed up by cylinder head airflow rigs, pump flow/pressure rigs, and rigs for evaluating valve gear dynamics. Apart from a set of weight scales and perhaps a chassis torsion stiffness rig, very little calibration was performed on the chassis. Component suppliers, such as those for clutches, brakes, dampers, and tires, tested their products for performance and durability. However, this discipline was carried across from their main line of business: road car components. Strength and fatigue testing of safety-critical components, such as wheels and suspension, was contracted out to test centers or a laboratory owned by an industry partner. Wind tunnel testing was in its infancy and was performed in aeronautical research establishments or universities, using small-scale model tunnels with fixed ground planes. All measurements were made on analog instruments and were recorded and post-processed manually. The time taken to acquire and analyze data was a major limitation on the use of test rigs.

The ability to acquire and analyze measurement data by computer has revolutionized the use of test rigs, enabling track conditions to be faithfully reproduced in the laboratory, and the rig test results to be correlated with simulations. As the cost of track testing escalates, and the available track test time becomes more limited, investment in test rigs will increase. To compete, the teams must have access to most of the following rigs.

Wind Tunnel

Fifty-percent scale-model wind tunnels are covered in the chapter on aerodynamics. They are by far the biggest single investment in test rigs and have become very specific to the task. Some teams have worked closely with research establishments that have suitable wind tunnels, investing in moving ground planes, force balances, and other test equipment to suit race car testing. Specialist tunnels for calibrating radiator matrices are also used.

Seven-Post Road Simulator

By mounting a complete car on a seven-post road simulator, it is possible to subject the vehicle to representative cycles of the vertical loads experienced on the track (Fig. 13.1). The simulator consists of four high-performance, vertical, electro-hydraulic actuators, one under each wheel. Three tension struts connected to downforce actuators pull down on the chassis, representing quasi-static aerodynamic downforce, roll, and pitching loads and the load transfer due to inertia loads in roll and pitch. The wheel actuators are able to excite the contact patch at up to 20g and 50 Hz. The downforce actuators are effectively very low-rate springs, with a load control loop to maintain the desired force, irrespective of the motion of the chassis (Fig. 13.2). The whole system is mounted on a massive seismic block to isolate it from other equipment, and power is supplied by a hydraulic system.

Step, sine, and sine-sweeps can be input by the control system, with different phase relationships for each wheel, to excite each of the four modes, while the chassis response is measured and characterized. Alternatively, track data can be used to tailor the input to simulate a particular circuit/speed profile. Actual road surface data are not needed because a model within the control software tunes the inputs until the response exactly matches that measured on the track.

Fig. 13.1 A car mounted on a road simulator. (Courtesy of Servotest Ltd.)

Fig. 13.2 Downforce actuator. (Courtesy of MTS Inc.)

Spring and damper settings can be set up prior to going to a race circuit, and certain durability schedules can be performed. Resonance or modal testing of the structure can be conducted, up to the frequency limits of the actuators.

Similar to the wind tunnel, this sophisticated piece of equipment measures only what is there. It does not tell the engineer what the result should be. Limitations to the faithfulness of the simulation are the absence of horizontal loads, which can cause stiction in the suspension and geometric effects, and the non-rotating tire. Rotating tires tend to have less damping than stationary ones, and this must be taken into account when setting up dampers.

Single-Corner Rig

Access to a complete test car can be difficult because there is often considerable call on complete cars for all types of test work. For more fundamental research into suspension characteristics, a single-corner rig is often used. Effectively a quarter of the car, the sprung mass is represented by an appropriate mass, which is constrained to move only vertically. The suspension is fitted to it, and, if required, a downforce actuator (or more simply, a low-rate mechanical spring) is used to simulate downforce. These rigs were essential tools during the development of active suspension, for the sorting out of both hardware and control algorithms. Engineers learned the value of the single corner rig as a development tool at that time and have applied them to passive suspensions since then.

Damper Test Rigs

Hydraulic dampers have been tested and calibrated on test rigs for many years, but the mechanical actuators (rotating crank) limited inputs to sine waves. The electro-hydraulic actuators now employed permit almost any waveform to be used and dynamic performance to be studied in detail. Damper modeling has created a need for individual valve characteristics to be measured, and dynamic (intermittent) flow rigs meet this requirement.

Kinematics and Compliance (K&C) Measuring Rig

A kinematics and compliance (K&C) facility is a test rig that undertakes quasi-static testing of vehicles, to measure the kinematic and compliance characteristics of the suspension and steering systems (Ref. 13.1, Fig. 13.3).

Kinematics is defined as the study of motion without reference to mass or force. Compliance is the deflection resulting from the application of force (i.e., the inverse of stiffness). These two words are particularly apt in describing a machine that measures kinematic characteristics due to suspension and steering system geometries, and compliances due to springs, anti-roll bars, elastomeric bushings (road cars only), and component deformations.

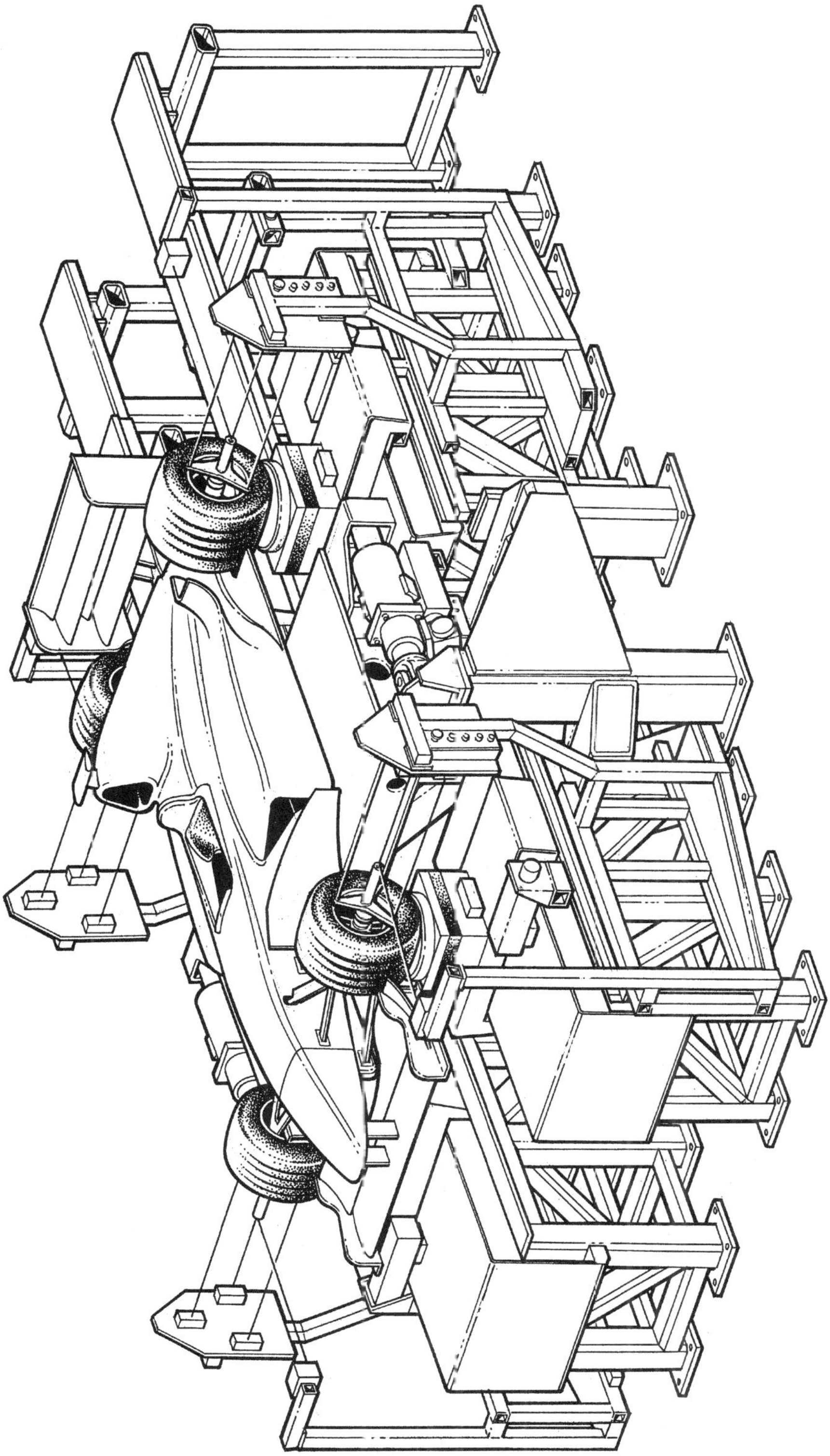

Fig. 13.3 Kinematics and compliance measuring rig designed and built by Anthony Best Dynamics Ltd.

Chevrolet pioneered an approach to K&C measurement in the early 1970s, with its vehicle handling facility. The body shell is held rigidly, while the tire contact patches are moved vertically. In the plane of the ground, the wheel platforms are servo-controlled to provide forces or moments if desired, or to provide zero forces and moments to allow unrestrained track, wheel-base, and steer changes. Later developments reversed this approach, by moving the body while keeping the wheel platforms in a fixed ground plane. To apply the correct tire loads, with the center of pressure in the correct position for a vehicle running on a level road, the relationship between the tire and ground should be correctly simulated. The simulated road surface beneath all wheels should lie in one plane, whatever the position of the body in roll or pitch. If the body is held stationary and the suspension is exercised by vertical actuators effectively moving the road, then each of the wheel pads must tilt about both a roll axis and a pitch axis and be controlled to lie in a plane. Conversely, if the body is moved, then the simulated road surface beneath each wheel does not have to be moved in the vertical direction, and all surfaces will remain in one plane. This effect is more important the wider the tires.

Typically, a K&C measurement rig has 19 servo-controlled axes of motion, enabling it to subject a vehicle to a wide range of motions and forces (Fig. 13.1). Up to 128 channels of data-capture simultaneously measure the forces on and deflections of the wheels, body, and suspension. The results may be formatted in several ways. However, the following lists a typical set of test results, describing the characteristics of a race car suspension.

Vertical test:

- Wheel vertical rate (contact patch vertical force versus wheel center vertical displacement).
- Tire radial rate (contact patch vertical force versus vertical compression displacement).
- Bump steer (road wheel steer change versus wheel center vertical displacement).
- Bump camber (road wheel camber change versus wheel center vertical displacement).
- Bump castor (road wheel castor rotation versus wheel center vertical displacement).
- Track change (contact patch lateral displacement versus wheel center vertical displacement).
- Wheelbase change (contact patch longitudinal displacement versus wheel center vertical displacement).
- Kinematic roll center height for each "axle."
- Kinematic anti-lift or anti-drive tangent for each "axle."

Roll test:

- Roll stiffness (roll moment versus roll angle).
- Wheel loads (road wheel vertical loads versus roll angle).
- Roll stiffness distribution (ratio of individual axle roll stiffness to total vehicle roll stiffness).
- Roll steer (road wheel steer change versus roll angle).

Steering system tests:

- Road wheel steer versus hand wheel steer.
- Ackermann curves.
- King pin offset, scrub radius, and mechanical trail.
- Hand wheel steer versus hand wheel torque for various control strategies of road wheel steer.
- Steering system stiffness.

Lateral compliance tests:

- Lateral compliance (road wheel center lateral displacement versus contact patch lateral force).
- Lateral compliance steer (road wheel steer change versus contact patch lateral force); this may be a family of curves with different effective longitudinal moment arms acting in the contact patch.
- Camber lateral compliance (road wheel camber angle change versus contact patch lateral force).
- Roll force center height for each "axle."

Longitudinal compliance tests:

- Longitudinal compliance (road wheel center longitudinal displacement versus contact patch longitudinal force).
- Longitudinal compliance steer (road wheel steer change versus contact patch longitudinal force).
- Castor longitudinal compliance (road wheel castor angle change versus contact patch longitudinal force).
- Force derived anti-lift or anti-dive tangent for each "axle."

All these parameters are important inputs to lap simulators and vechicle dynamics models. Although they may be derived from linkage models of the suspension, the stiffness of each component must be known if they are to be meaningful. The K&C measurements take account of every reality, including joint stiffness and friction.

Structural Test Rigs

The compliance components of K&C measurements on a race car are the results of structural deflections. Proof of design is required long before a complete car is built. Thus, a multitude of component test rigs are needed to measure stiffness and, on safety critical items, ultimate strength. The components include the monocoque (torsion) and all suspension and steering components and mountings, including the wheels. Likewise, the highly loaded aerodynamic assemblies are tested, especially the front and rear wings and their mountings.

The Technical Regulations require a number of static stiffness and strength tests to be performed on the safety cell portion of the monocoque. These will be conducted at the team's premises, under the supervision of an FIA representative.

Fatigue Test Rigs

The two main sources of vibration on a race car are the road and the engine. The road and wheel rotation induced vibration excites anything with a natural frequency up to approximately 50 Hz, and the engine almost everything else (the combustion frequency of a V-10 at 18,000 rpm is 1.5 kHz). Fatigue testing of components is critical to both safety and reliability, and there are many examples of a small design change resulting in a change of mode shapes, with disastrous results. Rear-wing assemblies are particularly susceptible to this problem. Fatigue testing is often subcontracted to specialist companies.

Resonance Test Rigs

Resonance testing up to approximately 50 Hz can be conducted on the seven-post rig. Above that frequency, an electro-magnetic actuator is needed to provide the excitation.

Control System Rigs

All electro-hydraulic control systems used on the car are rig tested to prove and calibrate the hardware and software. Each rig is purpose-designed to apply appropriate forces and motion to the system.

The level of integration of the electronic hardware used to control the systems, and therefore the level of integration of the software, is such that it is becoming necessary to develop and prove these parts of the system as a whole. A control system simulator maybe used, embodying all the control functions on the car, including engine management, and controlled by the central computer and the CAN network, via the actual car wiring looms. Apart from checking system integration, the simulator can be used to check system components in an almost real environment.

Gearbox Rigs

The performance of the lubrication system in the gearbox is crucial to the losses, durability, and cooling requirements, and is almost impossible to model. Minimizing the oil in the gear chambers at any time, to minimize churning losses and the heat to the oil, must be balanced against the need to lubricate and cool the bearings and high surface loaded gears. Simulation of the loads, speeds, and horizontal accelerations to which the gearbox is exposed on the track requires a special rig.

Pre-loading the output shafts to the input shaft at the maximum torque experienced by the gearbox obviates the need to drive the gearbox with the full power of the engine and absorb it with a dynamometer. Instead, the gearbox is driven at representative rpm, and the measured power requirement equates only to the losses. The whole rig must be capable of being tilted in two planes, to simulate horizontal acceleration. Although this is limited to 1g, it is possible to reproduce fluid level surfaces at higher g. As much of the gearbox case is machined away as is feasible, and the internal cavities are illuminated to enable the motion of the oil and oil/air mist to be visualized. More sophisticated gearbox rigs are now appearing. Hydraulic motors provide input power, and pumps absorb the output. The pump output provides part of the power to the motors, with the losses in the system made up from a hydraulic power supply. Full simulation of this transient loading on a gearbox is possible.

Impact Rigs

The front, side, and rear impact tests stipulated in the Technical Regulations must be performed by an accredited test house. Most teams have installed their own test rigs to facilitate the development of the impact-absorbing structures.

Engine manufacturers have their own specialist test rigs, of which the best known is the dynamometer.

Steady-State Dynamometer

Dynamometers have long been used to measure the performance of engines, throughout the range of speeds and loads. The development of computerized data acquisition enabled vast amounts of data on the engine and its systems to be monitored and collected. Ignition and fuel mapping, in addition to other variables such as intake trumpet length, can be performed only on the dynamometer. These processes can be automated to a degree, with computer-controlled test beds.

Fundamental research into gas flow, combustion, frictional losses, heat flow, fuels, and lubricants is conducted on single-cylinder engines on dynamometers. The role of the dynamometer is reversed when an electric motor is used to drive the engine, so that pumping, friction, and auxiliary losses can be measured.

Transient Dynamometers

To run an engine on the test bed, using recorded data of the driver inputs and speeds—either to prove durability or develop transient throttle response—a transient dynamometer is required. This takes various forms. A gearbox is attached to the engine (semi-automatic gear-change systems have facilitated the operation of the gearbox by the computer running the test rig), and either twin dynamometers are attached to the half shafts, or the final drive to the differential is bypassed, and the drive attached to a single dynamometer. Servo-control of the brake load controls the speed of the dynamometer, and hence the engine speed, according to the selected gear ratio at any instant. Thus, any inertia and resistance forces (aerodynamic drag, rolling resistance, gradient, and braking) can be simulated. The load-control loop must be high gain to provide the response necessary to control fast throttle changes with a low inertia system. A simpler alternative is to use a flywheel on the gearbox output shaft (or shafts) to simulate the inertia of the car. The only constraint is that the inertia cannot be varied while the engine is running, to simulate fuel usage during a race simulation. The load control must now respond only to changes that vary with speed and speed squared, which are much slower than throttle changes.

All dynamometers have temperature- and pressure-controlled air supplies to the engine intakes, to maintain standard conditions for power measurements. However, this does not represent the intake conditions on the track because cars are fitted with ram intakes and plenum chambers shaped to optimize external aerodynamics. Some manufacturers have fitted wind tunnels across the top of the engine on the dynamometer, and have fed the airbox from it to simulate the intake conditions, and to ensure that the distribution of air to the individual cylinders is representative. Uneven distribution can be corrected or compensated for, with individual cylinder fueling maps.

Engine Sub-Assembly and Component Rigs

Most subsystems are rig tested, to prove their performance and durability, early in the development of an engine. In most engine development facilities, the following rigs exist: fuel, oil, scavenge, water, and hydraulic pump rigs; alternator rigs; valve train rigs; fuel injector spray pattern rigs; variable intake trumpet rigs; drive-by-wire throttle rigs; air, oil, and water flow rigs; friction measurement rigs; and engine management system rigs. Component testing includes fatigue testing of dynamic components, and structural stiffness tests of the block and heads, at representative operating temperatures.

Of the major components supplied by specialist manufacturers, the two that require extensive performance and durability test rigs, are brakes and tires.

Brakes

Material development, disc cooling and attachment design, and caliper design must all be rig tested to establish the performance and wear characteristics under representative conditions of duty cycle, speed, and temperature. Durability must be proven, particularly at high temperatures and when near the limits of wear. Brakes are tested on dynamometers and are subjected to braking cycles typical of those used on the track.

Tires

The testing of road car and aircraft tires started in the 1950s (Ref. 13.2), in support of the development of vehicle dynamics theory. Early test rigs consisted of the wheel and tire under test, being mounted on a six-component force balance, which was mounted on a suspension system that provided a constant vertical load. The whole rig was mounted on a heavy truck and ran on real road surfaces, wet and dry. A range of loads, tire pressures, cambers, slip angles, brake torques, and speeds could be applied, and the loads measured. Tire testing is complicated by the wear and temperature varying as the test proceeds, both parameters having an effect on the measured characteristics.

Laboratory testing, durability proving in particular, was performed with the wheel and tire mounted on a similarly functioning load measurement rig, but with the tire running on either the outside or inside surface of a large-diameter drum, to which a variety of road-simulating surfaces could be attached. The curvature of the surface led to inaccuracies in the characteristics, and the first flat road tire test rig was developed in 1973 at Calspan—the TIRF facility. The flat steel belt developed for TIRF, which runs on a fluid bearing, forms the basis of the latest flat road tire test rigs, such as the MTS Flat-Trac III Tire Test System, capable of testing speeds up to 250 kph (155 mph) (Fig. 13.4).

Racing tires have been tested on road rigs, but the high side forces are so great that they tend to move the test truck laterally, affecting the slip angle. Pairs of tires have been tested on a dual rig, with the sign of the slip angles opposite so that the lateral forces cancel each other. However, the wear rates are so high that maintaining constant conditions is problematical. Race tire characteristics are now measured on flat road machines, but even then, the tires must be built with special hard compounds to achieve an acceptable wear rate. Extensive high-speed durability testing is also conducted on these rigs, and TIRF is still used for both this purpose and tire calibration.

Aerodynamicists are particularly interested in the change of rolling radius with speed and load, because this affects ride heights. Steady-state characteristics, combining longitudinal and lateral forces, are essential for vehicle dynamic analysis and lap simulations. Dynamic characteristics, especially the effect of suspension modes on maximum force generation and temperature, are yet to be fully researched, and will require significantly greater complexity in the design of test rigs, to simulate and measure the effects.

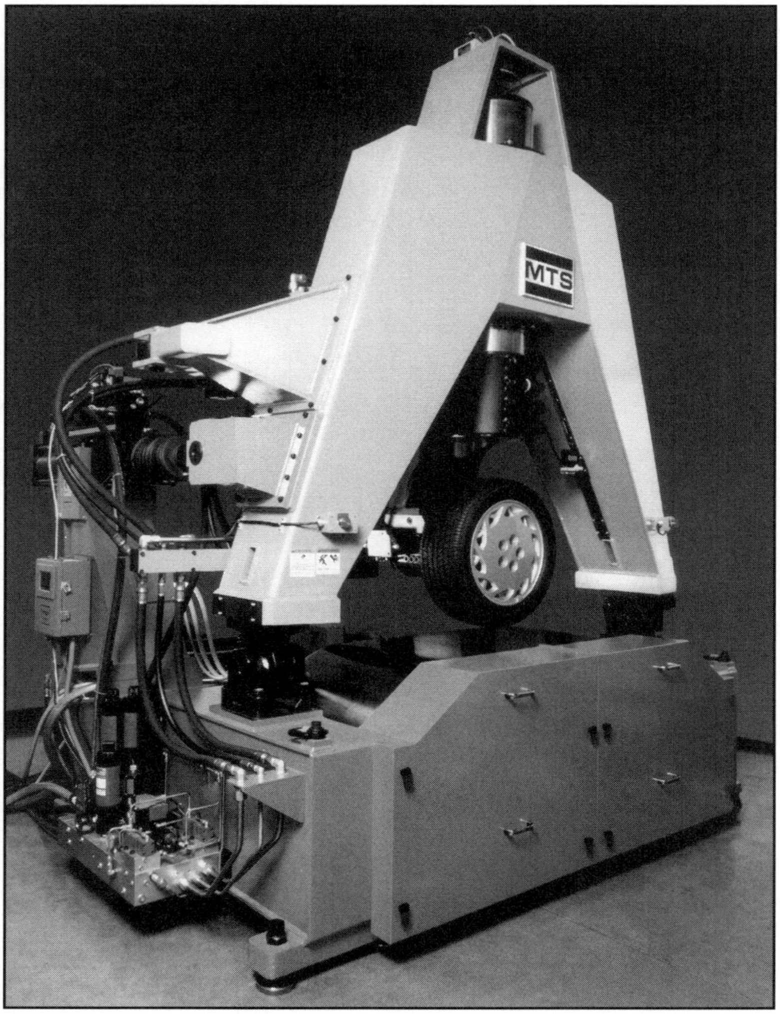

Fig. 13.4 Flat-Trac III tire test rig. (Courtesy of MTS Inc.)

The cost of track testing is so great and the time available for it is so limited that rig testing of almost every component, sub-assembly, and system on a car is a firmly established part of the process, from design to racing. As design calculation, simulation, and modeling become more widespread and more complex, the need to measure the performance parameters of parts, or the whole car under laboratory conditions, becomes more important. Much of track testing is proving durability under race simulations. The more development that can be conducted on rigs, the less pounding around a circuit the test drivers will have to do, and the greater the reliability will be in actual races. The R&D departments of Formula 1 teams have grown enormously during the last two decades, employing numerous people with graduate and doctorate degrees. Much of their work is connected to the variety of rigs employed in these tasks.

Chapter 14

Banned Technologies

The hardest thing for a driver to do is to control his desire to floor the throttle coming out of a corner, and spin the rear wheels all the way down the straight.
—Ayrton Senna

Formula 1 is not an unlimited formula. In fact, no automobile sporting endeavor is performance unlimited, except for the Land Speed Record. Grand Prix cars have been constrained in some way since their inception, either with a cylinder capacity limit and often a minimum weight, or, if unlimited engines are permitted (*formula libre*), a maximum weight. Laurence Pomeroy, author of *The Grand Prix Car* (Ref. 14.1), which was first published in 1954, wrote in the opening chapter:

> It will be observed that the favoured rule has been a limitation on cylinder capacity, which naturally had the effect of forcing the pace in the development of small high-output engines. Considerable technical advances have, however, taken place in the periods when restriction by weight was imposed. Despite the many theoretical arguments in favour of a fuel consumption rating, proposals on these lines have normally been out-voted or ignored.
>
> It is interesting to observe that entrants have never welcomed complete freedom in the construction of race cars. There has perhaps been a feeling that such liberty produces too many variables and makes an investment into a given design an unduly hazardous proposition.

Since Pomeroy wrote this, the development of performance, and the investment and running costs associated with competing to win, have involved more than power and weight. From the 1960s, tires and aerodynamics began to make an equivalent contribution to performance, and regulations to limit them were quickly devised. An awareness of safety also emerged at that time, bringing further regulations and the need to limit speeds to within the capabilities of the circuits to arrest the cars in an accident. The development of computers, and their ability to control functions that had previously been the sole province of the driver, resulted in a debate about the technical role of Formula 1 versus the sporting, commercial, and entertainment roles. At the end of 1993, virtually anything that could be defined as a driver aid was banned. As Max Mosley, president of the FIA, has said,

> The history of motor racing will show that 1994 was the year in which Formula 1 cars diverged technically from mainstream road cars.

The current Technical Regulations reflect a blend of influences—performance, safety, cost, sport, and entertainment—and are relatively stable. The history of power and downforce, as plotted in Fig. 2.11, shows periods of extreme instability, with technical escalation (steep upward slopes of the curves) subsequently curbed by regulation (steep downward slopes), followed by new technology which is then also curbed.

The development of major new technologies, such as turbocharging or ground effect, makes obsolete the existing cars and much of their established technology, as soon as their advantages become apparent. Banning that technology not only makes the new cars obsolete, but also the R&D that has gone into them. The magnitude of engineering investment within Formula 1 teams today makes this unpopular, and a new sophistication in the negotiation of regulation changes between the teams and the FIA has resulted. A much greater understanding of the parameters governing the performance of the car allows more precise control of performance with only small adjustments to power, downforce, and tires being needed to control speeds. Most regulation development is now associated with improving safety and limiting the escalation of costs.

The Formula 1 scrap yard is full of technologies that have suffered from the regulations. In some cases, the inventor or designer has sighed with relief at their ban; in others, automotive engineering in the widest sense has suffered from the killing off of a useful technology, its development under the stimulus of competition, and the lost marketing impetus it would have received from being showcased in Formula 1. The attempts by designers to find a "better way" have resulted in some of the most interesting, if not most successful, Formula 1 cars. The majority have been visibly different from the existing state of the art, and were eagerly anticipated prior to their debuts, due to often wild and inaccurate rumors about their specification. That has now passed, and it is even difficult to determine differences among cars, other than by their sponsors' paint schemes. Differences are buried inside control algorithms and the minute details of components, few of which escape outside the organizations employing them. Competitors are obliged to check with the FIA as to the legality of systems prior to racing with them. This is enforced by ever more sophisticated scrutineering methods and heavy penalties.

Chapter 5 described the key parameters governing power; that is, the rate of consumption of oxygen (engine cycle, swept volume, rpm, charge density, and oxygen content of fuel) and the energy released by the fuel when burnt with that oxygen. All these parameters have had to be regulated to take control of peak power. The early means of regulation, limiting swept volume, is no longer adequate. A number of interesting developments have fallen by the wayside.

Today, the only permitted engine is a normally aspirated (NA), 10-cylinder maximum, four-stroke, reciprocating (circular) piston engine, running on pump petrol.

Turbines

The 1967 and 1968 Indianapolis races had shown that a turbine engine car could be highly competitive in a racing environment, where high power settings were maintained continuously, in spite of USAC (the sanctioning body's) repeated attempts to throttle turbines. Colin Chapman gained firsthand experience of the pros and cons of turbines through the Lotus T56, which raced and almost won in 1968, except for a failed fuel pump drive. Graham Hill put one of the cars on pole for a road race at Mosport and enthused to Chapman about the potential of turbines for Formula 1. However, it was not until 1971 that Pratt & Whitney supplied a 3-liter equivalent turbine, with improved throttle response suitable for road circuits. The T56B (Fig. 14.1) was raced at circuits where high top speeds dominated, qualifying on the front row at Hochenheim and finishing second in the race, driven by Emerson Fittipaldi.

The turbine still lacked the throttle response a driver requires for car control on a twisty circuit, and it also lacked engine braking. The brakes suffered further when the drivers discovered that to maintain the gas-generator rpm for better response, they had to enter a corner under braking with the throttle open. The problems were too great to overcome at that time, and no other gas turbine cars appeared before they were banned along with other rotary engines such as the Wankel.

Fig. 14.1 The Pratt & Whitney turbine in the Lotus T56B. (Courtesy of Phipps/Sutton)

Turbocharging

When the FIA changed the engine regulations in Formula 1 from 1.5-liter NA to 3.0-liter NA for 1966, they also permitted 1.5-liter supercharged engines, as a sop to those who complained they could not build new 3.0-liter engines in time (Fig. 14.2). No one took up the challenge until 1977 because it was not considered a fair equivalency. Renault decided to enter Formula 1 with a turbocharged 1.5-liter engine, based on its successful racing sports car engine, to promote its range of turbocharged road cars. Although more powerful than the 3.0-liter engines, it required more fuel and presented the driver with driveability problems, due to throttle lag as the turbochargers spun up. It took intercoolers to increase the density of the compressed charge, variable geometry turbochargers to overcome the throttle lag, electronic fuel and ignition management to provide precise control of lean mixtures without detonation, and special fuels to realize the potential of the turbo-engines. It was not until 1981 that a BMW turbo-engine won a World Championship. Renault, the pioneer, never did.

Fig. 14.2 The Honda 1.5-liter turbo-engine in the Lotus T99.

As other major manufacturers joined the technical challenge by developing turbo-engines and supplying them to the teams, the escalation in power was unprecedented. The fuel chemists enabled the engine designers to keep increasing the pressure ratio of the engines, until the power went off the clock on 1000 bhp dynamometers. There were rumors of more than 1200 bhp from the BMW four-cylinder, as it trailed black smoke in qualifying trim. At 800 bhp per liter, these four-bar boosted engines, burning high energy-density fuels, did not survive for more than a few laps.

Boost limits were imposed from 1987, along with the alternative of a 3.5-liter NA engine, introduced in 1988, until turbocharging was banned completely in 1989. The intense development has undoubtedly benefited road cars, although the use of turbocharging for petrol engine cars has declined. However, the turbo-diesel has become more popular, as its economy has been matched by improved driveability. Turbochargers experienced accelerated development during their decade in Formula 1.

Fuel

As turbocharging turned Formula 1 back into a power formula again, after the ground-effect era, the major car manufacturers joined in the effort, and most have remained since that time. Alongside every engine development department is a fuel company, whose technical expertise is needed as much as its sponsorship funding. The near doubling of specific output in the six years between 1981 and 1987 can be partially attributed to fuel development. When turbocharging was banned and 3.5-liter NA engines were made mandatory, the rate of development was sustained, whereas the power increases, year on year, for the 3.0-liter NA engines in 1980 had tailed off. Relative freedom of fuel specification and the air valve springs developed by Renault helped maintain the climb in power until pump fuel was mandated, backed by sophisticated fuel analysis scrutineering.

The energy content of the fuel used is fundamental to the power released when it is burnt with the oxygen in the air that is pumped through any internal combustion engine. The oxygen content of the fuel must also be regulated, to give any meaning to a volumetric capacity limitation. The incentive to increase the oxygen content is enormous, and the rumors of competitors using nitrous oxide and other oxygenated fuels, in all forms of motorsport, have dogged the sport for years. The fuel specification can also be adjusted, to provide both charge cooling and internal engine cooling. The black smoke of the qualifying BMW turbo-engines was evidence of the need to use the fuel to cool the pistons when each was generating 300 bhp (Fig. 14.3).

The understanding of the role in the combustion process, of the range of hydrocarbons available to the fuel chemist, increased markedly with the ability to computer model both molecules and combustion. A measure of how this understanding had been used to increase the power of the 3.5-liter engines is that once the FIA had imposed and effectively policed pump fuel in 1994, more than 50 bhp was lost from the 820 bhp engines. Some of the constituents of these racing fuels were pretty nasty in their unburnt and burnt forms, and health and safety considerations led to their ban. Since then, the FIA has used Formula 1 as a high-profile example of environmental conscientiousness, and regulated the fuel to be used, such that it meets the most advanced proposed European regulations.

rpm

With swept volume, engine cycle, and fuel regulated, only rpm remains for the engine designer as a means of increasing power. Renault having removed the resonance limitations of the valve gear by replacing coil springs with mass-less air springs, the peak rpm is determined by resonance in other parts of the structure of the engine—the crank shaft in particular. Reducing the

Fig. 14.3 Black smoke reveals the need to fuel-cool the pistons of the BMW turbo in the Brabham during qualifying. (Courtesy of Sutton Motorsport Images)

mass and increasing the stiffness of the dynamic components provides the development path for increased rpm, and hence power. Two recent regulations have put limitations on the development of higher rpm: a maximum limit on the number of cylinders at ten, and restrictions on the use of materials. No part of the engine may use a metallic material with a specific modulus greater than 40 GPa/(g/cm^3), and neither the crankshaft, camshafts, pistons, cylinder heads, nor cylinder block may use carbon or aramid-fiber reinforced composite.

Volumetric Efficiency

Restrictions on volumetric efficiency, apart from the effect of the number of cylinders being limited to 10, are the limitation on the number of valves per cylinder to five, variable geometry exhaust systems, any form of charge or fuel cooling by, for example, water injection, and the banning of oval pistons. Honda developed oval pistons for Grand Prix motorcycles, gaining valve area; however, they were banned in Formula 1 before any engine designer was tempted to go to the expense of trying them.

Stored Energy

The opportunity for Formula 1 to make a major contribution to the development of road cars was denied, when the storage of energy and its subsequent use to propel the car were banned. (The regulation is written in such a way that the existing electrical and hydraulic energy storage systems are permitted.) With current cars up to 70 kg (154 lb) under the weight limit, the potential for energy storage systems replacing ballast is significant. It is doubtful whether recovering energy from the braking of the car would warrant the hardware necessary, but storing energy from the engine, when full power is not required, might have been worth the additional complexity. The stored energy then could be released when full power was being used.

Mercedes was rumored to have been investigating the possibilities in the late 1990s, working with engine builders Illmor. A high-speed flywheel would be necessary because batteries, while having approximately half the specific energy of a flywheel, are limited by a specific power of up to 0.2 kW/kg—a 50 kg (110 lb) battery pack would discharge only at approximately 10 kW. On a typical circuit, full throttle is used approximately 60% of the time, with the remaining 40% being at low-throttle settings. If the storage system were capable of delivering a 50% power increase on full throttle (from 800 up to the 1200 PS of the turbo-engines), an 80-second lap would require only 4.0 kWh of stored energy, provided by using surplus engine power during the 40% off-throttle periods. A flywheel system weighing 20 kg (44 lb) could store and deliver this, provided the electrical system was powerful enough. In reality, 400 PS would not necessarily be needed, when balanced against the weight of the system and additional fuel consumption. Also, the flywheel duty cycle would be less than a complete lap, so the whole system could be considerably lighter. The advantages in overtaking and at the start of the race, however, might have driven powers ever higher.

These rough figures show the enormous potential for stored energy systems in racing. Both Chrysler (gas turbine/electric/flywheel) and Panoz (piston engine/electric/batteries) have explored stored energy/hybrid race cars for the Le Mans race, but have run into technical problems with their systems, leading to their abandonment. However, the prospect of this sort of power escalation, and the costs of developing the flywheels, power, and control system led the teams and the FIA to agree to a ban. The opportunity to make a real contribution to the development of lightweight efficient hardware and control strategies for hybrid road cars has been lost.

Automatic Gear-Changing

Although the semi-automatic gear-change was not included in the ban on driver aids, on the basis that the savings in over-revved engines, due to missed gears, was too great to be discarded, full automatic control is no longer permitted.* The algorithms for automatic gear-changing are

* See author's note on page 290.

relatively straightforward, and road cars appearing with Formula 1 derived systems have this facility. A far greater loss to road car development was the limitation on the minimum number of gears, set at four, thereby banning any form of continuously variable transmission (CVT).

Continuously Variable Transmission

Step-less gear ratio changing permits the engine to operate within a narrow rpm band, or even at a discrete rpm, instead of requiring an operating band of approximately 2000 rpm. The engine designer can then tune the engine for that rpm, avoiding any compromises in intake and exhaust lengths or valve timing. The increase in available power (the integral of power throughout a cycle of accelerating through the speed range) must outweigh any loss in efficiency or increase in weight.

Williams started research into CVT's in the late 1970s, initially with a system based on the Perbury system. At the time, Williams was developing the Cosworth DFV and was seeking ways to achieve a power advantage over other DFV users. The Perbury proved to be too heavy and was encumbered with IPR problems. Some years previously, Van Doorne had raced a Formula 3 version of its rubber belt and cone CVT, based on the system sold in DAF road cars. It was reasonably competitive but unable to take the power of a Formula 1 engine. Williams decided to work with Van Doorne to develop a steel belt version, suitable for Formula 1. The issues of engine starting, engaging drive, reverse, and the hydraulic and electronic control integration with the engine all had to be solved. The gearbox that was mounted in a Williams FW15, powered by a specially tuned Renault 3.5-liter engine, weighed 6 kg (13 lb) more than the six-speed one it replaced and was approximately 93% efficient, compared to 96% for the spur gear variant (Fig. 14.4). This appeared in testing in 1993 but was banned as part of a package of control system regulations, at the end of the year, before its performance advantages could be demonstrated. History does not relate whether the ban was a result of the fear of having to develop competing systems, when Williams had been working on the principle for so long, or the thought of 24 Formula 1 cars, all with their engines continuously at peak rpm. Continuously variable transmissions are now becoming accepted road transmissions, competing with automatics and a variety of computer-controlled gear change systems. Audi launched a steel belt and cone-based CVT system in 1999, capable of accepting 280 Nm torque, which endows its A6 with better performance and economy than a manual five-speed transmission, equivalent version. Formula 1 could have promoted and hastened the development of what may become the transmission of the future.

Weight

A minimum weight limit, currently 600 kg (1323 lb) including the driver, was introduced to put a limitation on performance and discourage designers from paring down weight to such an extent that component strength and safety are compromised. It was also hoped that a minimum

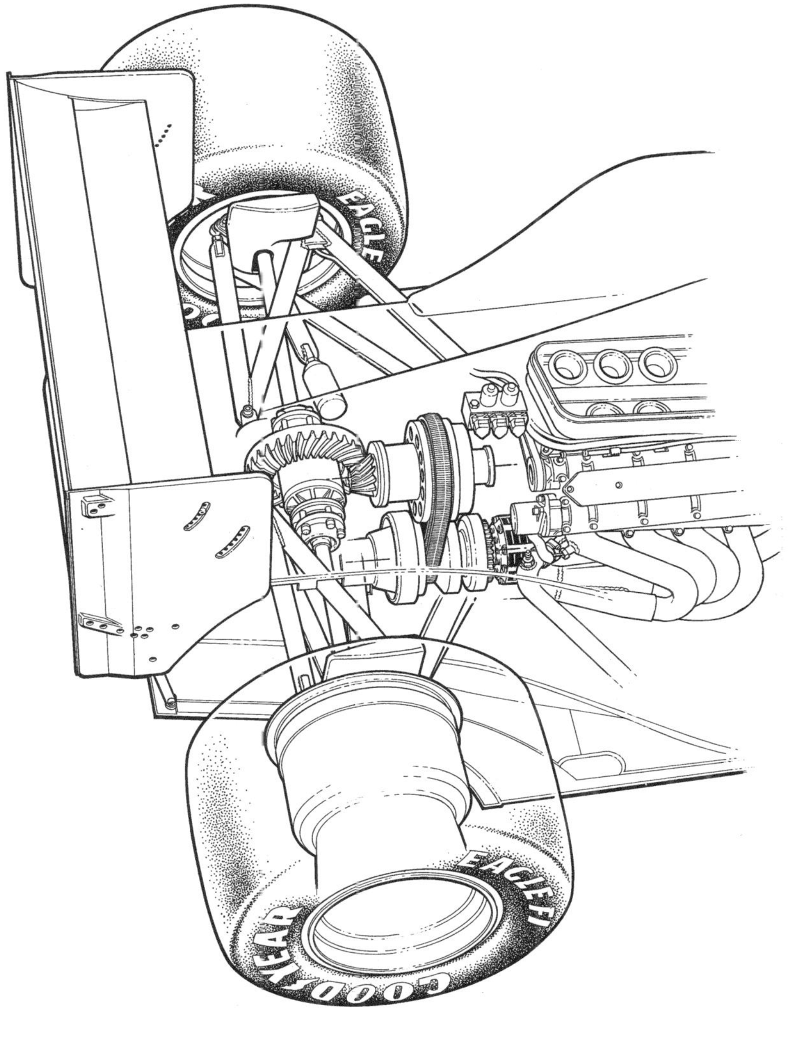

Fig. 14.4 A CVT gearbox in a development Williams FW15-Renault.

weight would prevent the use of costly, aerospace-developed materials and manufacturing processes. Unfortunately, it no longer achieves this.

The current cars are both narrow track and under-tired at the rear, so that a necessity exists to design for the lowest possible CG, to avoid overloading the tires. Added to this, the balance of the cars is sensitive to the weight distribution, and this may change from circuit to circuit, according to the aerodynamic setup and the tires chosen. As a result, the more the weight can be reduced below the weight limit, the more disposable ballast can be carried. Current cars weigh less than 450 kg (992 lb) empty, without driver and without ballast. Ballast, in the form of depleted uranium, compressed silicon carbide, or tungsten plates, is mounted on the lowest part of the undertray, and its position is selected according to the desired weight distribution. The quest for higher specific strength and stiffness materials is undiminished by the minimum weight regulation, and thus material specification regulations have had to be introduced, not on safety grounds but to limit costs. These are aimed primarily at the use of beryllium alloys, metal matrix composites, and the use of composites for dynamic engine components.

From the moment Chaparral brought to the notice of the Formula 1 designers the performance benefits of aerodynamically induced downforce, almost every device with the potential of generating a downforce at reasonable efficiency has been tried. The gains in cornering speeds and the reduction in braking distances have been so great, that the FIA has been stretched to keep a rein on developments and to maintain performance within the capability of the circuits to protect drivers and others from being injured by out-of-control cars. The history of downforce, plotted in Fig. 2.11, graphically illustrates the pace of development and the efforts to contain aerodynamics.

Wings

It took less than a year for wings to grow, from the first tentative devices that appeared in 1968, fitted low down over the engine (Fig. 14.5), to full span wings, strut mounted to the sprung masses, high above the car front and rear, and fitted with driver-operable feathering systems (Fig. 14.6). A year was not long enough to fully understand the aerodynamic and road-induced dynamic forces, to which these frail devices were subjected, and several of the more ambitious examples collapsed, causing many accidents. The FIA quickly intervened and placed limits on the height and width of bodywork, banned the fitting of aerodynamic devices to the unsprung mass, and banned movable aerodynamic devices. These last two regulations have survived, although their exact meaning has been severely tested several times.

> Any specific part of the car influencing its aerodynamic performance:
>
> - Must be rigidly secured to the entirely sprung part of the car (rigidly secured means not having any degree of freedom)
> - Must remain immobile in relation to the sprung part of the car

Fans

Again in 1969, Chaparral and Chevrolet R&D showed a new way to generate downforce, this time by using an additional engine to drive a pair of fans. The fans created a depression under the full plan area of the car, the edges of which were sealed by Lexan sliding skirts, operated by the suspension movement, to maintain a light contact pressure with the ground. Compared to wings, the advantages were that the full downforce was available from zero speed and the drag penalty was nil, the extra engine being permitted in Can-Am's free engine rules. It worked brilliantly.

Fig. 14.5 Driver-adjustable wing on a Ferrari.

Fig. 14.6 High wings, mounted to the unsprung masses, fitted to a BRM.

Formula 1 ignored this approach, as the "moveable aerodynamic device" regulation ruled it out. At least most designers thought this was the case. In 1978, faced with the ground-effect dominance of the Lotus T79, and the inability to follow suit with a flat-12 engine blocking the venturis, Brabham's Gordon Murray decided to test the interpretation of that rule. He argued that if the primary purpose of the fans is to draw air through the radiators, then any secondary effect is incidental. The argument pivoted on whether the function of a fan is to move air or to generate a pressure difference. If it is to move air, the consequence of which is to create pressure differences, then the radiator flow exceeded the leakage flow under the skirts, and thus it could be argued that it is legal. If the function is to create a pressure difference, the consequence of which is the movement of air, then the area of the car on which that pressure difference acted is far greater than the area of the radiators, and thus it is illegal.

The Brabham fan car (Fig. 14.7) was brilliantly executed, with a single fan driven by the Alfa-Romeo flat-12 engine. It won its first and only race in Sweden, with the Lotus T79 drivers complaining that it was faster in the slow corners although they continued to have an edge in corners taken at more than 160 kph (99 mph).

Lotus immediately designed a version of the T79 with two fans, one in each venturi. The theory was that the fans would power the venturis at low speed, providing "static" ground

Fig. 14.7 The Brabham fan car. (Note the lack of barriers between the track and the executive aircraft park.) (Courtesy of Phipps/Sutton)

effect, and that normal, car-induced flow would take over at high speeds, giving the best of both worlds. Politics took over from technical interpretations, and one of the most fascinating and irrelevant Formula 1 technologies was finally banned for good.

Advantage Lotus.

Twin Chassis

The *raison d'être* of the Lotus T88 twin chassis car (Figs. 14.8 and 14.9) is described in Appendix A on the development of ground effects. It took to new limits, the interpretation of the rule banning any aerodynamic effect being generated by unsprung parts of the car. Originally conceived to separate the suspension requirements for the chassis (soft) from those of

Fig. 14.8 Lotus T88 twin chassis car. Colin Chapman and I are on the left side. (Courtesy of Phipps/Sutton)

the downforce generating body (stiff), it might have been accepted, if skirts had not been banned in the middle of its development. As potentially the perfect solution to gaining the most from ground effect without skirts, the FIA and the other teams were not prepared to accept it. It was a classic case of the regulations being able to deal with only what had already been imagined, and Colin Chapman's fight to be allowed to race the T88 was long and bitter. He was fighting for the right to innovate and was never satisfied that the T88 was proven to contravene the regulations.

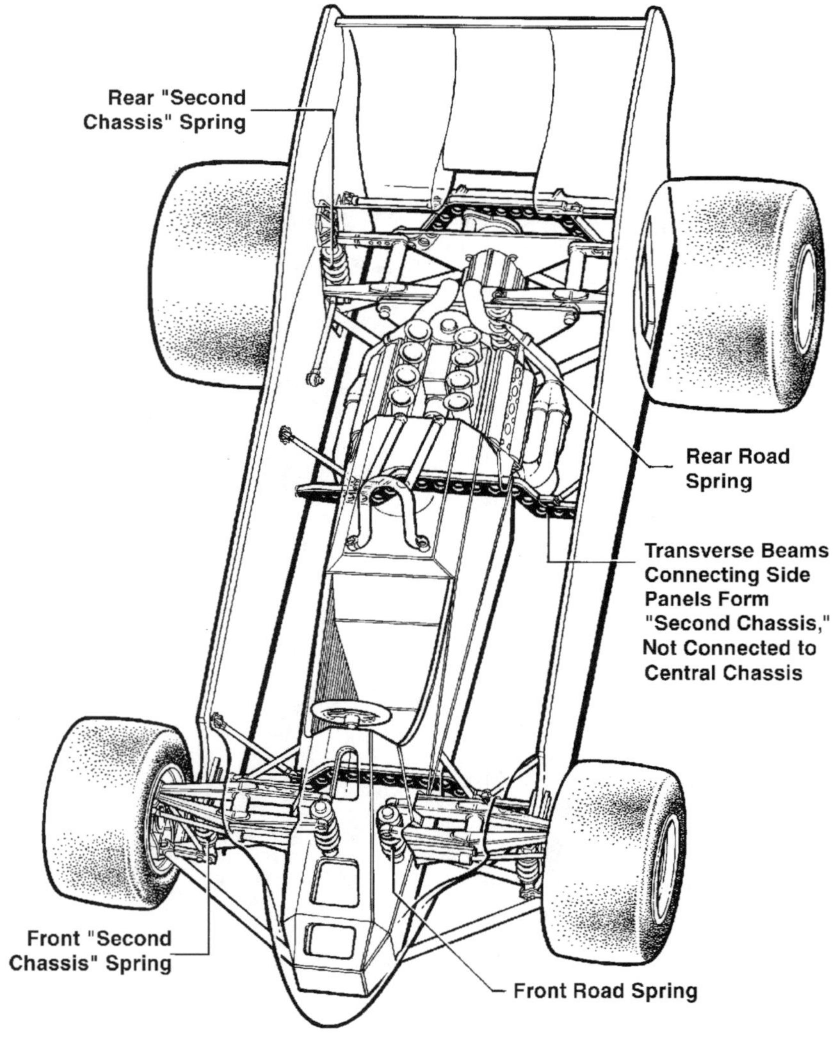

Fig. 14.9 The Lotus T88 with top and bottom body panels removed, showing the two suspension systems.

To make matters worse for Chapman, Brabham was able to race a system that lowered its cars to a running height, after the cars had left the pits and passed the 6 cm (2.4 in.) clearance test, and which raised them pneumatically on re-entering the pits.

Advantage Brabham.

Venturis and Skirts

It was always a bit of a mystery why skirts were permitted, in the light of the "movable aerodynamic device" regulation. When the FIA set out to curtail the level of downforce being generated by the cars in 1980, it chose to do so by stipulating a 6-cm (2.4 in.) ground clearance rule. Although this eliminated sliding skirts, it permitted flexible Vethane rubbing strips. By 1982, most teams had learned how to set up the cars stiffly, to minimize the skirt gap, and had regained some of the lost downforce. The cars were so unpleasant to drive that for 1983, flat bottoms were mandated, setting the stage for the much more pitch-sensitive aerodynamic configuration, that is still imposed on most race cars nearly 20 years later. Only CART and IRL have retained restricted venturis, without any form of skirt, to avoid pitch sensitivity on the high-speed ovals. There are advantages in enabling the cars to run closer together, thereby offering more overtaking opportunities. Eventually, other formulae, including Formula 1, may adopt this configuration.

Active Suspension

Active and semi-active suspensions were banned at the end of 1993, on the basis of the "movable aerodynamic device" regulation. A new regulation defining "sprung suspension" was introduced:

> The suspension system must be so arranged that its response is consistent at all times and results only from changes in vertical load applied to the wheels save only for movement permitted by inherent and fixed physical properties.

This effectively limited suspension systems to passive spring/dampers and denied the addition of energy to move the suspension according to algorithms that optimized ride heights at different parts of the circuit.

No fundamental technology regulations have affected tires, but limitations on size, tread patterns (grooved treads replacing slicks), and the number of tires used during an event have all affected their performance, as the FIA attempts to control speeds in the face of competition among tire manufacturers.

The way tires are used, however, has been the subject of limitations.

Four-Wheel Drive

In the early 1960s, Ferguson built a four-wheel drive Formula 1 car to showcase its transmission technologies. The car won a wet-weather, non-championship race, but managed only to inspire BRM to test four-wheel drive. They did this rather half-heartedly and succeeded only in convincing themselves that it was not the way to go in Formula 1.

Ferguson was also cooperating with STP on the development of four-wheel drive Indy cars. When Lotus built turbine, four-wheel drive cars for STP, Colin Chapman decided to explore the application in Formula 1 in 1969 (Fig. 14.10). At that time, racing tires were somewhat lacking in traction for the power of the Cosworth DFV, an engine that had a peaky torque curve. The combination of these two factors made it difficult for drivers to control the cars, because it was too easy to break traction when opening the throttle. Once the news was out that Lotus was building a four-wheel drive car, Matra and McLaren followed suit, and even Cosworth decided to build a car to defend the reputation of its engine (Fig. 14.11).

None of these cars was successful, even in the wet, and the drivers hated them. On high-speed corners, such as those at Indianapolis, four-wheel drive provides stability, without the loss of steering control that race car drivers abhor. On very low-friction surfaces, such as those found in rallying, the traction advantages outweigh the control disadvantages. On a race track, with a variety of corners and reasonably high grip, the loss of steering control, allied with the inability to steer the car on the throttle, far outweigh the traction improvement. The only circumstances where track race cars have benefited from four-wheel drive in recent times is when raced against powerful front-wheel drive cars. When a typical Formula 1 car accelerates at 1g, approximately 10% of the weight of the car is transferred from the front axle to the rear. With a static weight of approximately 40% on the front, the reduction to less than 30% when accelerating hard is too much, when combined with making the front tires transmit torque. Thus, excessive understeer results. Any attempt by the driver to induce oversteer with the throttle results in more understeer.

The complexity and weight of the transmission, and the need for a drive shaft to pass alongside the driver, added to the penalty. As the percentage of the torque transmitted by the front axle was decreased, in an attempt to reduce understeer, the advantages of four-wheel drive diminished, and all the projects were abandoned. The development of wings and downforce, as well as major improvements in the tires, finally killed conventional four-wheel drive experiments in Formula 1. There was no need to ban it at this stage.

Six-Wheelers

In 1975, Tyrrell built a car with two front axles and four front wheels—Project 34 (Fig. 14.12). The diameter of the tires was reduced from to 533 to 406 mm (21 to 16 in.), and the arrangement gave benefits in front aerodynamic downforce, braking, and tire performance. At that time, the front tires being used were poor compared to those of the rear. By requesting special

Fig. 14.10 The front axle of the Lotus T63 four-wheel drive. (Courtesy of Phipps/Sutton)

Fig. 14.11 The four-wheel drive Cosworth—its one and only Formula 1 car. (Courtesy of Phipps/Sutton)

Fig. 14.12 The Tyrell six-wheeler. (Courtesy of Phipps/Sutton)

tires from supplier Goodyear, Tyrell gained an advantage from new developments. The cars performed well during 1976, with a win and regular second places. However, after Goodyear responded to the pressure from the other teams for a better front tire and concentrated its development efforts on the needs of the majority, the six-wheel Tyrells fell off the pace. No one else followed this interesting initiative, and thus no ban resulted.

The novel idea that a race car could have more than four wheels set designers thinking. In 1977, March built a car with two driven rear axles, fitted with front wheels and tires—a four-wheel drive six-wheeler (Fig. 14.13). The concept was not developed, but was taken up subsequently by Williams in 1981, in its search for an advantage to enable its Cosworth DFV engines to compete with the turbos. The reduced drag of the four small rear wheels, better traction, improved underbody aerodynamics, and the retention of the driver's ability to steer the car on the throttle all added up to a potential performance gain. The car showed real promise in spite of the 55 kg (121 lb) weight penalty of the prototype (Fig. 14.14). Perhaps there was too much promise there, because before a racing version could be built, four-wheel drive and six-wheelers were banned from 1983.

The development of data systems, engine management, and subsequently active suspension in the 1980s introduced reliable and compact computers, sensors, and hydraulic systems onto Formula 1 cars. Soon these systems were required to control a whole range of systems to

Fig. 14.13 The six-wheeler March. Max Mosley (right) was the "M" in "March." (Courtesy of Phipps/Sutton)

Fig. 14.14 The six-wheeler Williams. (Courtesy of Phipps/Sutton)

enhance vehicle dynamics. It has long been realized in the aircraft industry that putting a computerized control system between the pilot and the control surfaces enables the aircraft to be given control and stability characteristics, which would be impossible for the pilot to handle without the computer. Equivalent systems have been developed for road cars, but with the prime purpose of making up for a driver's lack of skill or concentration (i.e., primary safety). When applied to Formula 1 cars, computer systems enhance control and stability, enabling the driver to use more of the potential performance envelope, more of the time. (Systems included traction control, ABS and anti-skid, four-wheel steer, automatic gear-changing, active differentials, and pit control by telemetry of the car setup, while on the track.)*

The sound of 22 cars misfiring around Monaco, as their traction control systems cut the ignition, and the lack of spectacle of the carefree 5g braking, which power brakes plus ABS afforded the drivers, was enough to convince the FIA that anything that assisted the driver in driving and controlling the car should be banned. The drivers, who felt that the "driver aids," as these control systems were termed, took away too much from the sporting aspects of Formula 1, supported them. At the end of 1993, all driver aids except semi-automatic gear-changing and drive-by-wire throttles were outlawed. Policing these regulations meant that all hardware and software used by the teams had to be inspected and approved. Rumors of cheating continued to abound, especially the application of traction control software within the engine management system; however, nothing has been proven. During 2000, the FIA tightened the parameters that could be controlled by the engine management computer, and the range through which they could be altered, to counter evidence that several teams were continuing to bend the rules. As is the case with accusations of the use of drugs in athletic sport, no one wants to believe that a competitor can be quicker.

Fiddle Brakes

In 1997 and 1998, McLaren raced one of the most interesting developments in recent years, differential rear braking or "fiddle brakes," before it was protested as being a rear steer system (which it is not). The finding of the race stewards was never appealed; therefore, the technical arguments for and against were never properly aired. The advantages of using the brakes to generate a yawing moment, and the ability to limit the torque transmitted to the unloaded, inside tire, are great enough for the principles to have been retained. The concept has now disappeared inside the gearbox and is no longer controlled by the driver. Instead, it has become a controllable differential, subject to the regulations governing these devices.*

Summary

If the FIA had not intervened to ban all these technologies, one can speculate on what Formula 1 cars would have become: 1500 PS turbo- or turbine engines; fan-assisted, skirted ground

* See author's note on page 290.

effect; variable geometry aerodynamics; active suspension; and full control of the individual tire slip ratios to provide artificial stability and control. Whether four driven rear wheels would be needed would depend on the level of downforce achieved with the use of fans and the performance of unlimited tires. One thing is clear. The performance of the cars would be beyond the capability of a human driver, with much more than 6g cornering and braking possible, and top speeds exceeding 450 kph (279 mph) on existing circuits. Of course, the circuits would have to change to accommodate these cars in the event of a failure. Most of the regulations that have been imposed over the years were inevitable, and only a few useful technologies have been lost in the process.

The pressures of the commercial and entertainment aspects of Formula 1 have dictated that the Technical Regulations define the cars closely. There is little scope for designing a car that differs significantly from the cars racing today, and similar performance and close racing are the desired consequences. However, this truism is limited by the imagination of the regulation writers. It is possible to regulate only what has already been imagined, and designers and engineers will continue to invent systems that test the spirit and meaning of the regulations. How many of these systems will work, in terms of endowing a Formula 1 car with more performance? How many will benefit road car development? Only time will tell. Formula 1 technology no longer pretends to be of any great relevance to road cars, other than in the marketing of brand names and the training of engineers. Different objectives for the premier form of motor racing will be needed before the regulations are freed, and invention and innovation rule again.

Chapter 15

Safety

Speed does not kill, but a sudden lack of it does.
—Henry Labouchere

Motorsport is slightly more than 100 years old, a few years younger than the automobile itself. For many years, apart from occasional steps being taken to protect spectators, little was done until the late 1960s to protect participants from fatality or serious injury as a result of accidents. Although drivers raced largely for glory rather than monetary reward, and automobile manufacturers participated as part of their R&D efforts rather than marketing, death and injury were accepted as part of the risk. In the same way that public attitudes toward road safety changed in the late 1960s, it was the drivers—now professionals earning substantial incomes from motorsport—who campaigned for greater safety.

The nature of motor racing encourages drivers to drive their vehicles at the limit of control and encourages designers to reach for the limits of the materials they use. Therefore, it is both inevitable and accepted that accidents will occur. Accidents are a necessary part of motorsport, providing spectator and television entertainment. However, the governing or sanctioning body is responsible for ensuring that accidents do not result in fatalities or serious injuries An adequate level of safety is required, not only for the drivers and spectators, but also for team participants and officials involved with staging the events.

In 1994, after 11 years without one fatality in Grand Prix events, two fatalities occurred during a single Grand Prix weekend. Because one of them involved triple World Champion Ayrton Senna and was covered worldwide by live television, the reaction of the press, public, governments, sponsors, and even the Vatican was so adverse that it became apparent that international motorsport could be threatened, unless even greater levels of safety were achieved. As the governing body of international motorsport, the FIA immediately took wide-ranging steps to improve safety, including new regulations for car construction and circuit design. It also initiated R&D programs to improve systems and standards for the future. Max Mosley, president of the FIA, has stated that the only acceptable safety objective is zero fatalities and serious injuries. Apart from the moral necessity of this objective, the loss of a top race car driver is a sporting loss. During the last five years of the century, Formula 1 racing would undoubtedly have been enriched by Senna-Schumacher-Hakkinen battles.

In the United States, the major automobile manufacturers participating in the prime racing series, such as Champcar, IRL, NASCAR, and IROC, are working closely with the sanctioning bodies, to research and improve safety. Oval racing, with its high speeds and rigid containment walls, provides a unique opportunity to research high delta-V (total velocity change during an impact) high-g impacts, and to investigate human tolerance of these conditions. Impacts generating 80g frontal, 160g lateral, and 135g rearward have been recorded, and are under detailed investigation, to determine why no g-related injuries were sustained by the drivers involved in many of these accidents.

Because every aspect of motor racing is so controlled, and the technology now exists to enable accidents to be monitored and analyzed in detail, the motorsport arena presents a valuable opportunity to research automobile safety *in extremis*. The physics and biology of racing accidents and injuries are almost the same as those of road car accidents. Motorsport safety can show what is possible in terms of human protection and tolerance, if no constraints are placed on the level of technology applied or costs involved. Although the lessons learned cannot always be directly or economically applied to the road system, much of the data and experience gathered is relevant.

Of particular interest to those American automobile manufacturers that are also involved in motorsport safety is human tolerance and its effect on legislative standards for road safety. The deceleration injury assessment criteria, derived from dummy data in crash tests, are based largely on the results of tests performed by Colonel John P. Stapp of the U.S. Air Force. He subjected himself to 40–45g sustained deceleration during rocket-sled tests in the 1960s, to determine criteria for ejection from aircraft. Further testing was conducted on cadavers to determine g-levels for internal organ damage. None of these results indicated survivable g-levels of the magnitude that race car drivers sustain in accidents. Motorsport is a unique biomechanical laboratory.

The deceleration forces involved in bringing a 600+ kg (1323+ lb) car to a halt from high speed in a short time, and the way in which the stored energy in the car is dissipated (i.e., by distorting and fracturing the structure of the car) make it a bad place for a human being to be located. The violence of an impact is on a par with an explosion and can lead to injuries or loss of life.

In a motorsport accident, the drivers of the cars involved may be injured or killed as a result of the deceleration incurred, or by striking the inside of the car structure (or occasionally the track or circuit features), or by being struck by another car or some part of it. Much less common today are fires caused by accidents. Fortunately, the days of ruptured aluminum fuel tanks are gone, as well as the horrendous conflagrations that accompanied a large proportion of serious accidents. It is rare for the CFRP/honeycomb tank, located in the center of the car, to be penetrated. When this does occur, the Kevlar-reinforced rubber bladder provides a second line of defense. Even if the engine is torn off the monocoque, self-sealing breakaway couplings prevent fuel spillage. Spectators and other non-drivers, such as team members and marshals, may rarely become involved in accidents and suffer injuries or death due to being struck by a car or part of it. Motorsport safety attempts to prevent each of these potential causes with regulations governing all aspects of an event. These regulations are inevitably shaped by the physics and biomechanics of accidents, and the resulting impacts.

When a car leaves the track and hits a barrier or other "solid" object, or hits another car on the track, three separate impacts occur: (1) the car hits the object; (2) shortly afterward, the driver hits the car; and (3) shortly after that, the driver's internal organs hit part of his skeleton. Prior to the impact, the car, the driver, and the driver's internal organs are all travelling at the same speed, but they are not rigidly attached to each other. When the car hits a solid object, it decelerates at a high rate, but the driver carries on at the speed at which the whole assembly was travelling prior to impact. When some part of the driver hits the car, that part decelerates quickly, but the driver's internal organs continue at unabated speed, until they hit the now-stationary or rebounding skeleton. The first impact, car to solid object, will not in itself cause injury. The driver hitting the car is likely to cause injuries to the flesh covering the skeleton, or fracture the skeleton itself. Neither of these types of injury is, in general, likely to cause death unless the impact is so severe or the part of the car so sharp that the skeleton is penetrated and an internal organ is also damaged. Michael Schumacher's accident at Silverstone in 1999 is a typical example of a relatively clean break in major bones. He was able to drive his Ferrari at competitive speeds only six weeks after his accident. Multiple bone fractures can take much longer to mend, as Johnny Herbert experienced after his Formula 3000 accident in 1987, in which his feet were severely crushed in a double impact. It was several years before he could provide full braking force on the middle pedal and he will always limp, but he can still drive competitively.

However, the impacts between the internal organs and the skeleton are, in many cases, the most serious and likely to cause death or long-term disability. The critical internal organs are located in the head and torso, and these bodily areas must be protected. Broken bones heal in time; damaged brains and spinal cords may never heal.

Unsavory as the preceding description may be, it is essential to consider the biomechanics of injuries, a massive subject in its own right, in order to understand the physics of an accident as a prelude to devising prevention systems. Ralph Nader alerted the automotive industry to the issues of road safety in the 1960s and, as a result, created a major automotive discipline that involves scientists, doctors, and engineers. The biomechanical side is symbolized by the development of anthropomorphic dummies, such as the Hybrid III and Euro-SID dummies, with their ability to measure loads, deflections, velocities, and accelerations that represent those to which the human body may be subjected during an accident. From these measurements, criteria that reflect the risk of injury or death are derived, enabling accidents to be reproduced and analyzed in laboratories and computer models. The criteria are hotly debated within the profession and among vehicle manufacturers and legislators. Motorsport safety research is beginning to make a contribution to that debate, being one of the few areas where actual injuries can be related to measurements recorded during an accident. Live volunteers who are prepared to take part in road accident research are rare!

Safety engineers can do nothing about the critical impact between internal organs and the skeleton because that area cannot be re-engineered. Thus, the deceleration and energy dissipation of the car, and the prevention or attenuation of the driver coming into contact with the structure of the car, are the areas that receive all the engineering attention. The two most important physical parameters in an impact are the energy to be dissipated and the deceleration rate during that process:

$$\text{Energy} = \frac{\text{Mass} \times (\text{Velocity})^2}{2}$$

$$\text{Deceleration} = \frac{(\text{Change of Velocity})^2}{2 \times \text{Distance}}$$

Both parameters influence the likelihood and severity of injury. The energy to be dissipated determines the amount of distortion and breaking that must take place in either the car or the object hit, in order to dissipate the stored energy. This influences the chance that parts of the broken structure of the car will intrude and injure the driver. The deceleration determines the severity of the subsequent two impacts: (1) driver to car, and (2) internal organs to skeleton. Both increase with the square of the velocity, while the deceleration rate decreases linearly with the distance over which it occurs. Four "golden rules" emerge from these simple equations:

1. Minimize the velocity of impact.
2. Maximize the distance over which the car is decelerated.
3. Maximize the energy dissipation in the object hit, rather than in the car.
4. Where the car must be used to dissipate energy, concentrate it in sacrificial parts, and protect the driver in a strong enclosure that will not distort or fracture.

The application of these rules is the basis of most of the regulations governing both car construction and circuit safety features.

Arrestor Beds

When a car goes out of control, either because the driver loses control, hits or is hit by another car, or something breaks, the car tends to start decelerating immediately while still on the track. The exception to this is the consequence of a stuck throttle. Also, the deceleration may be very low if there is brake failure, as with Schumacher's loss of rear brakes at Silverstone in 1999, or if wheels are missing. The trajectory of the car will roughly follow a tangent to the path the vehicle was following, and the car will depart the narrow track onto runoff areas. Various arrestor systems have been tried, to use these large areas to decelerate the cars by dissipating their energy. The essential requirements for an arrestor bed are:

1. It must decelerate vehicles without tipping them.

2. It must perform adequately with vehicle types that include: single-seaters, touring cars, GT/sports cars, trucks, historic cars, and motorbikes.

3. It must perform adequately if the vehicle has no brakes or no wheels, has a stuck throttle, or is inverted or rolling.

4. It must perform adequately, irrespective of vehicle heading.

5. Stranded vehicles must be easily removable.

6. It must not require refurbishing after an incident.

7. It must be weather resistant.

8. It must be economically viable.

The ideal characteristics of an arrestor bed are that it should generate a large horizontal force on the car, without making it rise up clear of the arrestor mechanism. A standing crop works well, as one or two drivers have discovered in the past, but it is somewhat impractical. Catch fencing was used in the 1970s and worked remarkably well with respect to Items 1–4 in the above list; however, it presented major problems with respect to Items 5 and 6, and it was not very good with respect to Items 7 and 8. Races often had to be halted to extricate cars from a tangled web of wire fencing. When incidences of drivers being hit on the head by fracturing fence poles began to occur, gravel beds started to replace catch fences (Fig. 15.1).

Gravel beds have been used successfully for many years to arrest brakeless trucks, and the mechanisms by which they dissipate energy are fairly well understood. A vehicle passing into a gravel bed transfers momentum from itself to the gravel, which "sprays" out of the way and dissipates its energy by friction. In addition, some shearing of the gravel takes place, according to the principles of soil mechanics. In many ways, the gravel behaves quite similarly to a fluid, with momentum and viscous shear properties. It also displays fluid-like properties, in that it causes the tires to "aquaplane," and any vertical motion of the car causes skipping over the surface, a motion similar to "ducks and drakes." Both of these effects are particularly prevalent with light cars, wide tires, and flat undertrays at low ride heights. However, they do provide a reasonable compromise, even if they have received bad press. They generally decelerate a Formula 1 car at slightly below 1g, whether the car has wheels or not, is spinning, or is traveling straight with brake failure. They work with a variety of vehicle types and

Fig. 15.1 A Surtees, caught in the catch fencing. (Courtesy of Phipps/Sutton)

require little repair or maintenance. They can be built anywhere in the world, although the characteristics of the available gravel vary significantly. No system with a better compromise has yet emerged (Fig. 15.2).

A classic example of the successful combined effects of a car decelerating on track and then in a gravel bed occurred when the rear tire on Mika Hakkinen's McLaren delaminated at Hochenheim in 1999. He was traveling at approximately 330 kph (205 mph) when the tire failed, and yet he hit the barrier head-on at such a low speed that the front of the car was hardly damaged. The worst scenario is when one car hits another, causing one or the other to take off into the air. No deceleration occurs, other than that due to aerodynamic drag, until the car lands.

Barriers

The type of barrier deployed on any given section of a circuit depends on the estimated angle at which the car may hit it and the likely speed. Rigid walls deal best with highly oblique impacts, typified by most impacts that occur on high-speed ovals. The energy component perpendicular to the wall (generally low) is absorbed by the crushable structures of the car, and the friction of the car sliding along it dissipates the energy component along the wall. If there is the possibility of more energy perpendicular to the barrier than the structure of the car can absorb, some potential for distortion of the barrier is designed into it. The best example of

Fig. 15.2 Gravel beds may not be ideal, but nothing that performs consistently better while being practical has yet been found. (Courtesy of Sutton Motorsport Images)

this is Armco. The steel rails and posts are able to bend plastically, thus absorbing energy, while the car is still able to slide along the barrier. Unfortunately, Armco is a one-shot system and must be replaced after an accident. Tire barriers, with conveyor belting attached to the front face, provide more perpendicular absorption, while retaining the ability to allow the car to strike the barrier at an oblique angle. For impacts perpendicular to the barrier, tires have been found to work well, with the number of rows determined by the predicted speed (Fig. 15.3).

How tire barriers work has been the subject of recent research by the FIA (Ref. 15.1), and General Motors and Wayne State University in the United States. The numerous parameters involved (e.g., dimensions, arrangement of the tires, fixing method, spacing between rows, and additional inserts in the tire stacks) make analysis extremely difficult because the energy is absorbed by a mixture of momentum transfer, friction, and tearing of the tires. However, if the tire barriers are built to the standards laid down, they are a versatile and economically viable barrier system, using raw materials that are available in all countries and requiring only relatively unskilled labor to build the many kilometers of multi-layered barriers that line circuits. They are also usually self-recovering, durable, and weather resistant.

Numerous proprietary barrier systems have been proposed, with every major accident generating a flurry of new ideas. Many are not economically viable, but even more do not perform in the way their inventors claim. Safety is first and foremost an experimental science, and experiments must be repeated with a range of initial conditions because small variations in the setup can have

Fig. 15.3 Michael Schumacher's Ferrari became embedded in a tire barrier at Silverstone, after hitting the barrier at approximately 110 kph (68 mph). He suffered only a broken leg. (Courtesy of Sutton Motorsport Images)

a large effect on the results. A major problem in selecting a barrier specification for a given section of the track is that it must perform with a variety of vehicle configurations and masses, as well as a range of impact angles and vehicle attitudes. A barrier that is impacted by the sharp, stiletto-like nose of a single-seater requires a different stiffness from one that is expected to absorb the same car traveling sideways. If it is correct for the first case, it may be too stiff for the latter case. Likewise, a barrier suited to a 1000+ kg (2205+ lb) GT car will be too stiff for a Formula 3 car of half the mass. Trucks and fallen motorbike riders are even more extreme examples of the range of objects that a barrier may have to absorb.

The dynamics of an out-of-control car spinning through a gravel bed and hitting a barrier are virtually unpredictable. At Le Mans in 1998, a GT2 Porsche spun into the gravel and started to bounce and roll. As a result, it landed on the top of the tire barrier and was catapulted over the fence into a spectator area, fortunately without serious consequences. It would be virtually impossible to repeat this sequence of events, and equally difficult to simulate it using a computer. It is difficult to imagine that such an accident could happen, and thus it is difficult to predict and guard against.

Cars

Barriers will absorb only a certain amount of energy. In the case of a wall, very little is absorbed perpendicular to the face of the wall. Therefore, the structure of the car must absorb a proportion of the total, without failing in such a manner that part of it intrudes onto the driver. The philosophy behind all race car technical regulations that govern the crashworthiness of the car is that the driver shall be enclosed within a strong survival cell that is surrounded by sacrificial energy-absorbing structures. The CFRP monocoques have revolutionized the ability to build lightweight survival cells; the tubular roll cage serves the same purpose in space-frame and production car-based formulae. Energy absorption is performed by every other part of the car—wheels, suspension, bodywork, radiators, engine, and gearbox—with the addition of purpose-designed structures front and rear, and at the sides.

Although the failure of the CFRP fan blades of a Rolls-Royce RB211 to survive a bird-strike initially gave carbon fiber a bad name as far as energy absorption was concerned, the development of sophisticated composite structures has progressed a long way since that time. Dynamic FEA of composites is allowing engineers to understand the failure and energy absorption of these materials, and it currently takes less than 4 kg (9 lb) of CFRP and aluminum honeycomb to absorb the energy of 780 kg (1720 lb) traveling at 50 kph (31 mph).

When high-speed impacts occur, the barrier, which is generally softer, aborbs the first part of the energy, followed by the crushing of the impact-absorbing structure of the car. If the car has not stopped by the time both the barrier and the nose have been used up, either the car comes to an abrupt halt, with massive deceleration, or some part of the structure of the car gives, such that the car is brought to a halt without excessive deceleration, but with potential intrusion injuries to the driver.

The car stops, but the driver does not. The purpose of the restraint systems in a race car is to decelerate the driver with respect to the car, in the small space available, before a part of the driver hits the car. Decelerating 70+ kg (154+ lb) in such a short distance involves high rates of deceleration, and hence forces. To prevent those forces from exceeding the structural strength of the driver's skeleton, they must be distributed as evenly as possible over the surface of the driver's body mass. Padding in the seat, headrest, and cockpit sides is extremely effective in performing this task for rear and side impacts, providing 50+ mm (2 in.) of deceleration distance and good distribution of loads into the torso, and via the helmet shell and padding, to the head (Fig. 15.4). This is so effective now that IRL and CART drivers, when hitting the walls on ovals, can survive rearward and lateral car decelerations of more than 150g for several milliseconds, without injury. During a frontal impact, the seat-belt harness restrains the torso. Made of 75 mm (3 in.) wide webbing to spread the load, the belts stretch significantly to absorb the energy of the moving driver's body and decelerate it before it reaches the steering wheel.

Fig. 15.4 The driver's head is protected against side and rear impact by 75 mm (3 in.) of energy-absorbing foam padding. (Courtesy of Sutton Motorsport Images)

Driver Equipment

The driver's head is a different matter. Restrained only by the driver's neck attached to his torso, the 6–7 kg (13–15 lb) mass of the head and helmet can easily generate forces in the neck that exceed the biomechanical injury criteria. If restrained by hitting part of the car (e.g., the cockpit rim or steering wheel), the decelerations imposed on the head can exceed the criteria for the brain. It is a delicate balancing act to limit neck forces while trying to prevent the head from impacting the car. Cockpit rim padding, collapsible steering columns, and the helmet itself all help to attenuate the deceleration, if the impact is severe enough for the head to eventually hit something. An additional serious cause of head injuries is glancing blows to the head, resulting in high angular acceleration of the skull. Angular accelerations in excess of 10,000 rad/sec^2 can cause irreparable damage to the torsional suspension system of the brain. The way in which damage occurs to the brain during an impact is not yet fully understood and is the subject of much research. Computer models of the head, which combine the physical, biological, and chemical properties of all parts of it, are under development and will eventually contribute to a better understanding of brain injuries and how to prevent them.

Until the 1950s, race car drivers used to drive with nothing better to protect their heads than a cloth or leather helmet. The early crash helmets were no more than a thick layer of cork protecting the top half of the head. They offered little real protection to drivers, who were not even strapped into their open-cockpit cars. In the 1960s, a change in safety philosophy led to drivers adopting five-point harnesses as used by aircraft pilots, in order to remain restrained inside the car during a crash. Head padding around the back and sides of the driver's head provides a means of decelerating the head without excessive neck loads, in rear and side impacts. Obviously, it is not feasible to put a padding system around the front of the driver's head to provide the same protection in frontal or three-quarter frontal impacts, because the driver would not be able to see where he was going! Mika Hakkinen's crash in Adelaide in 1995, in which he suffered severe injuries to the base of his skull, was instrumental in initiating a further research program to determine whether it was possible to protect drivers in this type of accident. It is not a simple problem to solve.

A human head weighs approximately 5.0 kg (11.0 lb), and racing helmets typically weigh in the range 1.2–1.4 kg (2.65–3.1 lb), giving an average total weight of 6.3 kg (14 lb) for both head and helmet. The head acceleration injury threshold, established for road car accident research and testing, is 80g. In reality, the factors contributing to a likelihood of injury are much more complex than a simple g-figure and depend on the time history of the linear and rotational accelerations the head experiences. Although the driver's torso is restrained by his shoulder harness, the only thing restraining his head is his neck. If the head is decelerated in a crash at 80g, such that there is little likelihood of head injuries, the neck must generate a force of 4.94 kN (almost exactly half a ton!) to provide the deceleration. It is established that the combined tension and shearing forces that the human neck, can sustain without likelihood of injury is only 3.3 kN, a force that would be exceeded when the head was subjected to anything greater than 50g. The only alternative is to relieve the neck by decelerating the head via an alternative mechanism. What happens currently is that the helmet hits the steering wheel, and the rim bends but absorbs little energy. To provide even more energy absorption, the steering columns of Formula 1 cars must collapse when struck by a test head form, traveling with a velocity of 7 m/s, and the head must not experience greater than 80g. The problem then occurs if the head does not strike the steering wheel squarely (as happened in Hakkinen's accident) and hits the side of the cockpit rim, which is now padded.

The situation is a balancing act. Take away anything for the head to hit, and the neck may be injured. Put something for the head to hit before the neck is injured, and the head may sustain too high a deceleration. The wearing of a helmet in this situation helps the latter case but makes the former worse.

These problems and their potential solutions are similar to those experienced by road car occupants, even when they are wearing safety belts. To overcome the head injuries sustained by the un-helmeted head hitting the steering wheel, or in the case of a passenger, hitting the windscreen, car manufacturers fit airbags—the soft, energy absorbing cushions that appear only when needed. This solution formed the basis of the initial approach by the FIA to the problem.

MADYMO-3D computer modeling and Hi Ge sled tests at MIRA indicated that an airbag would protect the head and neck, provided it was deployed at the right time and could be held stable when supported by the small diameter steering wheel. The work also created the problems of packaging the airbag module in the tiny, overcrowded steering wheel, and the problem of how to trigger it at exactly the right moment. Road cars have relatively long, soft front ends to absorb impact energy. They have much longer to deploy the airbag, typically 30–40 milliseconds or more, compared to perhaps half of that for a single-seater. Having determined that it was theoretically possible to protect the driver in a frontal impact, and that to turn such protection into reality was no simple task, the FIA approached Mercedes-Benz (now DaimlerChrysler) to see whether it could take on the development. Mercedes-Benz was the first manufacturer to fit airbags in road cars and also had racing experience of airbags in its DTM cars. There was no organization better suited for the task.

In the aftermath of Hakkinen's accident, the FIA was bombarded with "good" ideas about how to prevent injuries in similar accidents. One idea stood out as being a possible solution among all the crazy contraptions suggested. Moreover, it was already in use in the United States. The device had been called the HANS (Head and Neck Support) by its inventor, Dr. Robert Hubbard, a researcher and lecturer at Michigan State University. The device had been tested at Wayne State University and General Motors Research and had been shown to reduce both head deceleration and neck loads. A number of drivers in several classes of car and boat racing were using it, and it had already saved lives. However, the device was designed to fit a more upright driver-seating position than that used in Formula 1. It was clear that a great deal of development work would be needed, to adapt the device to the narrow confines of single-seater cockpits.

Mercedes-Benz agreed to take on the research and decided to take a broad-fronted approach, considering all possible methods of driver protection in a frontal crash. The possibilities were quickly narrowed down to only two: (1) the airbag, and (2) the HANS.

A novel airbag system for a single-seater was proposed, in which a bolster-shaped bag is stowed around the forward rim of the cockpit and pulled out over the steering wheel by an actuator based on a seat belt retractor. The non-inflated bag then lies across the steering wheel and is inflated in the normal way. This "sausage" is fully inflated when the head hits it, and a calibrated vent in the bag allows it to deflate at precisely the right rate to absorb the energy of the head. The timing of this process is critical. The system was tested at Mercedes-Benz on a Hi Ge sled with a Formula 3000 monocoque fitted to it. A Hybrid III dummy, suitably modified to accommodate a reclined seating position, was placed in the car and fitted with a six-point harness and full-face helmet. The sled was accelerated backward to simulate a 44g frontal impact with a change of velocity of approximately 55 kph (34 mph). Without the airbag, the critical criteria for the threshold of head injuries were exceeded by more than 50%, and for the neck by approximately 40%. With the airbag deployed, none of the criteria was exceeded.

The HANS works in a different way to decelerate the head and relieve neck loads. It uses the well-padded inside of the helmet to arrest the head, and feeds the deceleration loads directly into the driver's torso and hence into his shoulder harness, almost completely bypassing the neck as a load path. It comprises a CFRP yoke that fits around the driver's neck and sits on his shoulders, under the shoulder belts. Behind the neck, it forms a collar to which are attached two tethers that connect to the sides of the helmet. When the driver is seated in his car, the HANS is held in place by the belts, and the tethers are loose enough to give the driver full head movement. In a frontal impact, the torso and head start to move forward together, until the shoulder belts tighten. The shape of the HANS ensures that the tightening shoulder belts grip it and lock it firmly to the driver's shoulders and upper torso. The motion of the helmet is then arrested by the tethers pulling onto the collar of the HANS, now anchored to the torso. The crown of the head (the strongest part of the skull) bears onto the padding in the top of the helmet, and the head is decelerated. Because the relationship between the head and the torso is maintained, the neck is not stretched and does not have to generate any significant loads to arrest the head, as it does without the HANS. The motion of the head and torso is limited to such an extent that it is unlikely the head will reach the steering wheel, or any part of the cockpit rim, except in the most severe crash (Figs. 15.5 and 15.6).

Mercedes-Benz and Bob Hubbard developed the HANS to fit into a single-seater, reclined-seating-position cockpit and refined its characteristics until tested on the Hi Ge sled. It performed at least as well as the airbag, generating no injury-threatening loads in the head or neck (Fig. 15.7). Its performance in accidents where the car hits at different angles—three-quarter front, side, or rear—showed that HANS had no negative aspects to it, performing as well as or better than if it were not fitted (Ref. 15.2).

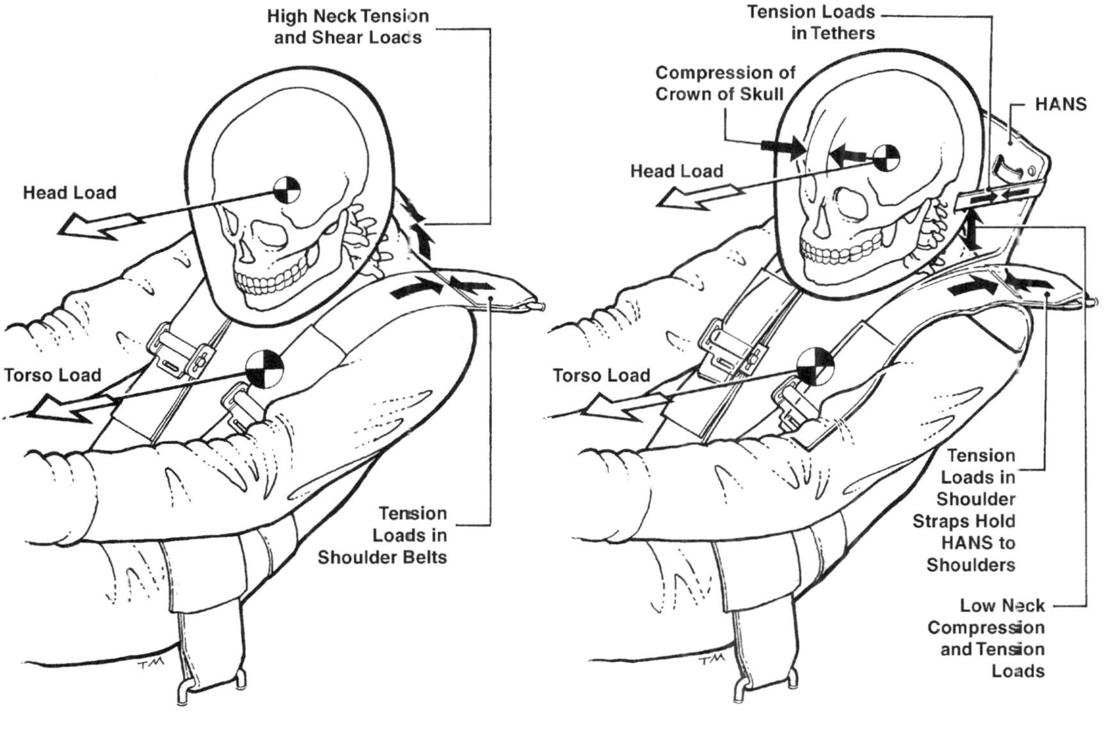

Fig. 15.5 HANS operation.

Fig. 15.6 The HANS yoke. (Courtesy of McLaren International Ltd.)

Fig. 15.7 The HANS test setup at DaimlerChrysler consists of a Formula 3000 monocoque fitted to a Hi Ge sled and a modified Hybrid III dummy. (Courtesy of DaimlerChrysler GmbH.)

The HANS is a simple but effective passive device, whereas the airbag is a fairly complex active device. To be acceptable in a race car, issues such as the fitting of two explosive devices in the cockpit (deployment actuator and inflator), sensing and triggering, double impacts (e.g., another car, followed by a barrier), and the effect of accidental deployment on the driver's ability to control the car must be addressed. Because the airbag would take a great deal of rig and track testing to perfect and guarantee reliability, and the HANS was proven to be at least as effective while being almost ready for track testing, it was decided to concentrate on the HANS. A number of changes were necessary to both the HANS and the seat/cockpit area around the driver's shoulders and neck, to create enough clearance for the driver to be comfortable. It soon became clear that the HANS could be worn without major problems.

In the year 2000, a number of drivers were testing and getting used to the HANS in Formula 1, Formula 3000, DTM2000, Champcar, IRL, and other North American series. The deaths of Gonzalo Rodriguez and Greg Moore in Champcar focused the minds of many drivers racing on the high-speed ovals where impacts into the concrete walls are so violent. It is a major step for a driver to accept a new piece of safety equipment to be worn in the car; it is easy to forget how long it took for helmets and seat belts to be accepted. The driver must not only be convinced that the device will enhance his chances for survival in all kinds of accidents, but he must also be convinced that it will not distract or inhibit him to such an extent that it slows him down. It is hoped it will be a long time before the HANS is tested in earnest in a real crash. Everything that can be checked in the laboratory has been checked (Mercedes-Benz conducted 53 Hi Ge sled tests during the program), but every accident is different, and racing is the ultimate laboratory for safety, as well as for so much else. Drivers will not like another piece of equipment to wear and look after. It will be a nuisance as they get in and out of the cockpit, and it will be more difficult for them to remove their helmets while strapped into the cockpit. However, one day, as several wearers of Bob Hubbard's original HANS have discovered, it might literally save their necks. This development program to solve the frontal impact problem is typical of how safety is now a subject that is amenable to a scientific approach, similar to the approach taken with the rest of the car.

If all systems work as specified, the driver will step from the car, probably bruised all over from his restraint system and perhaps suffering from a headache but with no broken bones and no internal injuries. The survival cell, which includes the fuel tank, will be undamaged, and there will be no fire. With the continually improving performance of car and circuit crash-protection systems, perhaps the greatest threat comes from the risk of being hit by another car, or part of one. The risk affects not only drivers, but also anyone whose job entails being within the confines of the circuit and, even in some extreme cases, someone in a spectator enclosure. The greatest risk is from loose wheel assemblies, which weigh approximately 30 kg (66 lb) and possess not only linear velocity energy but also rotational energy, making their behavior on hitting the ground or other objects highly unpredictable. In 1999, wheel tethers were fitted in Formula 1, and, after a wheel landed in a grandstand during a CART race, IRL and CART followed suit. Although not always successful in fully retaining the wheels, single and multiple tethers have undoubtedly reduced a wheel's energy after it comes loose. Avoiding the tethered wheel from swinging back into the monocoque or, even worse, into the cockpit is a major consideration. Such an event led to Ayrton Senna's death in what was a serious but probably otherwise survivable accident at Imola in 1994. Open-wheeled, open-cockpit race cars have always been an essential part of motorsport. The additional risk, although greatly reduced, will always be present (Fig. 15.8).

Fig. 15.8 Martin Brundle barrel rolls his Jordan over the pack at Melbourne in 1996. He landed inverted in the gravel bed but was uninjured. (Courtesy of Sutton Motorsport Images)

When things go wrong during an accident, usually due to a combination of small details, the first line of protection for the driver is his equipment: helmet, fireproof clothing, and gloves. The specification of these is laid down and controlled by the sanctioning body and will steadily improve as new materials become available, often as a spin-off from the defense and space industries. When these too are unable to protect a driver from injury, he is in the hands of the rescue and medical teams. They are equipped with, and trained in the use of, the latest resuscitation and extraction equipment (e.g., the Lear type of extraction seat) and are backed by emergency transport, to transfer an injured driver to specialized medical centers. The speed with which the right treatment is available often determines the eventual outcome of an accident.

In the 1960s, one in every eight accidents in Formula 1 events resulted in a fatality or serious injury (defined as an injury that prevented the driver from continuing to participate in the event or subsequent events), with some years as high as one in four (Fig. 15.9).

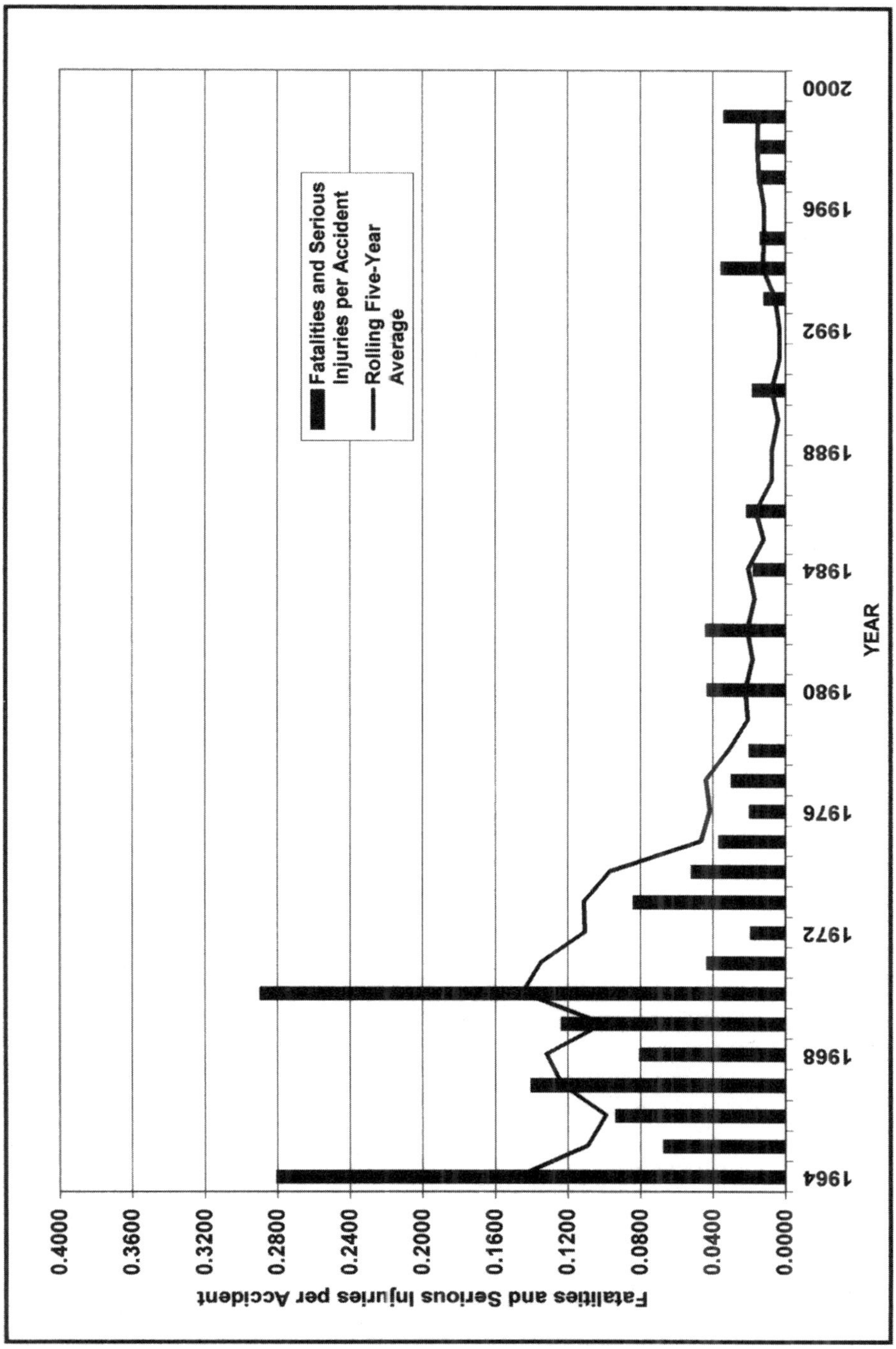

Fig. 15.9 Fatalities and serious injuries per accident in Formula 1.

The following passage from Tony Rudd's book, *It was Fun!* (Ref. 15.3), illustrates the conditions that gave rise to this:

> In the race it began to rain heavily on the far side of the circuit, so as the leaders came down the hill to the Masta Straight, Jackie [Stewart] aquaplaned off, striking amidships a large stone buttress of an even larger and very solid stone barn, and wrapped the car round the buttress like a banana. It then bounced off into a ditch with Jackie trapped in the car with a crushed tank leaking fuel all round him. Graham [Hill] and Bondurant went off in the same area for the same reason a few seconds later but without such disastrous consequences. Just as Graham was getting back on the road he saw Jackie's car and stopped to help him. It took nearly ten minutes to get him out of the car and another 20 minutes before the ambulance arrived. He had been sitting up to his waist in petrol, so the drivers helping him stripped off his fuel-sodden clothes and wrapped him in a spectator's rug.

This accident involving Jackie Stewart started the driver-led movement for greater safety in motorsport. The first steps taken were:

- Reduce the risk of fire.
- Establish circuit safety criteria, including arrestor areas.
- Six-point harness restraint systems and driver protective clothing.
- Standardization of signaling and marshalling procedures.

These were followed during the 1970s by:

- Structural specifications for the chassis of the car.
- Medical centers at the race circuit.
- Tire barriers.
- Graded driver licensing.

The effect of these measures was to reduce the number of accidents causing fatalities or serious injuries to an average of 1 in 40 accidents—a fivefold improvement in less than ten years.

Between 1980 and 1992, a further steady decline in fatalities and serious injuries per accident resulted in the rate falling to less than 1 in 250 accidents—a further sixfold reduction. To achieve this further reduction, all aspects of motorsport that influence safety were scrutinized, and measures were introduced to minimize the threat to participants and spectators.

The Technical and Sporting Regulations governing Formula 1 set the highest standard and provide a proving ground for safety measures prior to introduction into other international racing classes.

Car Construction

The prime objectives of car construction are:

1. To protect the driver from intrusion injuries
2. To generate a deceleration pulse, in any impact direction, that can be accommodated by the driver restraint system within his deceleration and load tolerances
3. To restrain the driver from impacting the inside of the car or any solid object, or, where that is not possible, to provide a means of attenuating the resulting impact to below a level that would cause injury
4. To avoid fires
5. To ensure that the driver can exit the cockpit quickly and easily

The Technical Regulations take the approach of defining a driver survival cell and stipulating a series of structural integrity demonstration tests, rather than laying down construction methods and materials.

Circuits

Closed-circuit racing tracks are licensed by the FIA, according to the classes of cars to be raced on them. Circuits are inspected according to the FIA Internal Guidelines for Motor Racing Course Construction and Safety, which lays down standards and guidelines for all aspects of circuit construction, including:

- Track design and construction
- Run-off areas and curbs
- Deceleration and arrestor areas
- Barriers—guard rails, concrete walls, tire barriers, and spectator protection fences
- Service and access roads
- Marshalling and signaling posts and equipment
- Track-side signs and all other potential obstacles

Drivers

Drivers are licensed to participate in racing classes according to their experience. The licensing system also provides a means of sanctioning drivers for actions that are considered prejudicial to safety. Standards for driver equipment, including helmets and overalls, are set and regularly checked.

Medical

The philosophy behind the medical service standards laid down by the FIA—largely the result of unstinting work by Professor Sid Watkins, the FIA Formula 1 Medical Delegate—is based on providing timely and expert medical attention within seconds of the occurrence of an accident, through to transfer by helicopter of an injured driver or other participant to an approved receiving hospital.

Much has been written about the critical "golden hour" in the treatment of trauma; however, it is intervention, reduced to minutes or even seconds in Grand Prix racing, that is deemed most important. Rapid response in the event of an accident is provided by fully equipped, rapid intervention medical cars, stationed around the circuit and able to reach the scene of any accident within 30 seconds. These are backed up by doctors and paramedics on foot, stationed at intervals around the circuit. At a circuit such as Monaco where access is difficult, three intervention cars, backed up by 100 doctors and paramedics, are required to provide this level of intervention performance. In several accidents in the last few years, such rapid intervention undoubtedly saved the lives of drivers who had obstructed airways.

At every circuit, there are also two or three trained extraction teams, which are equipped to extract a driver with possible spinal injuries from a damaged car. These teams practice on actual Formula 1 cars and familiarize themselves with potential problem features. Following extraction and stabilization, an injured driver will either be taken to the medical center by ambulance or will be taken directly to the receiving hospital by helicopter.

The FIA regulations demand a high level of equipment at the circuit medical centers. Staffing levels are a minimum of two consultant anesthetists, a consultant general and orthopedic surgeons, a spinal or neurosurgeon, and a burns specialist. Intensive resuscitation rooms and an operating theater equipped to university hospital standards are a minimum requirement. Only when the injured person is deemed stable enough to travel to the receiving hospital for more specialist treatment or care is that person transferred.

In addition, driver fitness is monitored, particularly post-accident. Random drug tests to International Olympic Committee (IOC) standards are conducted regularly.

Race Control

A permanent FIA Race Director controls the running of major international motorsport events. Race Control is in radio contact with all medical facilities, race officials, marshals, and signalers, and is able to survey the whole track and pit lane, via closed-circuit television. The Race Director can thus assess the seriousness of an accident, and either stop other cars from running or control their pace using the Safety Car. These measures enable medical intervention teams to gain unencumbered and safe access to the scene of the accident. If the driver is not injured, his stationary car can be removed from an exposed position without endangering track marshals.

Race Control also monitors the speed of cars in the pit lane and reports offenders to the Stewards of the Meeting, who administer penalties to those who exceed the limits set down on safety grounds.

Future Developments

In 1992, the safety record of Formula 1 approached zero fatalities and serious injuries per accident, a level where individual incidents have a disproportionate effect on the statistics. The events of 1994, when two drivers were killed and one was seriously injured, focused sharply on this effect. The FIA responded by initiating a series of R&D programs to investigate where further safety measures could be effective. The objective of meeting and maintaining the target of zero fatalities and serious injuries was confirmed. Many of these programs have resulted in new regulations covering many aspects of safety, and others are currently underway, including:

- Analysis of accident and impact data gathered using onboard accident data recorders (ADR) (Fig. 15.10).
- More severe crash-helmet test specifications are being developed.
- Accident reconstruction, using ADR impact data, television coverage, and helmet and cockpit damage for correlation, is performed on accidents in which injuries have occurred.
- Close cooperation with national motorsport safety researchers in different countries is being undertaken.
- In-cockpit signaling systems are being developed.
- Software for the analysis of safety features on proposed new circuits, and changes to existing circuits is being further developed.
- Protection of drivers in T-bone accidents.

Safety Regulation

The FIA governs international motorsport through the World Motorsport Council. This body is assisted by specialized sporting commissions. Those relevant to safety are:

- Safety Commission
- Medical Commission
- Technical Commissions of each class—F1, WRC, etc.
- Circuit Commission

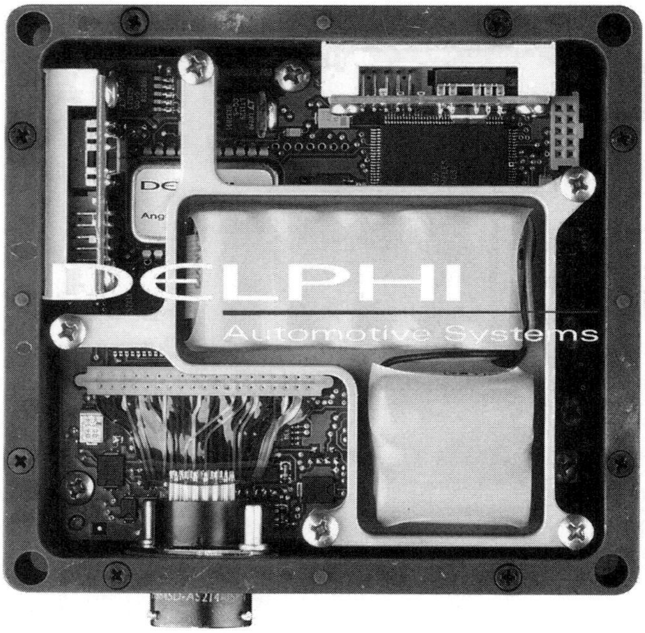

Fig. 15.10 The ADR2 accident data recorder. (Courtesy of Delphi Automotive Systems Inc.)

In addition, a number of working groups involve technical experts from the teams, circuits, the medical profession, and organizing bodies, such as the Research Group and the Equipment Approval Group.

Through this system of close cooperation among the FIA and all those involved in organizing and participating in motorsport, fast responses to new safety issues are possible. Safety measures can, if necessary, be introduced within weeks.

Not only is safety an experimental science, but it is also a statistical one. No two accidents are the same. However, by gathering data about those accidents, they can be grouped by common characteristics, and trends can be discerned. Tiny details can determine whether someone is injured or not, and also affect the severity of those injuries. Seat belt tightness is one such detail: if the belts are loose, the torso and head can build up speed before being restrained, and the velocity of the head at impact on the steering wheel will be significantly greater. Thousands of people participate in a variety of motorsport events throughout the world, and hundreds of accidents occur—they are part of the attraction of the sport for spectators. Only the seriously deranged want to see anyone hurt, and today the number of injuries and fatalities per accident is quite low. When a driver is hurt, it is a newsworthy event and has a negative impact on the sport, particularly if the driver is of the status of Ayrton Senna or Michael Schumacher. Sponsors do not like to be associated with the dead and injured, especially if their business is automobiles.

Although motorsport is not statistically one of the most dangerous sports, it is among the more dangerous of the high-profile ones. The quest for zero serious injuries and fatalities must continue until that target is achieved.

The value of research into human tolerance, which is a spin-off of involvement in motorsport safety, is now recognized by the motor industry. FIA president Max Mosley has stated that, due to the political and technical role that safety issues in motorsport play in FIA dealings with the European Economic Committee (EEC) over road safety, the safety approach used in Formula 1 has such a high profile that its influence in reducing the large numbers of road fatalities and serious injuries is, on its own, complete justification for the whole cost of Formula 1.

Chapter 16

Track Testing and Racing

In order to finish first, you must first finish.
—Anonymous

Approximately 250,000 km (155,000 miles) are covered by Formula 1 cars each year during practice, qualifying, and races. In addition, a further 150,000 km (93,000 miles) are covered during testing. Wearing out engine, gearbox, chassis, and tires at a rate of more than $500 per kilometer adds up to $75 million for testing alone. Add to that total the cost of drivers, engineers, and mechanics, plus their travel and subsistence costs, and it is easy to understand why most teams constantly push for some limits on testing in Formula 1. Only those teams with the largest budgets demand the right to unlimited testing, in the belief that they will have an advantage over their less well-endowed competitors. Major teams field dedicated test teams, fully equipped with trucks, pit equipment, at least two cars committed solely to testing, and up-and-coming or ex-Formula 1 test drivers.

The complexity of current Formula 1 cars is on a level with that of a light military training aircraft. While the aircraft may take five or more years from concept to introduction into service, a Formula 1 car will race within nine months of concept and be obsolete another nine months later. In that time, the full performance potential of the design must be realized, and all parts of it—the engine, transmission, electronics, etc.—must be raised to a state of reliability that enables the car to be driven at the limits of that performance, for a full race distance. Track testing is the primary means by which this is achieved.

An old motor racing saying claims, "In order to finish first, you must first finish." No points are awarded for pole position—only for completing the race in the first six places. For this reason, reliability testing accounts for most of the track kilometers covered during the year. Race distances on components are necessary before they can be released for racing, particularly the hot rotating parts such as the engine, gearbox and clutch, hubs and bearings, brakes, and tires. There is no way to simulate in the laboratory the exact conditions on the track, although an enormous amount of effort is going into simulation and rig testing using actual track-derived input data. The combination of speed, rpm, g-force, downforce, vibration, and driver inputs is unique to a given driver and track, and this is why a part of the car that has given no problems for half a season will suddenly exhibit a fatal weakness in a particular race. As a senior Boeing engineer once said, "There is no such thing as a single failure; it is always the start of an epidemic." Any failure in testing that does not receive immediate attention inevitably will reappear when it really matters—in a race.

After reliability, the next main reason for testing is performance. It is the most difficult type of testing to conduct successfully and from which to achieve firm, unambiguous answers. Few drivers are good at it, because it requires the car to be driven at a consistent level, slightly below the maximum, throughout the test session—if necessary, all day. The driver must resist the temptation to modify his driving style to overcome a problem and must be able to report changes in track conditions or tires that would invalidate a particular test run. Too often, after a series of changes that have improved the lap time, the car is returned to the original condition, and the driver, unaware of the nature of the change, improves his lap time again.

Complete honesty and objectivity about the results of a test, combined with the ability to drive the car at nine and one-half tenths with the regularity of a metronome, is the mark of a top driver. (The last half-tenth is driving around the remaining problems).

For the engineers as well as the driver, it is necessary to differentiate between changes that have corrected a particular deficit at the track where testing is taking place and fundamental improvements (Fig. 16.1). A change that corrects understeer, on a track where understeer is the limiting factor in improving lap time, will slow the car on a track where it oversteers. Any major new part must be evaluated on more than one track, and ideally by more than one driver, to fully understand the effect it has had on performance. Data systems provide a breakdown of where on the track the car runs faster and where it runs slower. The stopwatch reveals whether there is a net improvement or deterioration.

Reliability and performance are not the only reasons for track testing. Pure research and development—gaining answers to questions that are not directly related to making the car travel faster—can sometimes be conducted only on the car. It is difficult for R&D engineers to obtain track time with a test car to do their pure research. However, the fundamental questions can be answered only by conducting these experiments.

Calibration

When a Formula 1 car is designed, it is based on many calculated factors and data from the wind tunnel and test rigs. Early in the life of the car, all aspects of the car's design must be calibrated to confirm these theoretical and rig-derived numbers and to provide actual data on which the setup of the car is based. An example of this is aero-mapping, which will be discussed in more detail later in this chapter. Also, engine designers need to know accurately the performance under real conditions of the cooling systems, airbox, exhaust system, etc.

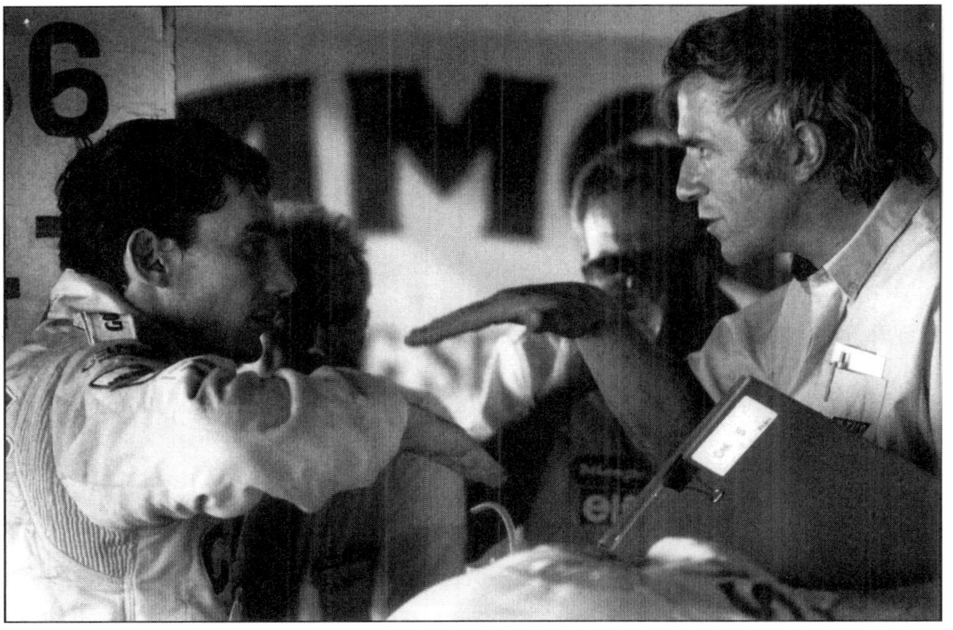

Fig. 16.1 Ayrton Senna (left) and I disagree about the attitude of the active suspension Lotus T101-Honda in 1987. (Author's collection)

Tire companies provide rolling radius data for their tires at various speeds and loads, measured on flatbed tire test machines. Using laser ride height sensors, the data, crucial to precise aero-mapping, can be checked on the track.

Modeling Validation

More and more of the car and its systems are being modeled mathematically, because the feasibility of doing so increases as computer power increases. Models allow engineers to make detailed studies of complete systems and to perform "what ifs," while monitoring parameters that cannot be measured in real systems. However, these models must be validated by comparing them with real measurements of those parameters that can be measured and the operating conditions (i.e., the inputs to the models) determined from track testing.

Driver Testing

Finally, testing is used to train and evaluate drivers. A driver's approach and feedback, as well as outright speed, are best assessed in the testing environment, under controlled conditions and away from the pressure of a race meeting. Even the most experienced driver may need track time to learn how to adapt his style to a new system. An obvious example is the use of the clutch during a standing start, operated by a steering wheel-mounted paddle. The lack of feel requires a totally different approach from that for a pedal-operated clutch.

The year can be divided into three main periods for testing, each with its own tempo and prime objectives: (1) the pre-Grand Prix season, (2) the actual Grand Prix season, and (3) the post-Grand Prix season.

Pre-season testing focuses almost entirely on developing new cars in the run-up to the first race. Both performance and reliability must be developed quickly, especially if the car has a new engine and/or gearbox. Ideally, calibration work also will be performed, and the car will be driven by both team drivers on as many tracks as possible. If the car is late, the time available is short and must be integrated with the period of maximum output of the factory, as a full team of new cars is built. The race results at the start of the season usually reflect who is ready and tested, and who is not.

During the season, testing objectives include completion of the work that would ideally have been done during pre-season and proving updates, both from performance and reliability viewpoints. New versions of engines must complete race mileages before being raced; aerodynamic developments must be proven at fast and slow circuits; and the process of adapting the car for a variety of circuits is worked on continuously. If the car is not competitive or is unreliable, emergency plans are put in place to rectify the situation, and test programs are changed to reflect the urgency of doing so.

Post-season provides the R&D department with its main opportunity for concept evaluation and validation work. With the next race being more than five months away, the imperative of racing recedes, and track time can be given over to less immediate matters. By the end of the race season, the design of the car for the following year will be almost complete, and if certain questions concerning the design—weight distribution, for instance—remain unanswered, or new features carry a high risk, they will be evaluated on a "mule" version of the current car, to ensure that the design can be finalized. If the new car is to be an evolution model, the current one can be updated to almost the new specification for early performance and reliability testing of as many components and systems as possible. Getting a new gearbox running early will pay handsomely in terms of race reliability in the new season. Recently, the FIA negotiated with the teams a ban on testing during November and December, severely limiting this phase of testing and requiring it to be brought forward into the end of the racing season.

The tools of the test engineer are varied: he has a stopwatch, which is of most use for timing competitors; the driver's comments; his own eyes, for watching the car on the track and for inspecting it on return to the pits; and the data system. Since control systems for engine, chassis, and gearbox were introduced in the late 1970s and early 1980s, reliable measurements and the logging of vast amounts of data have been possible, growing in importance until they virtually dominate testing, with numerous analysis engineers and an array of computers. Engines are monitored in real time, to avert a catastrophic failure by allowing the driver to be notified by radio if some parameter exceeds limits, and most teams telemeter chassis and gearbox data in real time also. Most of the data are stored onboard for downloading on return to the pits. There is virtually no limit to the number of channels and bandwidths of the system—only the bulk of memory chips.

Equipped with such sophisticated systems, every run provides several megabytes of data for either on-the-spot analysis or more considered investigation at the factory. It is unnecessary for all engineering specialists to be present at the track, because data can be transferred almost instantaneously to base. The difference that data systems have brought is that instead of running through a schedule of tests prepared in advance, and then analyzing the results later, it is now possible to run a dynamic test program, in which each test is based on the results of the previous one, varied according to the feedback from the computers and drivers.

One piece of data the test engineers would dearly like to have access to, but is not available, is the competitors' speed profiles around the track. To make an accurate assessment of where they are losing and where they are gaining relative to the opposition would remove much of the discussion and confusion, and it would focus everyone involved on the task to be accomplished.

Competitors' lap times are easy to obtain, but they can be misleading if they decide to create a lap that does not start at the Start/Finish line, or if they drive only individual fast sections and add them together later. Section times can be measured, as can local speeds, but every sector needs timing beams and a crew to operate them—they can be only sample measurements. What is really desired is data for the speed of the car at all points on the track. It would be possible to analyze this basic information and derive useful power, aerodynamic, and handling figures for a competitor's car. The process is the reverse of lap simulation, in which a team will input all performance parameters for its car, and then output a speed plot and lap time for the circuit. If the team has the speed plot for a competitor, and if the tires are the same, the weight is on the minimum, and the driver uses a similar line and effort, it is possible to make good estimates of power, drag, and downforce, and to gain some indication of the balance of the car.

Apart from placing a "mole" in each competitor's team, who then passes out data files on lap speeds and other information, there are a few legal ways to obtain this data. The legal methods are expensive to carry out all the time in order to obtain the data on that one lap, when the competitor is at peak performance. One possibility is to video a bird's eye view of the track from a helicopter. The images then can be analyzed by computer to give speeds all around the track. Another possibility is to fit ranging radar to the front of the car, and measure the relative velocity of the competitor while following him. Knowledge of one's own speed permits the speed of the other car to be calculated.

A more viable alternative would be to adapt a technique developed by P. Azzoni and D. Moro (University of Bologna, Italy), and G. Rizzoni (Ohio State University, Columbus, Ohio) (Ref. 16.1). Strangely, these authors did not put forward the speed profile of a competitor's car as one of the variables that could be analyzed by using the technique. From the proximity of

Bologna to Maranello, it is possible to surmise that Ferrari funded the work and that it did not publish this application because that is exactly what Ferrari uses it to determine!

The principle employed in the technique is the derivation of engine rpm, from the frequency analysis of the sound emitted by an engine. A V-10, four-stroke engine fires five times per revolution. As each cylinder fires and the exhaust valves open, it produces a pressure wave down the exhaust which, when combined with other cylinders and sound reflections within the exhaust system, generates a sound that is predominately composed of frequencies that are a function of the firing frequency. That fundamental frequency for a Formula 1 V-10 is

$$\left(\frac{5 \times \text{rpm}}{60}\right)\text{Hz}$$

Thus, with suitable analysis of the sound of an engine, it is possible to determine the engine rpm.

All that is needed to sense the sound is a good-quality microphone. If the microphone is stationary with respect to a moving car, the motion of the car will affect the frequency picked up—the well-known Doppler effect. Thus, it would be of no use to sense the sound of an engine using trackside microphones; the microphone must be placed on the car itself. Although most teams would object to a competitor mounting microphones on their cars, they cannot object to the microphones fitted to the video cameras they carry for transmitting onboard views for television programs. The researchers found that the sound track from digital pay-television channels produced a perfectly satisfactory sound source for their analysis.

Although rich in information about the engine and the way the driver uses it in corners, the analysis does not yield car speed directly. However, all the information is available. The speed of each car is measured at three places on the track during a Grand Prix event, by the TAG-Heuer timing system. A similar system can be set up easily by a team during a test session, and it usually is. By relating speed at a given point to the engine rpm derived from acoustic analysis, the overall gear ratio, in the gear selected at the measurement site, can be calculated. All other ratios can then be calculated from the rpm drop between gears during gear changes, derived from the rpm data. With the rpm and the gear ratios known, it is a straightforward task to calculate speed at all points around the circuit.

Each aspect of the performance of the car requires a different approach to testing.

Engine

It is not usually desirable or practical to change major engine components at the track, and thus engines of different test specifications tend to be complete units, which must be changed as a whole. The performance of the unit can be evaluated in a very short time—maybe after only one six-lap run. Measurements of temperatures and pressures (T's and P's) of all relevant fluids—fuel, oil, water, induction air, and exhaust—at each stage in their cycle will give a full picture of how each subsystem is performing. However, several runs may be needed to check the mapping of the engine, using Lambda sensors to analyze the exhaust chemistry and fuel consumption data generated by the engine management system.

The power output of the engine under known conditions will have been established on the test bed. If the effect of the installation is to be investigated—particularly the effect of the airbox or exhaust backpressure—the output shaft will have a torque sensor installed for the calculation of installed power.

There is much discussion about the driveability of engines, complicated by the adoption of the drive-by-wire throttle that is used to shape the response of the engine to throttle inputs. Although the basic shape of the torque curve can be established on the test bed, many test laps are expended on trying to improve driveability because of its influence on handling characteristics.

Aerodynamics

The aerodynamic characteristics of a ground-effect machine cannot be described with single lift, drag, and center of pressure numbers. Because the pressure distribution, and hence force coefficients (C_L, C_D, and C_M) are greatly affected by the ground clearance at each axle, and to a lesser extent by the roll angle and yaw angle, the characteristics must be described by an aerodynamic map. If the secondary effects are omitted, this will take the form of a contour map for each force coefficient, plotted with front axle ride height and rear axle ride height as axes.

Aeromaps are readily derived from wind tunnel testing. However, because of scale effects (Reynolds number too low) and imperfect ground plane simulation (model stationary, air and ground moving), they must be checked at full scale, in still air, and with the car moving over the road. This testing is best conducted on an airfield with a long runway. The surface must be very smooth to maintain constant car attitude, and wind condition must be low or zero (early morning or evening). The car will be fitted with at least three laser ride height sensors (to determine ride heights and roll angle), load cells to measure the load on each wheel, and air data sensors to measure air speed and yaw angle.

When the cars were fitted with active suspension systems, it was possible for the driver to select ride heights in the cockpit and thus perform a number of test points during one run. With passive suspensions, the ride height must be set to give the approximate desired height under loaded running conditions (governed by spring and tire deflection) and the data recorded. Smoothing and contour drawing software allows aeromaps to be developed from the results of numerous runs. If the track is smooth enough, it may be possible to replace the springs with struts and thereby gain greater control over the loaded ride height.

Drag measurement is much more difficult. Coast-down tests have been developed, in which the deceleration against speed characteristics are used to determine the coefficients of an equation for total resistance that includes aerodynamic drag, rolling resistance, and all friction effects (e.g., transmission, wheel bearings). With the development of accurate and reliable torque sensors on either the input shaft to the gearbox or the half-shafts, it has become possible to continuously measure the torque needed to drive the car under all conditions. This data provides many more points to calculate the resistance coefficients and hence potentially greater accuracy (Ref. 16.2).

Aerodynamic testing will also include pressure measurements all over the car, including inside the airbox and throughout the cooling systems, to determine flow through the radiators and their performance. Pressure sensors that sequentially cycle through 60 or more ports are available for mapping the pressure distribution on surfaces.

The interpretation of aeromaps in conjunction with suspension and tire characteristics is not advanced enough to enable engineers to determine handling characteristics. The braking and corner-entry instability of the current breed of Formula 1 cars is testimony to this statement. In consequence, substantial overlap exists between aerodynamics and chassis testing.

Chassis

Weight, power, lift/drag ratio, total downforce, and tire grip determine the overall potential performance of a race car. The chassis—with its weight distribution, structural characteristics, and suspension kinematics and compliances plus their influence over aerodynamics—determines how much of that potential can be realized by the driver. Systematic chassis testing releases the performance, step by step.

Vehicle dynamics models are becoming highly sophisticated, but the dynamic characteristics of tires, dampers, aerodynamics, and structures are not yet adequately calibrated to enable models to make a significant contribution to the development of transient handling. Thus, there is no substitute for an experimental approach to setting up the car. Although each circuit needs its own particular setup, tunable according to fuel load, qualifying requirements, weather conditions, and tires, testing at any circuit provides experience and knowledge for race engineers and drivers, as well as adding to the database of the team.

Chassis sensors, combined with DGPS and an inertial platform, provide the ability to reconstruct the trajectory of the car and analyze all dynamic loads, accelerations, and velocities, including the measurement of dynamic loads on the structure and suspension. Fully strain-gauged suspensions validate the calculated loads used to stress components at the design stage and to provide test rig load cycles for proving their fatigue lives.

Tires

Formula 1 is again entering an era in which there are competing tire suppliers. With a single supplier, every team has the same equipment. The need to be involved in tire testing, in order to ensure that the tires supplied are matched to your car (or at least vice versa) and not optimized for a competitor who is running on the same tires, quickly becomes among the most important reasons for carrying out testing. The need for a car that is adequately competitive, a driver who is not only fast enough but also able to provide good feedback, and sufficient resources for the extra testing are all attributes needed to qualify for tire testing.

Tire engineers seek data on tire temperatures, wear, and durability, as well as the effect of constructions and tread compounds on lap time. The full suite of chassis sensors will be analyzed for clues concerning grip and driveability.

Software

Engine management, gearbox, clutch, differential, and data systems all require elaborate software to control them. As is true with all software, it is difficult to ensure that no bugs exist either to cause erroneous actions to occur or the software to crash. However much time is spent in checking software on a simulator, only by subjecting it to real-life conditions will it qualify for race use. Every change or update must be tested.

Drivers

One of the most revealing ways of evaluating and comparing drivers new to Formula 1 cars is to plot their lap times, lap by lap, on a common graph, with the number of laps along the X axis. The resultant curve for each driver will be his learning curve. When the lap times stabilize, the quickness of the driver and the consistency of his lap times will be revealed.

Another characteristic important in a Formula 1 driver is feedback. If the team driver has set up the car and his comments are not reported to the new driver, a comparison between their impressions of the car can be made. The new driver is usually allowed to make small changes to the car and report the changes he feels, to assess his technical knowledge and sensitivity to changes.

Every test should set out to prove, disprove, or at least evaluate a hypothesis. The hypothesis may be simple (e.g., "This engine will last for 320 km [200 miles] when driven by Michael Schumacher,") or complex (e.g., "This front wing will reduce understeer in fast corners, but increase pitch sensitivity and thereby make the car more unstable under braking."). Testing a hypothesis focuses the experiment and allows a definitive conclusion to be drawn from it, positive or negative. Without it, the result may be inconclusive, subject to debate, and an inefficient use of $500 per kilometer ($805 per mile).

Racing

Preparing, setting up, qualifying, and racing the cars is the sharp end of Formula 1. Although the whole hierarchies of the team and its technical partners are present at a race, the race engineers supervise and run the cars. The job of a race engineer is one of the most difficult and involves the longest hours in the business. It may appear to be one of the most glamorous in Formula 1; after all, few have a position on the pit wall during the race, and even fewer have the undivided attention of the driver while his car is in the pits. With the authority to command half a dozen mechanics to descend on the car, and the technical ability actually to understand what all the computer screens are displaying, it is no wonder that up-and-coming designers and engineers and many mechanics aspire to become race engineers.

In truth, no position in motorsport requires longer hours, involves more paperwork, and carries as much day-to-day responsibility as that of the race engineer. On the other hand, it is the job that affords the greatest opportunity to gain hands-on experience in racing. Every engineering instruction from design and R&D passes through the race engineer on its way to the car. Every piece of data derived from testing and racing of the car feeds back through him. The race engineer must ensure that the car is ready on time and to specification. He is the front line of engineering, blending the skills of the driver with the brilliance of the designer's ideas (occasionally he may be heard using other terms to describe both driver and designer) to deliver on the day of the race. He must perform the "art of the possible."

At the Factory, Before the Event

In reality, preparation for a particluar Grand Prix commences at the end of that race the preceding year. All lessons learned from running on that circuit will have been filed away, to be brought out and reviewed in the lead up to this year's event. It is the beginning of a process that gets underway as soon as one race is over, and builds to a climax on the Sunday afternoon of the following race.

Up to a week before the race, the cars leave the factory, prepared to a specification compiled and supervised by the race engineers. Based on consultation with department heads and final approval by the technical director, the build specification is demanded by the chief mechanic before the cars arrive back from the previous race. In an ideal world, the build specification will be the same for all three cars.

Lap simulation computer models enable changes to the track—and major changes to the car caused by the Technical Regulations or significant developments—to be evaluated for their effects on wing setups and gear ratios, and as initial input into the development of the refueling strategy. If the track has been resurfaced and it has been impossible to run on it since the resurfacing, someone can only guess at the effect of bumps removed or introduced in the process, as well as changes in grip levels. A call to the tire supplier will provide early knowledge of any compound changes since last year, but the supplier will be equally in the dark concerning surface changes.

New specification parts, to overcome reliability issues or provide greater performance, may be available for the next race. The responsibility of deciding whether they are adequately tested, and whether the risk of introducing them into the race program is outweighed by the need to fix a reliability problem or increase speed, lies with either the technical director or the chief designer. The record books are filled with DNF's due to "better, faster" parts being rushed through for racing without adequate testing.

Everything must be prepared: lists of gear ratios to be taken, lists of spares, lists of wings, and lists of new parts to test if there is time. Have databases, spreadsheets, and word processors truly made the job easier, or have they simply made the task bigger?

As the build of the cars progresses, new parts must be chased so that they do not delay the build. Priority items on the job list of the last race must be checked, to ensure that they have been or will be completed. The engine supplier must be contacted to determine when the engine for Friday running will arrive. (Perhaps it will not be ready until after the cars have left and thus will be fitted at the circuit. The chief mechanics must be warned.) There is always a need to check that there are no subtle, untested changes to any of the software for the car, waiting to spring a surprise in the middle of qualifying. Rigging of the car must be supervised, the first of at least half a dozen times this will be done over the race weekend. List, chase, check, and supervise.

The driver should be informed about what the changes will be and what the plan for first practice is. This may take some time because the driver will probably have his own ideas.

After the cars leave the factory, there is no letup. Hand-carried items are chased and listed, running sheets are prepared, and paperwork is updated. For back-to-back intercontinental races, all this is done in the paddock at the next race, over a period of three or four days. The phone bills for communication with the factory are not small.

At the Circuit, Thursday

By the time race engineers arrive at the circuit on Thursday, the garage is set up, all equipment has been installed, and the mechanics will be completing car build, adding last-minute items brought there as hand luggage. By the end of the day or early the next day, the car will have been scrutineered, the legality checked on the FIA measuring and weighing bed, the car decaled with the latest sponsor deals, the engine run, the semi-automatic gear-change system exercised, all systems and software checked, and the car rigged again on the flat patches in the garage.

Meetings must be held with the engine supplier, the tire supplier, the FIA technical delegate and software checkers, component suppliers' representatives (particularly brakes and data systems), the team's tire and refueling personnel, the chief mechanic, the data analysis engineers, and the team manager. The plans, so thoroughly documented and communicated, can easily change when the technical director and/or the chief designer arrive.

It is always beneficial, whether or not the track has changed, and whether or not the drivers have driven on the track in the past, to walk around it with them, discussing lines, road cambers, and bumps, and generally starting mentally to note features of the circuit, ready for practice and qualifying. If the surface is new, it is extremely difficult to identify bumps that may disturb the car. If the surface has been used, long scuff marks and a darkening of the tire rubber ingrained into the tarmac show the crests. If the wavelength corresponds to that of the oscillations of the car, at the particular speed at that point on the circuit, problems can be expected. Wavelengths between 8 meters [26 ft] at (200 kph [124 mph]) and 12 meters [39 ft] at (300 kph [186 mph]) cause the most trouble.

Getting drivers away from the press also provides a chance to talk about the following day and to start focusing on setting up the car for qualifying and the race.

Thursday evening, after the paddock has emptied and close to dinner time, is the time to peer into opponents' garages, check on wing and suspension setups, and wander along the back of the garages where undertrays are stacked, with perhaps a chance to spot a new diffuser. Sifting the rumor and gossip of other race engineers and mechanics might yield a morsel of useful intelligence—or has it been planted merely to confuse?

Friday

With no qualifying on Friday, there is little point in striving for ultimate speed during free practice. The circuit will almost inevitably have changed by Saturday afternoon, and few others will bother to set times worthy of comparison. The first half-hour on Friday is useless for any setting up because the track will be dirty and slow. Only toward the end of Friday afternoon will enough rubber be down, and possibly the track temperature be representative of Saturday's qualifying period, for true performance comparisons to be made. Even then, teams and drivers who are confident of their capabilities will not waste useful track time revealing their true potential, and will continue to run with heavy fuel loads and conserve tire sets. Friday is about arriving at a fast and consistent race setup.

Tires are so important that it is not surprising to see cars going out on rain tires to do engine installation checks. Three (or even four) fresh, lightly scrubbed sets of tires must be conserved for the race, and two or maybe three new sets of tires for qualifying. This leaves only one or at most two sets of tires for free practice on Friday and Saturday. If a driver flat-spots a set of tires, there will be a shortage of tires to do the important part of the job properly. The details of which tire sets will be used, blanket temperatures, and hot pressures are decided by the race engineer and are arranged in advance with the technician in charge of tires.

Initial laps, as the driver builds up the speed and tries not to be confused by the changing track conditions, yield useful information on temperatures and pressures of all systems of the car, such as engine water, fuel, oil, and airbox air; gearbox oil and hydraulics; dampers; black boxes; brakes; and tires. Data analysis systems, both chassis and engine, start to pour forth streams of information concerning speed, rpm, aerodynamic loads and ride heights, gear-change points and suitability of ratios, accelerations, severity of bumps, wheelspin, brake lock, damping performance, and all system function parameters. Megabytes of sensor data, flowing into and out of computers, must be interpreted with the help of experts and then compared with the information being fed back by the driver.

If the driver reports that the car is awful, the race engineer must keep a cool head. Is it the track? Perhaps everyone has snap oversteer. Maybe there is a rigging error or a faulty damper. This is a good moment to stop, check the car, and let other cars clean up the track. Is the engine engineer happy with his charge? How is the gear-change? After all, 40 minutes remain in this session to sort it all out...

If the words, "It's OK, just like normal" crackle in the headset, it is time to start the setup program. Aerodynamics should be considered first—ride heights and wing settings to set the best possible balance everywhere. Perhaps later, when times mean something, it may be a good moment to consider overall wing to calibrate the tradeoff between downforce and top speed. It is definitely too early to be confused by other people's top speeds—they are probably in the opposite cycle of the same experiment.

Mechanical handling is next—springs, anti-roll bars, damper settings, toe-ins, and cambers, and then back around the ride height and wing loop again. Using driver comments, section times from the data as well as other data analysis, and the tire pressures and temperatures recorded by a tire engineer every time the car stops, the task is to develop a setup that is not only fast, but with which the driver can put in consistent laps. The working range of tire pressures is only approximately 2 psi. If the car comes in with the pressure in one or more of the tires being more than 1 psi different from the desired pressure, any information from that run must be ignored. It is necessary to filter circuit learning effects and everyone's over-sensitive reaction to the TAG-Heuer times. According to some drivers, the car can go from "good" to "hopeless" without leaving the pits, simply because someone else has put in a quick lap. Providing a reminder that they are probably running less fuel might have the right effect and calm him down.

A one-hour break is next, providing time to collect thoughts, change ratios, and perhaps change direction. If things have gone well and the driver feels he has a good setup, the afternoon session can be used to try a different one, with a back-to-back under identical conditions late in the session, to determine which is the best compromise—maybe one for the race and the other for qualifying.

In the afternoon session, the pace is a bit quicker. More people run low fuel with an eye to qualifying, but times are not yet indicative of absolute performance. By the end of the session, the car must be balanced, in both the quick and slow sections, ride the bumps well, and have the optimum wing settings. Likewise, the driver should be happy that he has a race setup. If not, time remains on Saturday morning to work on it, but by then the driver will want to be working up to qualifying, both with the setup and in his own mental state.

One of the hardest handling aspects to sort out, without the assistance of first-class driver feedback, is separating steady-state balance problems from transient ones. It may be possible to arrive at a good mid-corner balance and yet have a car that wants to swap ends while braking into corners, with power understeer turning to snap oversteer and poor traction in the exit. Most cars have very similar speeds through slow corners (second or third gear, 100 kph [62 mph] corners now dominate most circuits). However, the speed that the driver can carry into the corner and how soon he can get on the power are what determine lap times. The time spent at a steady-state speed around the apex is small compared to the time spent accelerating down the straights. A 10 kph (6 mph) increase in apex speed gains little time, whereas 10 kph (6 mph) superiority all the way down the straights and sustained to a late braking point is worth several tenths of a second. Section time analysis makes this all too obvious.

There are so many combinations of settings that hard and fast rules are dangerous. A setup that works at Silverstone may respond totally differently at Interlagos, where the low grip level and bumps are so different. The car will need a different setup for a circuit that is dominated by chicanes, from a circuit with long, fast, high-g corners. The driver's style and approach to a circuit will change according to the details he faces. If the fast line is over the curbs, the suspension must be able to cope with them without disturbing the car. It may be worth trading some understeer into slow corners for traction out—it depends on the arrangement of corner speeds and straights. Knowledge of the car is essential, backed by a good database from previous years and a rapport between driver and race engineer. It is all too easy to get into the situation where each change is putting a patch on some undesirable characteristic generated by the last one. The car simply goes slower and slower. Good cars respond to logical changes and are fast everywhere. Not so good ones shine at some circuits and are inexplicably slow at others.

While all this is occurring, an eye must be kept on the systems of the car—the engine, drive-by-wire throttle, gearbox, clutch, brakes, cooling, and electronics. Valuable setup time can evaporate easily as engineers pore over data readouts and mechanics change parts to try to cure a glitch or worse. There may be tests to perform on the systems that cannot be combined with handling runs, and they may or may not have been planned. A run-off into a gravel bed will destroy a session, even if the driver can get the car back to the pits. Hitting a barrier spoils the whole day, and only the best teams ever really recover from a crash and get back on program during the weekend.

At the end of the session, the car is derigged (the car is measured on the flat patch and the settings noted) to check that the differences from the original setup equal the sum of the changes. If they do not, the work begins. Has something broken or slipped? Did the race engineer or a mechanic make a mistake, or did communication break down? When did it occur, and what does it mean to the conclusions made during running? If the settings are logical, a calm analysis of the day's running and a thorough debriefing of the drivers is conducted. Comparisons between the two team drivers' comments and section times are made (if they are cooperating). When the TAG-Heuer section times and speeds come out, there is an attempt to puzzle about what others were doing and the significance of their times. Another tour of the track may be useful to look more carefully at a particular problem section.

Job lists are required by the chief mechanic, and problems discovered during strip down of the car demand attention. The best engines go in for qualifying. Brake wear is measured, and their temperature paints checked and renewed. Gear boxes are stripped and checked. Ratios are changed. Tire wear is measured and discussed with the supplier. Fuel is pumped out, and a first estimate is made of consumption, checked against the data from the engine management system. Computerized lifing sheets for components are brought up to date, and parts are changed accordingly. Setup sheets for race and spare cars are prepared, discussed with the driver, and issued in time for rigging. When rigging is completed, another legality check in the FIA weighing area is a good idea. It may be demanded by the scrutineers during qualifying.

As the evening continues, an intelligence sweep up and down the pit lane can yield clues about why one driver is quicker at the end of the main straight when he has less power. It is more likely to confuse, however.

Saturday

If things went well on Friday, the morning session on Saturday is a chance to do a full-tank run, determine how much the ride height must be raised, decide on the tire compound to be selected for both qualifying and the race, and obtain accurate wear and fuel consumption data. After that is out of the way, pump out the fuel and work up to a qualifying setup. Alternatively, if it is Spa, try a high-downforce wet setup to establish the balance because there is a reasonable chance it will be needed on Sunday. As the driver shifts from "test" into "compete" mode, he may need to be reminded that his chances in qualifying will be infinitely better if he does not crash the car in the morning.

All too soon the session ends, and preparation for qualifying begins. Another de-rig is carried out to double-check settings, unnecessary test equipment is removed, lightweight components such as brakes are fitted, ballast is adjusted to compensate, and another setup sheet is issued. If the driver has complained about the engine or there are any discrepancies in the data, a fast engine change might be done. The driver may request gear ratio changes, but any gain will probably be only psychological. If either of these are required, there is not much time for finesse or careful checking; therefore, the race engineer is wise to keep a close eye on everything that is happening to the car. The spare car is set up identically to the car of whichever driver has priority, in case it is needed.

All this leaves less time to discuss strategy with the driver, and to plan when he is going to go out. When will the track be quickest? When will the track be clear of other people? The driver wants to make his last run as late as possible without becoming caught in the last-minute traffic. The race engineer's job includes scheduling the whole process down to seconds, and then orchestrating it in the face of changing circumstances. It is satisfying when it turns out correctly, but frustrating when it does not.

The hour of qualifying usually starts slowly, with everyone glued to the TAG-Heuer times, waiting for an indication of track condition and what the others are capable of doing. The tempo suddenly rises to a crescendo, and the last few minutes turn into a blur. The effect on the mood of everyone in the garage is extraordinary—in a few seconds, it can go from elation to despair and back again.

The activities of that evening (and probably night) are governed by the routine of race preparation. Scheduled parts are changed, or they are cleaned, inspected, and crack checked; wear is measured; systems are exercised; the race engine is installed and run up; race ratios are fitted to the complete race-prepared back end of the car, pre-assembled at the factory; the gear change is operated up and down through the box a few times while the engine is running; and the car will be rigged again before the night has ended.

A race strategy meeting to discuss refueling and tire use will fill an hour or more. Computer simulations must be run, and every possible combination of circumstances must be evaluated. It may rain, so how does that change things? Not all of the data are available yet, so the final choice of strategy cannot be made until after the warm-up in the morning.

Sunday

First on the agenda is pit stop practice, using the spare car in case a wheel nut cross threads on a hub. Although race engineers take little part in routine pit stops other than setting their schedules, the responsibility for pit stops lies within their province.

Warm-up is a half-hour scramble to check all three cars, with engine installation laps on each, followed by a short run to check everything. To fit it all in, the driver jumps from one car to the other and back again. A sudden preference for the spare car will throw the plan into disarray, but drivers will typically prefer to race the car they have worked up through the weekend, and try a long run of 10 or more laps to obtain a feel for the potential pace at the start of the race. If the balance remains bad, a last desperate attempt can be made to sort it out. However, there really is no time for more than two or three runs with changes to the car between the runs. Nobody knows how much fuel anyone else is running, so times are not much of an indication of race form.

If no problems exist, the four hours prior to the race are spent checking and rechecking everything. The mechanics lock up anything that has any possibility of loosening, using locking wire, tie-wraps, and silicon adhesive. Driver drink systems are installed, and data systems are set up for logging race data. A final rig and legality check, and the car is as ready as it is ever going to be. If problems exist with the engine, gearbox, or any of the systems, it will be one long, controlled scramble until the time the pit lane opens.

While the drivers eat pasta and have their shoulders and necks massaged by their physiotherapist, a final race strategy meeting takes place. Race tire sets are prepared, and refueling instructions are passed to the chief mechanic and refuelers. Quantities for each car at each stop are calculated, based on the morning's final consumption check on the race engines. Possible changes in the plan are briefed, and every contingency is thought through and discussed with those who might be involved.

As the race approaches, a countdown starts. Various operations and procedures must be completed by certain times to ensure that the car leaves the pit lane within the 15-minute window in which it is open. If the driver is still not happy about the handling of the car, it is possible to try one change by leaving the pits and doing a lap back into them before a final lap to the grid. The timing must be right, or the car may be stuck in the pits when the pit lane is closed.

The countdown continues on the grid, leading up to removing the tire blankets one minute before the start and engine starting at approximately 30 seconds. The grid is one of the few opportunities that engineers and designers have to inspect the cars of the opposition. As bodies come off for final checks, suspension adjustments are made, and even whole corners of the car are changed, security tends to fall by the wayside. The sight of mechanics struggling to work on cars, covered by "security blankets," is one of the more bizarre sights of Formula 1.

The race engineer might try to ensure that the race plan and any last-minute instructions are briefed to the driver, but it is already too late. The look in a driver's eyes will tell you he is elsewhere. When the engine is running, there is not much else to do except wish the driver a good race, and then follow the stream of people and equipment leaving the grid as the cars drive off to begin their parade lap. The time has come to think about the regulations for a false start, or a race stopped after a lap or two. It is too easy to receive a penalty in the chaos of these situations, by working on some part of the car or topping up fuel when such action is not permitted.

When the race is underway (Fig. 16.2), the race engineer's primary task is to record all information concerning the team's cars: lap times, gaps to other cars, fuel consumed, any problems appearing in the telemetry data, laps to go to pit stops, tire wear, etc. He must also communicate the information to those who need to know (i.e., the driver, chief mechanic, refuelers, and anyone making the final decisions on race strategy).

Even when the race has ended, crucial tasks must be performed. The first, if the car has completed the race, is to be present when the car is post-race scrutineered, to deal with any queries that emerge and to check that software uploads are handled smoothly. Next, race data and any information such as fuel and oil consumption, tire and brake wear, and obvious failures and damage must be noted. If accident damage or a failure halted the car, information about any parts needed for the next race are faxed to the factory to give the manufacturer a head start on replacements. A full analysis of failures and any other problems will come later when the car is stripped down.

Finally, the driver must be formally debriefed, after he has been reclaimed from the clutches of the press corps, to complete the record and to extract from his memory any nuances concerning

Fig. 16.2 The standing start of a Formula 1 race is one of the more exciting moments of motorsport. More than 17,000 horsepower is unleashed. (Courtesy of Sutton Motorsport Images)

the performance of the car and its systems over a race distance. Only then can the race engineer relax, grab a beer, and head for the airport.

On Monday the whole process begins again.

Timing and Information

Any sport in which speed is the measure by which the winner is decided requires precise timing to one-thousandth of a second to determine the differences among drivers and cars.

For years, the human eye and hand, linked to clockwork chronometers, were used to time cars in all forms of motor racing, including Formula 1. Splits of one-hundredth of a second could just about be read, but it is doubtful whether the push of the button as the car crossed the line was accurate to the same degree. The advent of quartz timing mechanisms enabled time to be read to one-thousandth of a second, but it required electronics (light beam and photoelectric cell) to determine precisely the moment the car crossed the line and complete a system that really achieved that accuracy. With time in an electronic form, it did not take long before computers were added to the system to calculate speeds and sort the order of all competing cars. Identification of the cars was still done by visual and manual input until transponders on each car provided unique identification of that car as it crossed the line. These transponders enabled the position and time of every car to be determined at multiple points around the circuit, as well as in the pit lane. A complete and automatic analysis of the progress of the cars could be made and displayed widely throughout an event. This is the basis of the TAG-Heuer timing and Siemens information systems used in Formula 1.

The two parts of a Formula 1 event during which the speed of the cars has official standing are qualifying and the race. At other times (i.e., the practice sessions), speed is officially irrelevant, but it is of great interest to competitors and spectators. In qualifying, the fastest lap-time of each competitor is the only thing that matters. During the race, the total time taken to complete the race distance and the order in which competitors complete that distance are what count. These pieces of information and the timing process to determine them are essential to a Grand Prix so that the official results can be declared. Before computers and electronic timing, the race officials and each team equipped themselves with sufficient numbers of people (often wives and girlfriends) who used stopwatches and lap charts to gather this data. Certain individuals among the teams gained reputations for their ability to keep accurate lap charts, even during a race confused by rain or multiple accidents, and they were often consulted by the officials to help confirm or at worst disentangle the official results. Disputed race results, with the winner being argued about days after the race, are now fortunately a thing of the past.

The timing system times each car not only as it crosses the start/finish line, but at two other places on the circuit as well, enabling three sector times to be measured which, when added together, make up a complete lap. At each point, the speed of the cars also is measured. The time of entry to and exit from the pit lane is also recorded, enabling the system to determine the whereabouts of each car and to count the total number of laps started and completed, even when the car crosses the start/finish line while in the pit lane. All this data is transmitted to the central computer that calculates a number of useful parameters and assembles lists of times, speeds, and car orders, which then are transmitted to television monitors all over the circuit complex—race control, pits, pit lane, press, and television. The data are assembled into "pages," along with other information such as weather data, track and air temperature plots, and the transmitted television picture, to provide a complete information package that the competitors, organizers, and press can use to base decisions on and make reports. At the end of each practice and qualifying session and the race, sheets of sector and lap times, speeds, and the resulting order for each car are printed and distributed to the teams. The summary sheets of the order of the cars are made widely available to the press.

As the information provided to the teams has become more complex and more complete, the teams have come to rely on it more for developing their strategies for qualifying and the race, particularly since the resurgence of pit stops (Fig. 16.3). All teams now have multiple television monitors on the pit wall, and each page of data is shown separately, so that situations can be assimilated quickly by several key personnel monitoring them. The data shown varies according to the session taking place (Fig. 16.4).

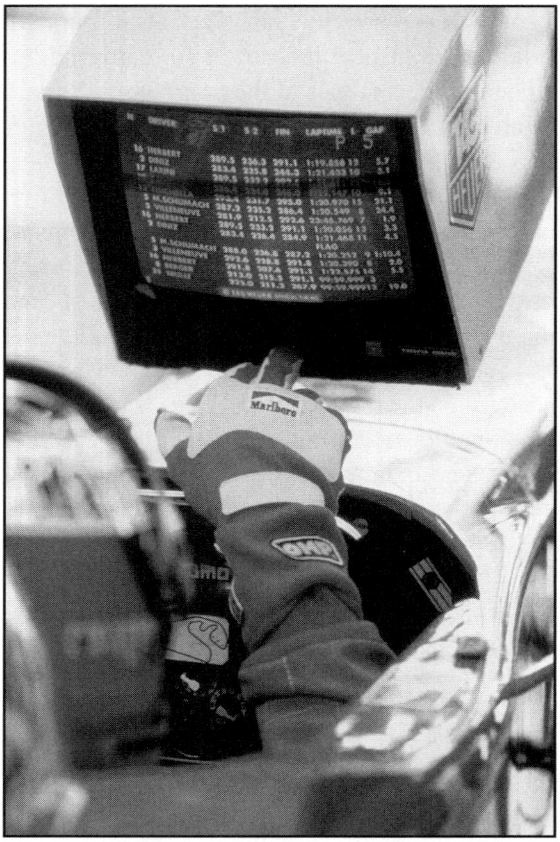

Fig. 16.3 Drivers have their own timing screens, accessible to them while sitting in the car in the pits. (Courtesy of Sutton Motorsport Images)

For the practice and qualifying sessions, the most relevant data are:

- Each driver's last and best lap time (if the last lap is a new best time, it is tagged)
- The difference between each driver's best lap time and the quickest driver's lap time
- Segment times and top speeds; individual and overall best times and speeds; and individual "idealized" lap times
- Which drivers are on the track
- Number of laps completed by each driver
- Time of day and time to the end of the session
- Race control information and messages, including notification of any penalty incurred by a competitor and the red flag

Color coding is used to highlight latest, individual, and overall best times and speeds in a multi-page display that takes practice and concentration to master.

For the race, the data displayed on the television monitors are changed. The main data are now as follows:

- Number of laps completed
- Elapsed time
- Race order
- Individual last lap times (marked to indicate whether it is the driver's quickest lap and/or the best race lap)
- Gap to preceding and following cars, and to the leader

Fig. 16.4 A typical "command center" on the pit wall. (Courtesy of Sutton Motorsport Images)

- Cars in the pits, time spent there, and the number of pit stops completed by each driver
- Cars stopped on the track
- Race control information and messages, including red flag, stop/start penalties, the safety car on the track, and the black flag
- Segment times and top speeds, although this information is not a great deal of use during the race

The row of team personnel scrutinizing the television monitors on the pit wall will try to analyze this ever-changing mass of data and use it for two basic purposes: (1) to inform the driver by pit board and radio of the race situation as it affects him, so that he can modify his tactics according to what is happening to his rivals, who may be out of his range of sight; and (2) to modify the pre-planned pit stop strategy according to circumstances (e.g., tire degradation, traffic, and competitors' strategies as they unfold).

For the driver, who may be running on his own with no other cars in sight except for back markers, knowing the gaps to his rivals is all important. Once pit stops have started, and until everyone has completed their last stops and are back on equal footing, there is almost no way that the driver can know how the race is developing. He relies entirely on information transmitted to him by his team, and the team must extract that which is relevant from the displayed data. The data must be carefully analyzed and filtered, so that the minimum essential facts are sent to the driver concerning potential threats and opportunities. The final decision about what to do with the information lies with the driver—only he can know whether he can speed up to catch the car in front without putting his own car, engine, or tires at risk.

The timing of pit stops is something that only the team can really assimilate. The details of the stops made by other cars and the effects on their speeds are too complex for the driver to process while driving. His lap time variation will indicate whether he is being held up by the car in front or by back markers as he comes up to lap them. It may be worth bringing him in a few laps early, rather than lose time behind a slow car. A quick calculation will provide an estimate of where the car will exit the pits again, relative to other cars, and whether it would provide the driver with a clear track on which he can make up time on a rival who is held up. If the lap times lengthen without traffic, it may be the first indication of a problem with the car or tires. Checks of the telemetry and a radio consultation with the driver might lead to an unplanned or early pit stop to correct the problem. The knowledge that a rival has slowed or stopped is equally valuable, because it may allow the team to advise its driver to conserve his car.

A change of track conditions, such as rain or a drying track, forces a decision to be made about changing tires, which may turn out to be a race-winning or, conversely, race-losing decision. The right time to change to wet tires, or to change back to dry tires, is crucial. Therefore, the lap times of the first car to do so will be watched very closely. The first sign that the change has yielded faster lap times than possible on the alternative tire results in frenzied radio messages and a stream of cars heading for the pits.

The system is also used by race control to communicate with the teams. Any incident that causes the safety car to come onto the track, or for the race to be stopped, will be announced on the monitors. Likewise, any penalty exacted on a competitor is displayed for all to see. Race control is provided with a display showing the positions of all the cars, derived from the triggered presence of the cars in a sector. From there, race control can spot an incident and seek further details from the television screens linked to cameras covering the whole circuit.

The system now produces so much precise and instant data that it takes an army of personnel in each team to monitor and analyze it. The data is transmitted as RF video data, rather than as a digital bit stream. This prevents the teams from inputting the data into a computer and performing further analysis on the data, using custom software. They would like to do this, and some teams even employ staff at the factory to input the printed data manually into computers for further analysis (it takes several people several days to do the job). Suitable expert system software could perform "what if?" analysis, as well as monitoring the data for threats and opportunities. Different pit stop strategies, on both the team's cars and those of the competitors, could be updated continuously with the lap times actually being achieved and then used as a guide to decision making. Graphical plots of gaps and lap times would enable trends to be spotted at a glance—the possibilities are almost infinite, and the facility will undoubtedly become available in the future.

Not only will the teams benefit from being able to post-process the data in virtually real time, but it could benefit the television viewers just as much. As viewers change to watching digital television and gain access to the TAG-Heuer pages, similar analysis would be invaluable in enabling the complexity of a Formula 1 race to be unfolded and explained.

It is difficult to believe that 25 years ago, the whole process of timing and monitoring a grid full of cars consisted of manually operated analog watches and a handwritten lap chart. Undisputed accuracy and instant information are the prizes won by this new technology.

Chapter 17

The British Motorsport Industry

Science finds, industry applies, man confirms.
—Anonymous

In the year 2000, 14 of the 22 cars on a Formula 1 grid were designed and built in Britain. Likewise, typically 26 of the 28 cars competing in the CART series are British-built, as are all Formula 3000 and Indy Lite cars. It is estimated that 75% of the race cars in use today were built in the United Kingdom. Although Britain's independent motor industry has declined to virtually zero since World War II, for that country to have built such a successful high-technology industry in that same period is remarkable and certainly worthy of investigation. The nearest parallel is the growth of Silicon Valley in California, and many business analysts would like to understand the factors leading to such phenomena.

More and more, it is being recognized that business communities are incredibly complex systems, prone to unpredictable and unexpected outcomes. The best product does not always win in the marketplace, nor does the best company survive. For example, Apple and Applesoft lost to the PC and Microsoft. Analysts are beginning to liken business communities to biological ecosystems and model them accordingly, applying evolutionary principles in an attempt to determine likely courses those communities will take. "Survival of the fittest" is a cornerstone of evolution theory, and few environments are more competitive than that of motorsport, except perhaps wartime. If the British motorsport industry is regarded as a species within a highly competitive ecosystem, it is easier to understand why it has succeeded, displacing the previous dominant species—German and Italian race car companies. It is also easier to understand why America and Japan—two of the most successful industrial nations today—have not displaced Britain.

World War II initiated the opportunity for Britain. The war years caused a cessation of motorsport and a leap in engine, materials, and design technologies. The playing field was completely leveled. Britain emerged from the conflict with certain key features, which by chance were to prove crucial in the development of a new and different way of motor racing:

1. A pool of designers and engineers skilled in aeronautical sciences: engines, structures, flight dynamics, and aerodynamics
2. The establishment of new aeronautical engineering colleges by a government (temporarily) committed to a strong indigenous aircraft industry
3. A diverse components industry that had grown to support the war machine
4. A country literally covered in airfields that had no further uses but were ideal for holding motor races
5. A nation confident in its engineering ability to win an ever more technological struggle

These were the "genes" that would come together to form a human motorsport "species," but another decade would pass before the sport had grown to the point that it would really be noticed on the world stage. Meanwhile, the dominant species that had existed before the war had not died out.

Pre-war motor racing had been dominated by major motor manufacturers, each competing to demonstrate the superiority of the industrial might of its respective nation. Mercedes and Auto Union, supported by the Nazi party, had swept away the Italian and French companies that had led the way in motor racing for decades. After World War II, Italy re-emerged at the forefront with Ferrari, Maserati, Alfa Romeo, and Lancia. Britain competed with BRM (part of the Owen Organisation) and Vanwall (part of Vandervel), and in sports cars with Jaguar and Aston Martin (part of David Brown)—all major industrial companies. Mercedes returned in 1954, swept all before it, and withdrew triumphantly at the end of 1955—the year marred by the tragic accident at Le Mans.

German motor racing had always had national support, and Italy has always had the full backing of the whole of the Italian motor industry (arguably more influential than the numerous varieties of government that have attempted to rule Italy since World War II). In Britain, motor racing was only half-heartedly given official recognition. Jaguar's successes at Le Mans were duly applauded, but the efforts of BRM to overcome the problems of a technically ambitious project were often publicly derided.

While the traditional approach to top-level motor racing in Britain continued to fail, in spite of producing some World Championship-quality drivers (e.g., Moss, Hawthorn, and Collins), a new species was forming at the club motorsport level. Individuals, rather than organizations, were designing specials, often in their back bedrooms, and building them in their garages. On weekends, they took the cars to disused airfields and raced each other on makeshift racetracks. The competition was intense, and newly learned engineering skills were brought to bear on every aspect of the design of the cars and engines. Under the auspices of the 750 Motor Club, which had started as a club for people wanting to race Austin Seven-based specials but later evolved to encompass cars based around small Ford engines, this racing thrived. A few small companies formed to manufacture copies of their founder's successful designs, so that others could race competitive cars. Those fit enough to survive during the early days are typified by the following:

- Lotus—Colin Chapman
- Lola—Eric Broadley
- Cooper—John Cooper
- Cosworth—Keith Duckworth

Progressing through Formula Junior, Formula 2, and sports cars to Formula 1, these small companies provided employment for other engineering aspirants who lacked the business acumen to run successful companies. In the late 1950s and early 1960s, very little money could be made in motorsport, and not being the racing departments of automobile manufacturers or components companies, budgets were extremely limited. These companies were not in a position to create Grand Prix engines nor to fabricate gearboxes and suspension parts in the way that Mercedes had done and BRM attempted to do. Instead, they took proprietary parts and modified them using ingenuity, innovative flair, and total disregard of established practices. This approach also helped create the British motorsport component industry, because component manufacturers were forced to recognize that their products were being raced. It generated a wide-ranging network of race departments and specialist engineering companies that started life by modifying standard production parts for racing. The most famous examples are:

- Coventry Climax—Started by modifying a production fire-pump engine and eventually produced a Championship-winning Formula 1 engine
- Hewland—Started by modifying Ford Anglia gearboxes and developed into the major supplier of racing gearboxes for all classes
- Cosworth—Tuners of Ford engines, now a major race and road car engine design, development, and production company, the racing side owned by Ford and the road car side owned by Audi
- AP Racing—The racing department of Automotive Products, producing racing brakes and clutches
- Specialized Mouldings—Pioneers in using composites for race car bodywork

A new species will not always prosper or even survive in an existing ecosystem that supports a dominant species. Why should Lotus, Lola, Cooper, and Brabham necessarily be able to enter and succeed in Grand Prix racing? Portions of the British Empire (or what remained of it) such as South Africa, Australia, and New Zealand also had an environment that was right for the development of grass-roots motorsport, and many of the best engineers and drivers migrated to the United Kingdom to advance their ambitions. The answer lies in a detail of the environment in which the design and engineering skills were developed: the technical regulations of the 750 Motor Club.

To keep costs low, the regulations tightly controlled what could be done to the engines but gave great freedom to the chassis. With limited power, the best chassis design, aerodynamics, and setups usually won. In Grand Prix racing, the opposite was true: the best engine at any given time dominated the results. This emphasis on chassis technology rewarded those with the requisite skills in the aeronautical sciences (i.e., graduates from the new technical colleges).

The evolution of the chassis moved quickly toward lightweight, rear-engined, softly suspended, low-drag cars that, given an adequate engine, could take on the heavy, front-engined, stiffly suspended Ferraris and Maseratis. The Coventry Climax engines were just adequate, and the new species gained a foothold. Next came the battle for dominance of the motorsport ecosystem, and a few chance events ensured the emergence of the newcomer as the victor in the evolutionary struggle.

In the early 1960s, Lola and Lotus had taken their new approach to chassis design to Indianapolis and inflicted on the front-engined behemoths what they had already wrought on the traditional Grand Prix cars—they killed them. In the process, Colin Chapman formed a business and racing relationship with Ford and discovered sponsorship.

In 1966, egged on by Chapman, Walter Haynes of Ford U.K. persuaded Ford to give Cosworth, the most successful of the Ford tuners, £100,000 to design and build the Ford Cosworth DFV—the best bargain in motor racing history! This incredible power plant gave the British constructors the next "adequate" engine (it was never as powerful as the Italian engines) and enabled them to consolidate their position by breeding the next generation: McLaren, Williams, March, Tyrell, etc.

This ever-growing species needed more food (i.e., money) to sustain it than was available within motorsport. Chapman brought back the concept of commercial sponsorship from the United States and in 1969, uninhibited by tradition, painted his Lotuses to resemble cigarette packets. It took Ferrari another 28 years before it would even change the shade of red in which its cars were painted. That inability to adapt is one of the fundamental characteristics that explains Ferrari's lack of success, through a period of continuously changing technical and commercial influences.

The increasing size of the motorsport business necessitated the development of structure and leadership to ensure that it adapted as it grew to the equally fast-changing outside world. The creation of the Formula One Constructors Association (FOCA), led by Bernie Ecclestone and Max Mosley (who later became president of the FIA), organized Formula 1 into a $1+ billion-per-year worldwide business, based on television revenues. This approach has been a model for all classes of motorsport since then.

The final event that influenced the success of the British was their grasp of the importance of aerodynamics. Downforce was the ingredient that their superior chassis needed to remain ahead of the more powerful Italian engines. Through the development of wings and downforce-inducing body shapes (Lotus T72) to full ground effect, the speed with which the British teams adapted was so much greater than that of Ferrari (the only non-British team that was able to win World Championships in this era). Ferrari is rumored to have tested a Lotus T78 look-alike model in its wind tunnel in 1977 and declared that it did not work! Perhaps it was because the flat-12 engine interfered with the underbody shape and airflow. Likewise, perhaps the belief at Ferrari that the engine was everything carried all opposing opinion before it. The V-8 DFV was exactly the right shape for a ground-effect car, and the engine continued to power British cars to victory.

By the 1980s, the chassis and aerodynamic technologies, which gave the British dominance at the top level of motorsport, had filtered downward into the varied range of race cars produced by Lola, March, Ralt, and Reynard, and all the smaller constructors evolving from the large ones. Cosworth gave birth to Illmor, Hewland to Xtrac, and so forth, evolving steadily and improving at almost every stage. The unsuccessful companies failed to survive, in the same way that genes often fail to survive in biological ecosystems. For a species to survive threats from other new species, it had to be able to adapt to changing environmental conditions. One such threat was the turbo.

The emergence of the turbocharged 1.5-liter engines to challenge the 3.0-liter NA engines led to the re-entry of major automobile manufacturers into Grand Prix racing. Renault started the movement, choosing to race its own car. Other major manufacturers decided to form partnerships with established teams, which significantly proved to be the better approach. If it had not, Renault and Ferrari might have proved superior to the British teams and ended the British dominance by encouraging other automobile manufacturers to enter Formula 1 in their own right.

At the same time as the turbos came carbon-fiber composites and electronics, first in engine management systems and then in chassis control systems. These technologies are difficult to copy because their secrets are invisible. The continued success of the British teams through the composite and electronics age of the late 1980s and 1990s can be attributed to a particular characteristic on which much of the success of the British motorsport industry is based. The major automobile manufacturers are experienced at protecting their intellectual property and maintaining loyalty among their employees. They use their involvement in motor racing both to develop in-house technology and expertise, and to motivate and train their engineering staff. Little cross-fertilization exists among competing companies, each carrying out parallel time- and money-consuming R&D programs.

In contrast, the British race car manufacturers swap and exchange staff with almost gay abandon, although not always willingly. An advantage developed by one is quickly adopted by competitors. The information is exchanged through the movement of staff and the social exchanges that inevitably occur within an industry based in a small area. This diffusion of successful genes ensures that the species as a whole benefits from the qualities that prove success in competition, thus improving the chances of both survival and expansion of the species.

Isolated in Italy, Ferrari has only its own pool of genes on which to rely. It cannot adapt easily to a changing environment, because mutations do not occur often enough to provide new and potentially successful characteristics. Ferrari has tried "artificial insemination" from time to time: Harvey Postlethwaite introduced composite technology, and John Barnard brought up-to-date design techniques. Today, Ross Brawn and Rory Byrne have brought with them technical management and state-of-the art aerodynamics and chassis technology, while Jean Todt, who joined Ferrari from Peugeot where he had led the company to win World Rally Championships and Le Mans, has forged the team into a single entity, rather than fragmented groups.

Britain's dominance of Formula 1 and Indy cars has ensured that it also dominates the lesser formulae, in which production race cars prevail. Forty years of steady growth has seen the consolidation of an industry in which around 1,200 firms participate, with a turnover around $4.2 billion per year, employing approximately 25,000 people. (Exact figures are extremely difficult to determine, because motorsport does not spend much time or effort on self-analysis.) These figures represent only the engineering industry and do not include peripheral activities such as promotions, circuits, press, publications, and all the service industries that have grown up around the sport, which account for a further $2.7 billion per year turnover and employ further 15,000 people in another 1,200 firms. Of the whole, Formula 1 probably represents 20% (Ref. 17.1).

This phenomenally successful story has taken place without government support or intervention. The ingredients were the right ones back in the 1950s, and a number of chance events occurred which gave the emerging species the opportunities to establish itself and finally reach dominance. The process by which this industry has established itself is now being studied by governments to determine how it can be emulated. Ecosystems are unpredictable, whether they are biological, business, or social. Trying to repeat the process, orchestrated by officialdom, would inevitably be doomed to failure.

There is no reason to believe that the British will cease being Number One in motorsport. Now that their economic and technical strength is so great, they should dominate the industry for some time to come. However, other countries would dearly love to have the *kudos* and the exports that such an industry brings. Likewise, changes in the environment, such as the loss of tobacco sponsorship, greater involvement by the motor industry (both technical and financial), investment by financial institutions, and technology changes can have serious effects on the development of the industry. One such self-imposed threat is the lack of British investment in promoting motorsport engineering in universities and colleges (though this shortcoming is now being addressed), while others (notably the United States and Japan) further motorsport through education. From where else will the new designers and engineers come as the industry continues to expand?

Chapter 18

Formula 1 Now and in the Future

Prediction is very difficult, especially about the future.
—Niels Bohr

A little more than 100 years ago, Daimler and Benz made their first motorcars. A little more than 50 years ago, the FIA created Formula 1. Since then, Formula 1 has become one of the greatest sporting, entertainment, and business phenomena. The features that make Formula 1 so popular globally and set it at apart from its main sporting rivals, which are soccer and the Olympics, are that it alone involves real danger, high technology, and an activity in which hundreds of millions of people throughout the world regularly participate—driving a car. In spite of currently the greatest motor industry involvement since the 1950s, Formula 1 cars bear less resemblance to road cars, and have less technical relevance to the direction in which road cars are going, than ever in the past. Applying normal automotive engineering logic to the prediction of how Formula 1 cars will evolve in the future is fruitless, because the forces that influence the direction in which the sport/business will go are not primarily technical.

Those who make predictions must usually wait a few years before they are proven right or wrong. If they are lucky, their predictions are forgotten by the time those predictions are proven to be way off target. Making predictions about the future of Formula 1 is a much shorter cycle. In the past, new technologies have developed to make existing cars obsolete overnight, likewise making nonsense of any predictions about what might happen next. The months between writing this chapter and publication of this book are long enough to potentially render any predictions in this book completely wrong!

The specifications of a current Formula 1 car, epitomized by the latest cars from McLaren and Ferrari, embody several key technologies that have either been developed within the motorsport industry itself or adapted from road car or aerospace technologies. The FIA Technical Regulations have rejected certain technologies and limited the use of others, in an effort to curtail speed and costs, ensure an acceptable level of safety, and maintain the sporting spectacle. Some of those rejected were irrelevant to road cars, while others might have benefited them.

It is inevitable that the technologies essential to a current car (i.e., the ones that have contributed to the expansion of the performance envelope and maximize its access by the driver) would be invented and developed because they are so fundamental. However, some individual or group of individuals within a racing organization were always the first to realize the potential of a new technology, apply it, and sometimes gain a temporary advantage over their competitors, although such an advantage never lasts long.

The current layout of a Formula 1 car was established by 1978 and is unlikely to change suddenly as it is enshrined in the Technical Regulations. The driver sits reclined in a monocoque chassis (Lotus T25, 1962) ahead of a single fuel tank on the center of gravity (Lotus T79, 1978) to which is bolted a mid-mounted (Cooper, 1958) V-configuration engine as a structural member (Lotus T49, 1967). (Ferrari and BRM had explored the concept of structural engines previously, but Lotus established it as the definitive arrangement.) The transmission, bolted to the rear of the engine, also carries structural loads. Wings mounted to the nose and above the gearbox (Chapparal, 1967) generate downforce that is additional to ground effect (Lotus T78, 1977) now produced by the flat undertray (FIA Technical Regulations, 1983).

The monocoque is now made wholly in CFRP composites (McLaren MP4/1 and Lotus T88, 1981). The engine is a V-10 (Honda, 1989) revving to more than 18,000 rpm, thanks to pneumatic valve springs (Renault, 1986). Electronic engine management controls all aspects of the engine, including fuel and ignition (Bosch, 1981). The transmission is operated by a computer-controlled, semi-automatic clutch and gear-change (Ferrari, 1989). Computers and electro-hydraulic control systems measure, optimize, and control several functions to the extent permitted within the regulations.

Belted, radial ply tires are fitted (Michelin, 1977) almost slick-treaded (Firestone, 1971) but with circumferential grooves (FIA Sporting Regulations, 1998). Carbon-carbon brake discs and pads (Brabham, 1976, originally aerospace) and clutches are universal.

Several useful technologies have been banned, and thus their potential contribution toward the development of road cars has been lost. These include turbocharged engines (Renault, 1977); active suspension (Lotus, 1983); driver aids (traction-control, ABS and yaw-stability, rear-steer, active differentials—all from road cars, late 1980s); CVT (Williams, 1993—never raced; developed from Van Doorne road car system); stored energy, i.e., hybrid (McLaren/Mercedes/Illmor, 1998—never raced); and a variety of means of using the tire, including four-wheel drive (Ferguson, 1961) and fiddle brakes (McLaren, 1998).

The basic layout of the cars was set by the end of the 1970s, and all the successful technologies were applied by the end of the 1980s. Since that time, the only major new technologies to appear have been the application of road car systems, and these have all been banned. The enabling technologies have long been established. These include desktop and on-car digital computers (1970s), carbon fibers (Royal Aircraft Establishment, Farnborough, 1964), electro-hydraulic servo-valves (Moog, 1948), and the principles of ground-effect aerodynamic testing in moving ground plane wind tunnels and the description of the flow around a rotating wheel (both from the Imperial College of Science, Technology and Medicine, London, late 1960s). One of the strangest twists was that several of the most effective technologies were not pioneered in Formula 1, but in Can-Am racing in the United States. Chaparral, in conjunction with Chevrolet R&D, showed the worth of variable incidence wings mounted directly to the unsprung masses, fan-induced downforce, skirts, hands-free gear-changing, composite monocoques, data systems and simulation, and even under-car aerodynamics—all in the 1960s.

The last decade has been one of optimization and adaptation to changing and ever more restrictive regulations. Almost nothing revolutionary or fundamental has been introduced by Formula 1 during this period. The single technology that has brought this process to the current state of refinement is the computer, along with its associated applications to measurement, analysis, and process control. It is the ability to model and understand, design and analyze, manufacture and inspect precisely, and test and measure that have taken the key new technologies from the 1970s and 1980s and turned them from black arts into sciences.

In the face of the foregoing analysis, it would be rash to predict that any revolutionary technology will burst onto the Formula 1 scene. Indeed, most chief designers have been saying each year for several years that their new cars are "evolutionary, not revolutionary." It would be equally rash to predict little change while Formula 1 is such a high-tech part of a rapidly changing world. There are a number of prevailing influences that shape Formula 1 and thus will determine the technical direction it takes. The most important influence on the design of the cars is, of course, the Technical Regulations, and how these are determined in the future will be the most significant factor.

Many vested interests are involved in the development of the Technical Regulations. On the one side, the technical departments of the teams and engine suppliers would like more freedom to innovate and show their technical prowess. The better-funded departments (i.e., those with large resources, staffs, and budgets) have an advantage whenever the regulations change significantly. They are able to do more research quicker and arrive at the optimum solution faster, gaining a temporary advantage until others that are less well off can catch up with them. Pulling in the opposite direction are the team owners and the Formula 1 promoters, who seek rule stability and the competitive equality that generates close racing without excessive costs. In the middle is the FIA, which must try to satisfy all parties while ensuring Formula 1 exhibits the highest standards of safety for all participants and spectators, so that the threat to motor racing that appeared after Ayrton Senna's fatal accident never appears again. These opposing forces make it difficult to implement the major technical changes that sometimes seem so necessary to outsiders. The running of Formula 1 today is more democratic than it has ever been in the past, but democracies seldom encourage revolutions. However, major changes to this apparently stable situation are taking place. The major players themselves are changing.

Since the early 1980s, one person has effectively controlled the running of the FIA World Championship—Bernie Ecclestone. The Formula 1 teams, especially Ferrari, have had a strong influence at times, as has Max Mosley, president of the FIA. However, Ecclestone's long-term vision and single-minded promotion of Formula 1 have turned it into the globally successful business it has become today. Now, the ownership of the business and many of the teams is changing or will soon change. As Ecclestone heads toward retirement, he has found new investors to purchase shares in the $4+ billion Formula 1 business he built. The principals of several of the leading and most influential teams have also found investors to buy into their businesses. In some cases, these investors are motor industry partners, such as Ford, DaimlerChrysler, and Renault, which join Fiat as major or even majority shareholders in Formula 1 teams. Toyota is almost ready to join as a wholly owned team, as Honda almost did and may still do. Likewise, BMW will probably not be able to sit on the sidelines.

Overshadowing even these changes is major rationalization within the motor industry. General Motors is poised to take over Fiat, owners of Ferrari. Ford, among others, would like to purchase BMW. Can Honda and Peugeot survive independently? Some people within the motor industry have predicted that eventually, only five or six major companies will make the vast majority of road cars. How many Formula 1 teams will each own? It is unlikely that they will want to spend the R&D sums involved to do the job more than once, even if they want to promote more than one of their brand names via Formula 1.

The whole business depends on the man in the street wanting to watch Formula 1 on television. The Formula 1 "product" is the digital television signal that is piped out of the paddock down a fiber-optic cable. Without the following of millions worldwide, the television revenue will decline, and the sponsors (particularly tobacco sponsors and the motor industry) will gain less from involvement. Formula 1 must excite ordinary people in the face of competition from other sports and activities. NASCAR manages to sustain massive brand following with negligible technical interest and high-speed close racing. Formula 1 currently offers the opposite and appeals to a different audience. A shift in the public's preferred motorsport entertainment would force Formula 1 to change. At the moment, Formula 1 holds the Number One position in terms of technical interest and driver status, but it is criticized for lack of on-track racing entertainment.

The enormous complexity of the business model of Formula 1 means that the uncertainty about its future is so great that predictions would be unreliable, particularly because it is not known who will be the main players in the next five years. It is possible that Formula 1 may continue along its present growth curve. More countries may become enamoured of it as car ownership increases in their populations, and the remaining motor industry giants may vie with each other to gain the marketing benefits of winning on the track. Alternatively, the public may finally embrace environmental issues and turn its back on the pollutant-emitting, gas-guzzling dinosaurs, sending the motor industry running, or demanding a totally new formula based on environmentally acceptable and relevant transport technologies. The decline of tobacco sponsorship may leave Formula 1 short of other major funding, although telecommunications companies are becoming large enough to take over at the level of the former tobacco sponsorships. Or, Formula 1 may have reached a peak and will stabilize with slow evolutionary change, adapting to the desires of the motor industry and the public. So far, under Ecclestone's promotional and Mosley's regulatory and political stewardship, Formula 1 has proved to be adaptable and able to survive, willing to change quickly when necessary in response to outside pressures. It is, of course, not known how their successors will perform.

The effect of all this on the Technical Regulations is uncertain. The FIA has stated that it is determined to stabilize speeds around those achieved in 1996, even in the face of the inevitable escalation in power and the improvements in lift/drag ratio that the ever better-equipped aerodynamic departments achieve each year. Competition among tire suppliers is one of the biggest threats. Therefore, in the absence of any better suggestion from the teams, the FIA has provided itself with a reasonably low-cost but unpopular means of controlling speeds—tire grooves, which may evolve to treaded tires.

The physics of overtaking, considered by many to be the essence of racing, is now understood to the extent that the effect of changes to the cars can be predicted reasonably in this respect. For a marginally faster car to be able to overtake one ahead of it, the faster car must be able to corner closely enough to the car in front, to benefit from the reduction in drag afforded by its wake. It then can accelerate in the slipstream and, if the straight is of sufficient length, overtake prior to the next corner. Critical aerodynamics, low-grip tires, and low-drag configurations prevent a car from getting close enough in a corner to benefit from the wake. On the high-speed ovals in the United States, a car with equal or less performance can slipstream past a car ahead of it, but this level of contrived overtaking entertainment is considered by many to be inappropriate to Formula 1, analogous to the difference between the dozens of points scored in basketball and the one or two goals scored in soccer. Only time will tell which the public prefers and how much the regulations must take overtaking into account.

Some continue to consider that a reasonable risk of being killed or severely injured should be a part of motorsport, but their numbers are diminishing. The public's and sponsors' reactions to Ayrton Senna's death showed that in the preceding decade without a fatality in Formula 1, attitudes had changed. Max Mosley has tried to ensure that motor racing is never threatened in the way it was after the Mercedes accident at Le Mans in 1955 or after Ayrton Senna's accident in 1994. An adverse political reaction to a major accident at the highest levels of motor racing could easily be the biggest threat to the survival of the sport. Technical changes to improve circuit, car, driver equipment, procedures, and medical safety will follow steadily in the wake of increased research into the subject. Motor industry interest in motorsport safety should ensure that.

As the regulations increasingly limit the opportunity to gain a technical advantage, the cost of improving by a few percent each year becomes greater. The short life cycle of a Formula 1 car design—only 18 months from concept to the museum—demands a higher rate of expenditure to gain the most from the time available. Constraints on the use of exotic materials, and on the amount of expensive track testing, are already in place and will probably be extended.

Ross Brawn, technical director of Ferrari, has stated that the nearest equivalent to the conditions under which Scuderia Ferrari operates is a state of war. If he is right, we should look to the techniques of strategic planning, personnel training, intelligence gathering, propaganda, and political maneuvering, as much as to the technologies of the hardware deployed in battle, to understand how Formula 1 may develop as the stakes become greater and the competition becomes more intense. Computers will be employed to make decisions more quickly. The last of the black arts (e.g., tires, dampers, transient vehicle dynamics, and aerodynamics) will be turned into sciences by analysis and modeling. Optimization routines will continuously provide the best setup solutions, responding to changes in conditions. The hardware design of individual cars and engines will converge, and differences narrow down to differences in control software and the performance of the optimization routines. Race strategies will benefit from data gathered about the performance of competitors, and game theory will narrow the options in the "battle" on the track.

Technical progress will be made, if permitted by the regulations. Fiber-reinforced composites are a technology that is relatively in its infancy. Fibers with higher specific strength and modulus, such as hollow carbon fibers, will be developed, and matrix materials with higher temperature characteristics will enable composites to be used in applications currently denied to them. The problems associated with CFRP gearboxes will be solved. Higher-temperature carbon-carbon materials for brakes and clutches will come from the aerospace industry.

Furthermore, rpm and hence power will rise. However, with the number of cylinders restricted to 10, it may be necessary to reconfigure the engine to overcome torsional dynamic problems. Better modeling of all aspects of the engine will yield steady improvements in volumetric efficiency, combustion, friction and windage losses, and lower weight. Auxiliaries will benefit from modeling and rig testing.

The ability to model complex separated and rotating flows with computational fluid dynamics codes is the key to major steps forward in aerodynamics. Until then, better wind tunnel simulation using more detailed models at larger scales will sustain aerodynamics as a largely experimental technique.

Measurement and simulation techniques to improve understanding of the forces involved in vibrating chassis and engines structures will lead to more optimized designs and lower weight. The performance benefits will be small due to the minimum weight limit, but designers will always strive to lower the center of gravity.

The dynamic characteristics of tires and their interaction with the suspension will be understood, such that it becomes possible to optimize tread temperature under all conditions. Compounds with peakier characteristics can then be used for the treads.

The most difficult safety problem to solve is a function of the fact that Formula 1 cars have open cockpits and open wheels. Retaining the wheel and upright during an impact is difficult, although enormous progress is being made with the use of tethers. The driver's head is inevitably vulnerable in an open cockpit to any loose object or airborne car. Open-cockpit, open-wheeled cars are perceived as the ultimate in racing, and a major catastrophe would have to occur to change that.

Whether computers will elevate the techniques of optimization, systems integration, and strategic planning to levels where they can do a better job than a human in the time available is not yet clear. If and when they do, the need to pay driver-sized salaries to designers and engineers of the caliber of Adrian Newey, Ross Brawn, and Rory Byrne will end. In the same way that driver aids detract from the driving side of the sport, designer and engineer aids have taken something away from the technical interest. In the 1960s and 1970s, aspiring race car engineers, and indeed anyone with a technical interest in cars, hurried to the newsstand to collect the latest magazine with the announcement of a new Lotus Formula 1 car. The anticipation of discovering what Colin Chapman had created this time was seldom disappointed. The announcement of new cars now tends to cover only new sponsorship deals, with bland statements about how many new parts are in the car, and how it has a certain percentage of improvement in the lift/drag ratio. The loss of designer and engineer stars, who are able to influence the competitiveness of a car by their instinctive feel for what is needed, would be a great shame and would detract from the human aspects of the technical interest in Formula 1.

Perhaps the most intriguing possibility for the future would be caused by a sea-change in public awareness of environmental issues. Inevitably, the car-buying public will one day become concerned about the rate of consumption of fossil fuels, the emission of their locally polluting products of combustion, and the global atmospheric buildup of carbon dioxide. The motor industry has already developed many of the technical solutions, or such solutions are well underway in development. Electric cars, various hybrid powertrains, hydrogen-fueled cars, and fuel cells are all running, and some are entering production. Costs will come down only when the infrastructure to support them is in place, and the demand for them increases. Formula 1 would be a powerful showcase for these technologies, helping to build awareness and changing public opinion about them, as well as being a major contributor to their development.

More than 20 years ago, Keith Duckworth, co-founder of Cosworth Engineering and designer of the Ford-Cosworth DFV, suggested an engine formula based on the use of a fuel flow-rate valve. The valve would control the regulated maximum rate of fuel consumption, and the engine that most efficiently turned that fuel into power would succeed. The chassis designer and the driver would strive to use as much of that power as much of the time as possible. The proposal had none of the disadvantages of a total fuel consumption formula, because it did not discourage drivers from racing as hard as possible all the time. It also provided the regulating body with a simple means of controlling maximum power: as new technology emerged, the maximum rate of fuel flow could be lowered to maintain the peak power. It also made no differentiation among different engine cycles, thereby encouraging the most efficient technology. If the concept was extended, so that, by analyzing and calibrating the valve according to the energy content of fuels, it became a rate of energy consumption valve, any type of thermal engine could be used. The formula would encourage the most efficient means of turning transportable energy into work.

Today, it is possible to build such a valve that, with closed-loop flow control, would impose a flow limit to better than 0.5% (i.e., 4 bhp on today's 800 bhp engines). There is currently no interest in such a formula. However, if and when public opinion changes, Formula 1 could pitch all the proposed environmental technologies against each other in the intensely competitive arena that is motor racing, with the most efficient emerging as the winner. The public would be encouraged to buy that technology based on a valid demonstration of its efficacy and the image given to it by Formula 1. That the motor industry does not encourage any motor racing formula to go in this direction tends to indicate that, at the moment, selling their existing products is far more important to them than doing anything about protecting the environment.

Chapter 19

The Technical Regulations and Policing Them

The Theory and Practice of Gamesmanship,
or the Art of Winning Games Without Actually Cheating
—Title of a book by Stephen Potter

The FIA Formula 1 Technical Regulations (Appendix D) and those parts of the Sporting Regulations (Appendix E) that affect the technical side of racing are not worth the paper on which they are written if they are not enforced. Formula 1 depends enormously on the credibility of the scrutineering and enforcement system that the FIA applies to its regulations. If the major participants in Formula 1, including some of the largest companies in the world, did not believe that they compete on a level playing field, they would leave the sport to spend their marketing budgets elsewhere.

When the FIA Technical Delegate responsible for scrutineering arrives at a Grand Prix on the Wednesday before the race, he is accompanied by several full-time assistants and a number of technical specialists in computers, software, and fuel analysis, plus nearly three tons of sophisticated measuring equipment costing more than $0.5 million. Thus, it would be logical to assume that the FIA is trying to catch Formula 1 teams cheating. It is not. Rather, the intent of the FIA is to persuade the teams not to even try to cheat.

Nothing devalues a sport more than the specter of participants cheating the regulations that govern it. Among the responsibilities of the FIA are not only ensuring that it is impossible to get around or deliberately contravene the necessarily complex Technical Regulations, but also convincing the world that no competitors are doing so. Since the abolition of driver aids and active suspension in 1994, the temptation to gain the proven advantages of these systems is enormous.* The job of designers and engineers is to gain the maximum possible performance within the regulations. Although no car has been shown to have raced with deliberately illegal systems, some have been extremely close to the line, and one or two have inadvertently crossed it.

At the end of 1993, the FIA put together a set of carefully worded changes to the Technical Regulations, procedures to carry out regular monitoring and random checks on the eligibility of the competing cars. The FIA also invested in the hardware, software, and expert personnel to implement those changes. In the first year of these changes, irregularities in the software of the Benetton were identified. In the subsequent year, analysis indicated that Elf fuel samples used in Brazil did not match their prior-approved test sample. These discoveries may not have generated good publicity for Formula 1, but they served clear notice that the FIA can and will catch anyone who puts even a toe over the line.

Of course, with the sophistication and complexity built into the current breed of Grand Prix cars, it would not be difficult to temporarily configure one car to be outside the regulations and provide some small advantage in performance. Twenty liters of special fuel for a qualifying series of laps, or modified drive-by-wire software downloaded for only one run, might be things a competitor could do without getting caught. However, for the team principal to sanction such actions, he would be betting the company on gaining a few tenths of a second. With draconian penalties for premeditated breaking of the rules, and the Technical Delegate armed with the powers to dismantle any car or take parts or samples to ensure compliance with the regulations at all times, the risk is probably not worth it.

There are two sides to scrutineering—safety and performance. The bulk of checking to ensure that the cars meet the extensive safety criteria laid down in the regulations is conducted away from the circuits. Witnessed frontal, rear, and side impact tests and a number of static tests, including roll-hoop and nose push-off loading, are used to qualify the monocoque design. Individual moldings are subsequently weighed and subjected to various load-deflection checks to ensure that they meet the same standards achieved in the full inspection. An FIA "smart label" is molded in, incorporating a transponder that can be interrogated so that records can be maintained on each monocoque.

At the circuit, safety checks are limited to confirming that the cars to be raced have been through these procedures and that vital apertures through which the driver must pass to escape from the

---

* Author's note: The problems of checking software and the need to eliminate all suspicion of bending the rules has led, in the spring of 2001, to major changes in the Technical Regulations: (1) Complete freedom of electronic engine controls, including traction control; (2) Complete freedom of electronic control of gear-changes and clutch, including automatic gear-changes; (3) Complete freedom of electronic control of the differential, except the use of "torque-steering" systems; and (4) No powered or electronically controlled systems in suspension, brakes and steering. These changes should permit simple policing of the regulations. Where electronic control is permitted, no software checks are needed. Where it is not, the hardware can be inspected for the absence of actuators and power supplies.

car have not been encroached upon by the latest black box to be added inside the cockpit. New car-driver combinations must be checked for an egress time of under five seconds, with another five seconds for the driver to replace the removable steering wheel. Mandatory safety equipment is inspected, and the rearward view through the rear wing is checked. It has been a long time since anyone in Formula 1 has been caught deliberately braking a safety regulation; however, regular checking sometimes catches items that have been overlooked in the effort to get to races. Scrutineering of these features occurs on the Thursday before practice, in an atmosphere of cooperation among teams and scrutineers.

To ensure cooperation in the checking of features that affect performance, a useful article in the Technical Regulations is Article 2.7—Duty of Competitor:

> It is the duty of each Competitor to satisfy the FIA Technical Delegate and the Stewards of the Meeting that his automobile complies with these regulations in their entirety at all times during an Event.

With all those areas that determine the performance under continual pressure of development, the main part of the Technical Delegate and his team's work during a race weekend is to give the competitors the opportunity to satisfy him, at times that are not always convenient to the competitors, that they comply with the regulations concerning:

- Weight
- Aerodynamic devices—dimensions and rigidity
- Engine capacity
- Fuel specification
- Control systems, especially potential driver aids
- Tires
- Refueling systems

Weight

Any time during qualifying and immediately after the race, cars can be weighed and must not be less than 600 kg (1323 lb) including the driver. All drivers are weighed at the start of the season, and the finishers are reweighed at the end of a race. Each driver's weight is entered into a computer to allow the total to be calculated when the cars are checked without the drivers at the end of the race. Previously, when the minimum weight did not include the driver, their recorded weight was used to calculate the total weight of the car when the driver was aboard, as in qualifying. It was desirable to tip the scales as low as possible, and there was no real way to cheat this. Now the reverse is true, and thus a high official weight for the driver would allow him to race under weight by the amount equal to the difference between his actual weight and his official weight.

Each of the four weighing platforms is supported on four cantilevered, strain-gauged load cells, with full temperature compensation. The output is summed electronically to give a total weight, rounded to 0.5 kg (1.1 lb). They are calibrated to a gravity reference of 9.80471 m/sec/sec at a latitude of 43.56°N. Why? When weighing was performed by counterbalancing a known mass, the local value for gravity did not matter. Since the introduction of electronics, mass is no longer measured. Modern weighing systems measure the force exerted by gravity on the mass and thus are sensitive to local variations in the gravitational field. Because the earth is orange-shaped, gravity varies with latitude to the extent that at Interlagos, Brazil, at latitude 22°S, it equals 9.78778 m/sec/sec; however, at Silverstone, latitude 52°N, it equals 9.81266 m/sec/sec. This is a variation of 0.25% or 1.5 kg (3.3 lb) in the weight of a car, quite enough to disqualify a car. Therefore, the scales are recalibrated at each circuit to compensate for changes in the strength of the gravitational field.

Cars to be weighed during qualifying are selected randomly by computer. Eight to ten cars are weighed, and their key dimensions are checked during the session. The process takes approximately two minutes and thus does not significantly delay drivers, in spite of the occasional gesticulations that emanate from the cockpit.

Aerodynamic Devices—Dimensions and Rigidity

The old flat patch at the entry to the pit lane, on which the cars were weighed and measured, has been replaced by a new, purpose-built weighing and measuring platform. Made almost entirely of CFRP and aluminum, the 2.3 × 4.8 meter (7.5 × 15.75 ft) structure was designed and manufactured in the United States. For transportation, it breaks into sections, which are doweled and clamped together. The platform is leveled on adjustable feet, using a laser leveling system. The four weighing platforms are built in and can be slid sideways independently on linear bearings, being locked in place pneumatically.

The car to be measured is pushed up ramps onto the platform and placed on the weighing scales, and the weight is recorded. When this is complete, the weighing platforms are unlocked, and the front and rear of the car are moved sideways until the centerlines of the car and measuring platform coincide. The car is next aligned longitudinally, so that the front axle centerline lines up with reference marks on the platform.

The area of the platform approximately in the shadow of the sprung parts of the car can be raised on pneumatic cylinders and is free floating in pitch and roll. Mounted on it are five fixed aluminum bosses, 30 mm diameter × 14 mm high (1.2 in. diameter × 0.55 in. high), which engage in the 50 mm (2 in.) diameter holes in the skid block (also known as the "plank"), the positions of which are laid down in the Technical Regulations. A sixth, 40 mm (1.6 in.) diameter boss is mounted on the rearmost section of this floating platform, which is free to slide longitudinally. The sixth hole in the plank is referenced to the front edge of the rear wheels. When the boss engages in it and is pulled fully rearward, the rear section of the platform is also is referenced to the rear wheels.

The platform is raised until the top surfaces of the six bosses come up against the lower surface of the monocoque, which is the reference plane, as defined in the regulations. The floating part of the platform is now parallel to and 14 mm (0.55 in.) below the reference plane. For 1995, the regulations were changed so that all height measurements are now relative to the reference plane rather than to the ground. This was done at the request of the chief designers to make their jobs easier, and it enables this new equipment to be used for all measurements.

The car is now ready for checking. Marked on the fixed and floating platforms are longitudinal and lateral grid lines relating to all critical length and width dimensions, referenced, where appropriate, to the front and rear wheels. No measuring scale is needed to check the car dimensionally. A series of machined aluminum or CFRP templates and Go/No-Go gauges are used to check every dimension covered by the regulations, even plank thickness, and the results can be witnessed by the team's engineer (Fig. 19.1).

Fig. 19.1 Rear wing dimensional legality checking template. (Courtesy of Sutton Motorsport Images)

The rigidity of aerodynamic devices has had to be checked, because many designers have applied the reasoning that as everything deflects under load, it is worth seeing how far one can push the interpretation of the word "rigid" and gain an advantage by allowing a wing to feather under high-speed loads (Fig. 19.2).

Fig. 19.2 A load-deflection test is applied to a rear wing. (Courtesy of Sutton Motorsport Images)

Engine Capacity

Swept volume checks on engines are not performed on a regular basis and are usually conducted at the manufacturer's premises on an engine sealed at the circuit. However, the Technical Delegate is equipped with a cunning device that can be inserted through the spark-plug hole and which expands to measure the bore of the cylinder. The stroke is measured using a depth gauge, also inserted down through the spark-plug hole, while the engine is rotated slowly.

Fuel Specification

In 1995, the news that Elf fuel samples had been analyzed and found not to match prior-approved samples caused the company and Elf's teams using the fuel considerable embarrassment and surprise. The discovery was made because the FIA turned up in Brazil with a new mobile fuel testing laboratory and a fuel analyst. It is possible that this was not anticipated by everyone.

The concept of holding the potentially lengthy discussions, away from the circuits, on whether a particular fuel meets the specification laid down in Article19 of the Technical Regulations, and then only comparing the fuel used at a race with an approved sample, leads naturally to the type of tests that must be conducted at the circuit. To approve the fuel, a full analysis of its constituents must be made, and this requires the use of a mass spectrometer. Measurement of the mass of the individual molecular ions permits the precise determination of the structural formula of the organic compounds that make up the fuel sample. Additional tests of the type that motor manufacturers perform are applied, to ensure that the fuel used in Formula 1 is suitable for use in road cars. However, it would not be practical to repeat this process for all samples taken from cars during a Grand Prix weekend, nor is it necessary.

Instead, the sample is tested and approved away from the circuit. It is then "fingerprinted," and the "print" is put into a data base. At the race, three one-liter samples of fuel are taken, usually during qualifying, or immediately before or after the race, and then sealed. One sample is "fingerprinted" using the same equipment as used for the approved sample, and the results are compared with "prints" in the data base. If no match is found, the Stewards of the Meeting are informed, and the saga begins. The sample tested is put through the full mass spectrometer analysis in the United Kingdom, and one of the other samples is sent to an independent laboratory for analysis. The third sample is for the competitor, from whose car it was taken, to have analyzed. The results pass into the hands of lawyers, and the ensuing argument bears remarkable similarity to those heard in criminal courts over actual fingerprints and will be heard even more vehemently in the future, over human DNA matching.

Two tests are performed on each sample: (1) density measurement, and (2) gas chromatography. Density is measured to 0.01% in a density meter. Two milliliters of the fuel are injected into the machine and held at a constant 15°C (59°F) in a glass U-tube. This is vibrated by a piezo actuator, and the natural frequency is measured. The natural frequency is directly proportional to the mass of the U-tube, with the fuel sample in it. As the volume is known, the density can be calculated. Calibration is checked by injecting a reference sample of Dodecane, a density calibration fluid.

For fingerprinting the fuel, the FIA uses gas chromatography. The PC-controlled equipment runs the test automatically, records the data, and displays the gas chromatograph. Only one microliter of fuel is injected, vaporized, and mixed with the mobile phase or carrier, helium. The helium is filtered to remove all traces of oxygen and hydrocarbon that would otherwise modify the analysis. The mixture is split so that one-fiftieth is carried by the helium along a 30 m × 0.3 mm (98 ft × 0.01 in.) diameter glass capillary tube, internally coated with the stationary phase, substituted polysiloxane, to a thickness of 3 micrometers. The capillary tube starts at 30°C (86°F) and is raised to 280°C (536°F) over 60 minutes in a precisely controlled profile. The various compounds present in the sample emerge from the tube at time intervals dependent on their components and are detected by a flame ionization detector (FID). The hydrogen/air flame of the FID ionizes the hydrocarbons present, and a highly sensitive ion detector signals their presence. The timing of any given detection signifies the component, and the amplitude versus time (area under the curve of the plotted trace) signifies the quantity. Analytical software lists the characteristics of each detection event and provides a highly detailed "fingerprint" of the sample.

The whole process takes slightly longer than an hour. Thus, it is a full-time job for the FIA fuel analyst during and after the race, in order to complete the checks on samples from three cars and enable the Stewards of the Meeting to confirm the race results.

Control Systems, Especially Driver Aids

In the interest of an ordered life for all concerned, the procedures for checking that the control systems and software on the cars cannot perform any illegal functions have been formalized. Although the process does not require elaborate hardware, it does involve a great deal of hard work. The task has been entrusted to specialists in checking software integrity and requires a team of engineers armed with powerful notebook PCs.

Teams are offered two options for having their systems checked and approved. The first option is that the team can have the whole system inspected and the software checked, line by line, and a copy of the machine code kept by the FIA for comparison at any time during a Grand Prix. In the second option, the system is inspected prior to the season, but the software is not. However, an upload can be demanded, and it is then held by the FIA for a full check if deemed necessary. The deterrent effect of knowing that software may be uploaded from the car and read at any time is enormous. Because of the pace with which changes are made to software and the sensitivity to issuing copies of commercially valuable algorithms, most teams and their engine partners have opted for the second option. The FIA maintains a set of dual-key safe-deposit boxes, one for each team, to protect sensitive material.

Hardware inspections focus on three prime issues. First, does the wiring lead to and from where it should, particularly that connected to sensors and actuators? (Wheel speed sensors and the drive-by-wire throttle are closely scrutinized.) Second, it must be established which of the units (i.e., black boxes, dash display, or steering wheel) are passive and which are active (i.e., able to output signals in response to inputs). Finally, the architecture of the active units is analyzed to ensure that when an upload of code is performed on them, all the code is accessed. The whole task has been made more difficult with the introduction of local area networks, using CAN protocol, in which the various units are linked by one or more data buses, in the interest of reducing the number of wires snaking around the car. To establish what goes where, it is now also necessary to read the software.

Any piece of hardware that plugs into the car when it is stationary must be checked, as must any sign of radio frequency links from the pits.

System software is inspected to ensure that uploading, decompiling, and comparison software do not eliminate or modify the control software in any way. Especially important are checks to ensure that it is impossible to download an illegal program into RAM, which is overwritten and lost when the system is switched off, and then switched on again.

Control software for the engine and chassis systems takes a minimum of three man-days to interpret and verify. It is an eye-straining, brain-fading exercise that must be performed so that the world will eventually accept that winning teams are not cheating in Formula 1. All the control system checking must be done away from the circuit and involves experts from teams, engine suppliers, and electronic system suppliers, as well as those from the FIA. The main activities at the circuit are following up on new features and changes, and random software uploads for comparison with approved code or for later inspection.

Tires

To police the restricted use of tires as laid down in the Sporting Regulations, the Technical Delegate and his assistants must monitor and record the identity of every tire on every car as it leaves the pits during all the free and timed sessions, qualifying, and during the race. The cooperation of the tire companies is essential to make this work, and each tire is bar coded to speed the identification process.

Refueling Equipment

Much has been written about the missing filter in the Benetton refueling rig at Hockenheim in 1994, and whether it was the cause of the major conflagration that occurred in the pits, which led to the discovery of the missing part. The basis of the refueling regulations is that each team shall use the FIA-specified, Intertechnique-designed and built refueling equipment. The flow rate of the system is governed by the pressure difference between the pressurized reservoir in the rig and the tank of the car, and the flow losses in the valves, filters, hoses, and nozzle. It is set at a nominal 12 liters/second (3.17 gallons/second) by setting the reservoir pressure to a prescribed and regulated value. Any attempt to speed the fuel delivery time would involve either modifying or removing any flow restriction, such as the famous filter, or increasing the pressure.

Pressure is controlled by a high-flow pop-off valve and can be checked easily and at any time by attaching a gauge to an access point on the front panel of the rig. Scrutineering of the equipment consists of occasional inspections to ensure that no unauthorized modifications have been made, and spot checks on pressure are made during the race.

The work of the FIA Technical Delegate and his team of experts is, quite simply, to ensure that competitors and public alike believe that motor racing at its highest level *is* a sport.

Appendix A

The Development of Ground Effect

Ground effect is something for nothing.
—Colin Chapman

Ground effect has been around for a very long time. If one watches a swan taking off from a lake, one will see the swan applying ground-effect aerodynamics as it flies close to the surface of the water, gaining speed before climbing into the sky. Nature has provided a phenomenon that enables large, heavy birds to take off. Without it, they would be unable to leave the ground.

Ground effect can be defined broadly as the effect the ground imparts to the free-air aerodynamic characteristics of a body moving through the air. In the case of the swan and airplanes, ground effect increases the lift at a given incidence and reduces the induced drag. This increase in aerodynamic efficiency enables the swan to accelerate in ground effect to a speed where flight away from the ground is possible.

Race cars apply ground effect in the opposite way, by exploiting the fact that as a curved surface approaches the ground, the speeding up of the air flowing between the surface and the ground reduces the pressure and creates downforce—a sort of "attraction" between the surface and the ground. There was an awareness of this effect in the 1930s, but it was not until 1949 that Milne-Thompson calculated the theoretical effect of the ground on the aerodynamic characteristics of a sphere (Ref. A.1). Far away from the ground, the lift coefficient (C_L) is zero because it is a symmetrical body. Milne-Thompson showed that C_L became negative as the ground clearance reduced, reaching a $C_{Lmax} = -0.375$ when it touched.

More than ten years later, people in motor racing started to become interested in using aerodynamics to generate downforce, increasing the load on the tires without increasing the mass of the car. In 1962, Jim Hall experienced lift on his Chaparral 1 sports car. Chevrolet R&D took up the problem, and Bill Mitchell, the legendary chief stylist of General Motors in the 1950s and

1960s, suggested an inverted airfoil body shape. Mitchell knew little of aerodynamics (looking at the wonderful Cadillacs of the late 1950s, he obviously cared even less), and General Motors did not have its full-size wind tunnel at that time. At speed, the front wheels of the 1963 Chaparral 2 lifted off the ground until it was fitted with a bib spoiler.

Chaparral went on to develop downforce-generating wings with such effect that the whole motor racing world was alerted to the enormous increase in cornering, traction, and braking grip that downforce provided. Also around this time, the speed and size of sports cars led to a tendency for them to flip over backwards—a particular brow at Mosport gave one or two drivers some very exciting moments. The same problem was exercising the minds of designers of cars and boats intended for attempting world speed records.

John Stollery's involvement with Donald Campbell's Land Speed Record car Bluebird led to a key fundamental investigation of the effect of the ground on car-like shapes. The results of this research were reported in a paper titled "Forces on Bodies in the Presence of the Ground" by J.L. Stollery and W.K. Burns, published in 1969 (Ref. A.2). Using the Imperial College wind tunnel, with moving ground plane, they tested a streamlined body of typical car proportions. The tail could be cut back to blunt it (as per the cutoff tails of race cars), and various nose shapes were tested. The lift, drag, and pitching moment characteristics were fully calibrated through a range of incidences and ground clearances. The tests on the "whale," as it was known, show all the characteristics of a skirtless, ground-effect venturi car (Fig. A.1). Problems with separation of the underbody flow and moving centers of downforce are highlighted—the same problems that plague race car aerodynamicists today.

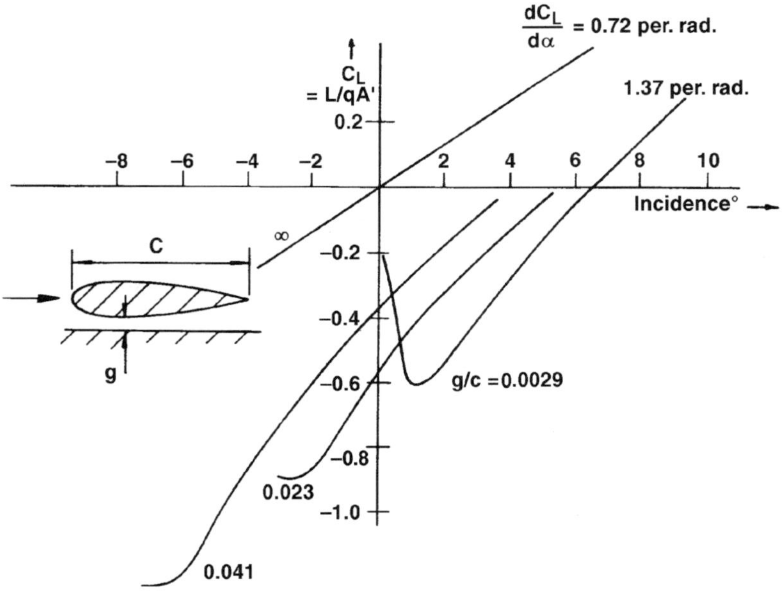

Fig. A.1 C_L *versus incidence for a streamlined body near the ground. (Ref. A.2)*

One component of a truly effective ground-effect car was missing—skirts. It would be another eight years before the whole package came together and appeared on the race track as the Lotus T78. The lack of skirts limited the maximum C_L of the "whale" to approximately -0.24 based on plan area. A typical race car of this shape would produce approximately 1.1 kN at 160 kph (100 mph), ignoring any degradation due to wheels, etc.—somewhat less than a good wing setup of the same era.

Almost simultaneously with Stollery's work and totally unknown to either of us, I had been working on some similar concepts for BRM in the same Imperial College wind tunnel. My boss, Tony Rudd, had been appalled at the explosive growth of wings in Formula 1 and their disastrous effect on the cars when they failed, epitomized by the twin failures on the Lotus T49's at Barcelona in 1969. He asked me whether it would be possible to generate downforce without the use of separately mounted wings. I did some quick calculations on my slide rule, which showed that if we could use the whole plan area of the car to generate downforce, including filling the permitted area between the wheels, only a $C_L = -0.35$ would be required to do the same job as a rear wing with $C_L = -1.0$. Tony Rudd agreed that a wind tunnel program should investigate the possibilities in greater depth.

Working with chassis designer Alec Osborne, I drew an inverted airfoil-shaped body, with a slender central fuselage for driver and engine. Internal radiators were fed by NACA surface intakes, while nose wings and a trailing edge flap permitted the overall downforce and its distribution to be adjusted. Wind tunnel tests at the Imperial College indicated that overall downforce would be similar to that of a winged car and that L/D would be significantly better. Cooling might be a problem because the NACA intakes in the top surface were very sensitive to details of the front suspension. A secret workshop was set up away from the main works to build cars for the 1969 Italian Grand Prix.

At the time, BRM was running the P126 V-12 cars (Fig. A.2), and one was fitted with stubby, airfoil-section side panniers, bolted to the sides of the monocoque to try and replicate the new car. Tests at Snetterton were inconclusive due to a lack of measurement equipment. Before any further work could be conducted, there occurred one of BRM's periodic "palace revolutions." Tony Rudd left BRM to join Lotus, and all work on ground effect halted.

I left BRM shortly afterward and joined Specialised Mouldings, pioneers of composites in motor racing. One of the first tasks I was given was to work with Robin Herd, chief designer of the new March company, in the design of bodywork for its Formula 1, 2, and 3 cars. The Formula 1 car, the March 701 (Fig. A.3), needed additional fuel capacity in certain races, and I suggested inverted airfoil-shaped side tanks. Robin Herd agreed, and they appeared in several races fitted to the cars. I doubt that they generated much downforce, but they housed the fuel at no drag penalty—not really ground effect.

Fig. A.2 P126 BRM fitted with wing section side pods. (Author's collection)

Fig. A.3 March 701. (Courtesy of Phipps/Sutton)

During the four years I spent at Specialised Mouldings, I designed and ran a one-quarter-scale wind tunnel. Most of the work was conducted on sports race cars for Lola, March, and GRD, among others. I carried out a limited amount of basic R&D into sports car aerodynamics, investigating the effect of undertray incidence and shape on the pressure distribution under the car. I also carried out a series of tests on OMC (makers of Johnson-Evinrude outboard motors) Formula 1 catamaran racing boats and some outboard motor Water Speed Record boats. These were all lifting, ground-effect craft that had a tendency to flip over backwards at speeds greater than 160 kph (100 mph). I now appreciate how important was the insight into the potential of ground effect, both lifting and downforce, that these various test programs provided.

In 1974, I rejoined Tony Rudd at Lotus to run the Lotus Group's composite R&D facility. Team Lotus was struggling to replace the phenomenally successful T72, which was made obsolete by understeer as Goodyear developed tires to suit the new generation of McLarens, Tyrells, and Ferraris. Toward the end of that year, Colin Chapman asked Tony Rudd to assemble a team, separate from the day-to-day racing, to rethink the Formula 1 car from first principles. Tony asked me to plan and conduct a wind tunnel program in support of this effort, working closely with chief designer Ralph Bellamy.

Ralph Bellamy had drawn the slenderest possible monocoque around the driver and the Ford-Cosworth DFV. We agreed that any additional bodywork or aerodynamic devices would be included only if they met a certain L/D in the wind tunnel. We knew there had to be some additional bodywork, because no provision existed yet for radiators or around half the required fuel. Because of the lack of front end grip in the T72, Chapman had set the target of making the new car, which was to become the T78, aerodynamically more pitch sensitive than the T72. He believed that if the downforce on the front increased under braking, it would improve turn in. We decided that all testing would be conducted in the Imperial College one-quarter-scale wind tunnel and that we would always use the moving ground plane. The first task was to make a model of the T72 and fully calibrate it in the wind tunnel, with special emphasis on measuring the pitch sensitivity.

Testing started in earnest in 1975, and by late summer the overall configuration of the T78 was defined—slim nose with oil cooler and front wings (the full-width nose had been rejected on the basis of L/D), airfoil side pods to house fuel and accommodate the water radiators in the leading edge, similar to those on the de Havilland Mosquito aircraft. Again, the side pods had been found to generate little downforce, but they packaged fuel and cooling radiators with no drag penalty. A large conventional rear wing completed the aerodynamic configuration.

Through the winter of 1975, we continued to refine the aerodynamics and tune the pitch sensitivity. By the end of a week of tunnel testing, the strong wooden model would have been so modified with card, modeling clay, and sticky tape (none of the CAD-designed, CFRP model bodywork used today) that it had usually lost most of its structural integrity. Toward the end of one of the weeks in the tunnel, I noticed that it was becoming almost impossible to obtain consistent balance readings. Something was wrong. Looking carefully at the model, it became clear that the side pods were sagging under load and that as the speed of the tunnel increased, they sagged even more. That indicated two things: (1) that the side pods had started to generate downforce, and (2) that it had something to do with the gap between their edges and the ground.

Thin wire supports restored the side pods to their correct position and stopped them from sagging—no downforce and consistent balance readings. Next, we taped card skirts to seal the gap between the edge of the side pods and the ground, leaving only approximately 1 mm (0.04 in.) gap. The total downforce on the car doubled for only a small increase in drag! Ralph Bellamy looked somewhat startled, and I do not think either of us believed the results, so we repeated the tests, with and without the skirts. Ground effect was real, powerful, and efficient, if the outer edges of the side pods were sealed to the ground.

I contacted the Lotus consultant aerodynamicist at the Imperial College, Frank Irving, to discuss with him the theory behind what we had found. The model was not pressure tapped, so we were unable really to investigate the effect of skirts and side pod shape. A new model would be required to do this. Ralph and I also had a lengthy discussion about skirts, their legality, and their practicality, before calling Tony Rudd to give him the news. According to the Technical Regulations, movable aerodynamic devices were not allowed. However, toward the end of 1974, Gordon Murray had fitted a transverse skirt underneath the Brabham BT44 to generate a low pressure behind it. A precedent had been set.

The T78 was built and shown to the press, fitted with nylon brush skirts similar to big draft excluders. Initial testing in autumn 1976 indicated that the brushes were quite incapable of sealing the sides. For winter testing at Kyalami, polypropylene skirts were tried, but they too sucked inward and became distorted. Andretti and the race engineers wanted to abandon messing about with all this trick new stuff that did not appear to work and to concentrate on developing the car. Colin Chapman would have none of that and insisted on continuing to work on the skirts until the potential of the car gradually became apparent. On returning from that test, Chapman instructed me to spend 100% of my time on skirt development.

The gradual improvement in skirt performance through the development life of the T78 is reflected in its race performance (Figs. A.4 through A.7). It was fortuitous that, at the beginning, its performance was nothing spectacular. Other designers, looking at the wing-shaped side pods, dismissed them as a gimmick. When the T78 started to go quickly, an elaborate deception concerning the differential put many people off the scent. It was not until the dominance of the T79 the following year, that it became obvious to all that ground effect did work.

To create a test rig for skirts, I commandeered one of Team Lotus' Renault 4 parts delivery vans and attached a skirt mounting rig onto the rear bumper. While it was driven around the Lotus test track at Hethel, I lay in the back, watching the skirts from the rear doors. We tested many designs and rounded off a test by driving back to Ketteringham Hall, along the bumpy

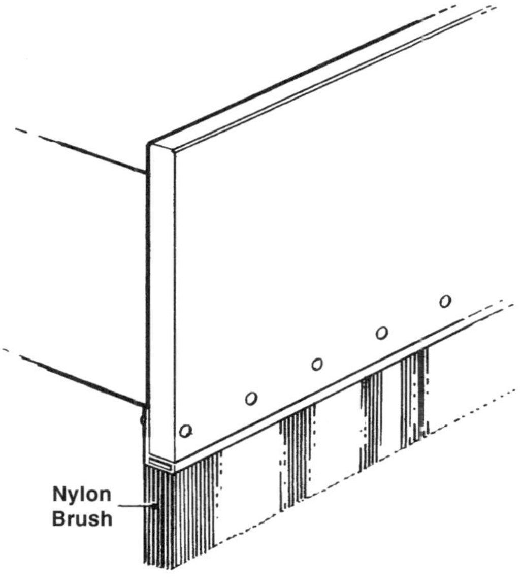

Fig. A.4 Lotus T78 brush skirt.

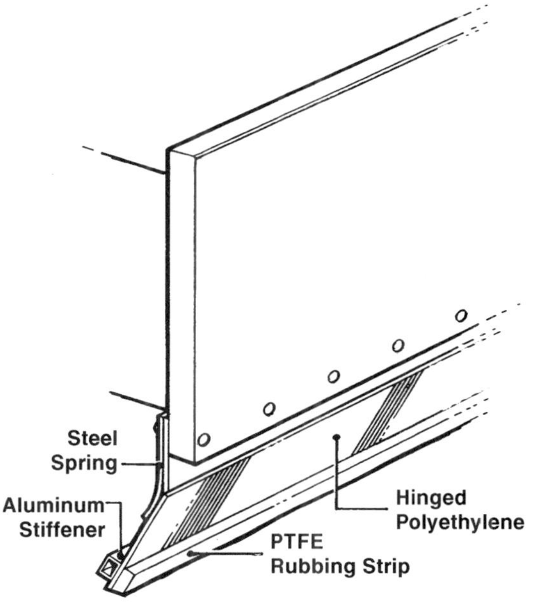

Fig. A.5 Lotus T78 "suck-up" skirt.

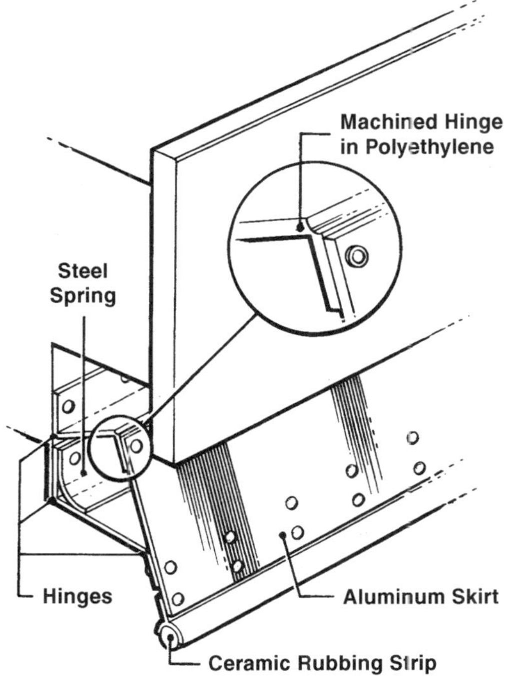

Fig. A.6 Lotus T78 "suck-down" skirt, with ceramic rubbing strip.

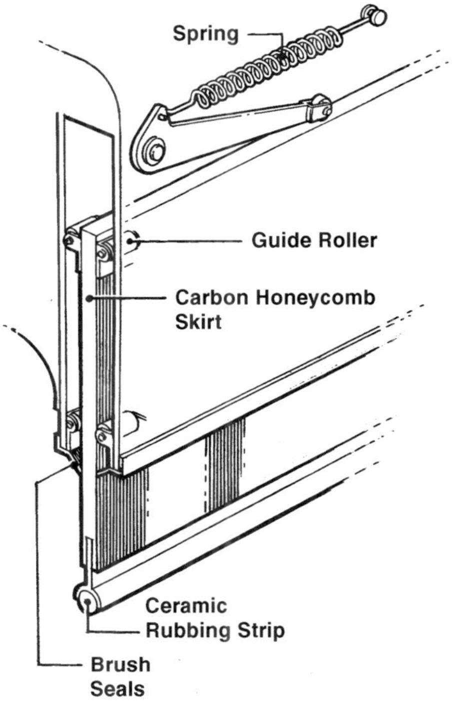

Fig. A.7 Lotus T79 "board-in-a-box" skirt.

and potholed Norfolk lanes. To prevent excessive wear, the skirts had to be sucked up by the low pressure beneath the car. Steel shim springs loaded them down, and there was a compromise between performance when new and how long they would last before they became so worn as to be ineffective. PTFE rubbing strips were fitted to their lower edges to limit wear and reduce friction.

One day, while further investigating and refining underbody shape at the Imperial College, I visited the ceramics expert in the College of Mining. He suggested various ultra-hard materials we could try to prevent wear. Tipping the skirts with ceramic transformed the wear rate and allowed the skirts to be spring loaded harder onto the road. So successful were these tests that I decided it might be possible to devise a skirt that was sucked down instead of sucked up (Fig. A.4). Designed similar to a twin, unequal link suspension, it took up little width and had a small area for the suction beneath the car on which to act, to minimize the wearing load. The Renault 4 tests appeared good, and a T78 was modified for Gunnar Neilson to try at Snetterton. After only a few laps, he was 1.5 seconds per lap faster and came into the pits with eyes bulging and a big grin on his face, saying, "I go quicker than I ever go before, and I look over the side of the car and there is a meter of track left!"

From that point onward, the T78 was the car to beat. However, only Harvey Postlethwaite built a ground-effect car—the Wolf WR5—to compete with the T79 the following year. Rumor had it that Ferrari tested a ground-effect model in its wind tunnel, but that the model did not work. Perhaps Ferrari did not use a moving ground.

Such was Colin Chapman's vision that, even before the T78 showed its true potential, he returned from an Easter holiday at his villa in Ibiza with sketches of the T79, a completely new car designed to optimize ground effect.

The T78 suffered from the downforce generated by the side pods being located too far forward, with the result that it was always necessary to run a lot of rear wing to balance it. As a result, the T78 had a reputation for poor top speed. The brief for the T79 was to move the center of pressure of the underneath of the car rearward and to further increase the pitch sensitivity. In tunnel tests, the cleaning up of the suspension and engine ancillaries that Chapman had achieved was immediately apparent in better downforce figures. I extended the throat of the venturi as far rearward as possible and worked on profiles that maintained attached flow with as sharp a recovery as I dared, bearing in mind that testing was only one-quarter-scale and at a low Reynolds number.

Postlethwaite's Wolf was fitted with sliding board skirts when it appeared at Monaco. We had not considered that they would be legal. However, once accepted, they were the only way to go (Fig. A.8). From its first appearance, the T79 was a sensation and totally dominated the 1978 World Championship, except for a short time when the Brabham BT46B fan-car appeared and won the Swedish Grand Prix before being banned.

Fig. A.8 The first two generations of ground-effect cars: the Lotus T79 heads the field, while the Lotus T78 flanks the Wolf WR2 (sixth and seventh cars). (Courtesy of Phipps/Sutton)

The story of fan cars is covered in Chapter 14 of this book, including the Lotus response to the threat by Brabham to the Lotus World Championship campaign. While Chapman concentrated on the politics of getting the Brabham fan banned, he tasked me with developing a better fan-car. A model showed that suction at the exit of the venturi caused sufficient airflow to generate significant downforce without the car moving through the air. All that was needed was to blend the back of the venturis into a pair of fan ducts, and a drive train from the gearbox to power them. Martin Ogilvie, Team Lotus's Chief Designer, had drawn it all within a week, and we found some ideal fans. The target for the car was in two races time, only four weeks later.

Meanwhile, Chapman took Polaroid photographs of the model, with its twin fans, to a FOCA meeting. He always claimed that he showed the photographs to Brabham's owner, Bernie Ecclestone, after which the constructors agreed to ban fans. I have always wondered whether the Lotus fan-car would have worked.

Although the T79 was not a good car, it possessed a special ingredient—ground effect. The monocoque structure was torsionally weak (too narrow), and neither the brakes (magnesium calipers) nor the exhaust system (inadequate cooling inside the bodywork) would last a race without deteriorating. Instead of pausing to refine the technology and sort out the new car systems that ground effect brought upon a race car, Chapman, in true Lotus spirit, demanded more—more downforce and more pitch sensitivity. While Williams built the FW07, the definitive ground-effect Formula 1 car, Lotus built the T80 and ran into big trouble (Fig. A.9).

Fig. A.9 Lotus T80. (Courtesy of Phipps/Sutton)

The T80 tried to dispense with conventional front and rear wings, and derive all its downforce from ground effect using the full length of the car. Although the downforce and L/D figures were excellent, the pitch sensitivity rocketed. The "if some was good, more was even better" approach—combined with sticking curved skirts running from just behind the front wheels, around the inside of the rear wheels, to the extreme rear of the car—caused that driver-frightening phenomenon known as porpoising. The car would hit a bump and bounce up and down, pitching at the same time, until the driver applied the brakes. I vividly recollect Andretti reporting the front wheels leaving the road at high speed at Silverstone. Ultra-stiff springs made the problem worse—the car bounced, under-damped on its tires.

While Team Lotus stepped back one pace and raced the T81 in 1980, I set out to investigate and attempt to solve the porpoising problem. The T80 proved that more downforce was possible; however, unless that downforce could be controlled, it was unusable. Since 1977, I had been working with David Williams, head of the Flight Instrumentation Department at the Cranfield College of Aeronautics. He had designed and built a digital data system for use on the T78, when it became apparent that it would be absolutely essential to gather data from the chassis to progress with the development of ground effect. When the T80 porpoising started, I discussed the phenomenon with David Williams, and he offered to model it and validate the results with the data we had. He established that it was an aero-elasticity problem, akin to flutter in an aircraft wing. The changing aerodynamic loads as the car bounced and pitched excited the pitch and heave modes of the sprung mass on its springs and tires.

One day, David said to me, "What you need is an active suspension," and he took me into the laboratory to demonstrate an artificial-feel control column he was developing for a fly-by-wire research aircraft. He demonstrated how the control computer could provide any spring rate, mass, and damping for the system, and that the technology could provide a suspension system with unique properties. We already had a Plan A—twin chassis—and decided that active suspension would be Plan B.

By the time we had struggled with the T80 for a year, we realized that while the body and associated aerodynamic forces required a very stiff suspension, if the remainder of the mass of the car was attached to that suspension, it would porpoise and the driver would have an unacceptably rough ride. The main masses and the driver wanted a soft suspension, so why not build a car with two chassis and two suspensions? At the end of 1980, we built the T86 test mule to try out the idea. No sooner had testing started, in the greatest secrecy, than skirts were banned along with the imposition of a 6 cm (2.4 in.) minimum ground clearance to be measured in the pits. In the twin chassis concept, Lotus had the perfect answer to the new rules! Chapman realized that coming out with such a car would not be popular because it drove a coach and horses through the new regulations, but knowing its potential made it almost impossible for him not to use it.

Security measures to protect the project were stepped up, but such were the rumors concerning the car that Ferrari persuaded FISA (the then sporting authority of the FIA) to add a new regulation banning cars with no suspension. At Lotus, we were not worried. We were building a car with two suspensions!

The T88 twin chassis car never raced, although it attended the Long Beach, Brazilian, Argentinean, and British Grand Prix. Chapman fought tooth and nail, not only for the car but also for the fundamental right to innovate. When he finally conceded defeat, he lost much of his interest in Formula 1, putting his efforts into developing an ultralight aircraft business for the last two years of his life. The potential of the twin chassis solution for controlling ground effect was never really tested. Every time it appeared on a racetrack, it was black flagged.

Meanwhile, Brabham pioneered a hydro-pneumatic, self-lowering suspension that allowed its car to run close to the ground and yet pass the 6 cm (2.4 in.) rule in the pits. By the start of the European season, most of the cars had driver-controlled, hydraulic, or pneumatic raising and lowering systems and flexible skirts (Fig. A.10). Full ground effect had returned. It all made the T88 appear straightforward and completely legal.

Proceed to Plan B. Active suspension is a whole story in its own right, covered in Appendix C. However, the reason for using it in motor racing is to control the attitude of the car relative to the ground, without having to use very stiff springs. Computer-controlled hydraulic, synthetic spring/dampers are used, which are insensitive to loads applied from above (aerodynamic and inertia loads—i.e., cornering, braking, and accelerating) and yet respond conventionally to loads from below (due to road irregularities and bumps). Lotus, Ligier, and Williams all took different approaches to applying this technology to ground-effect cars, and Lotus appeared at

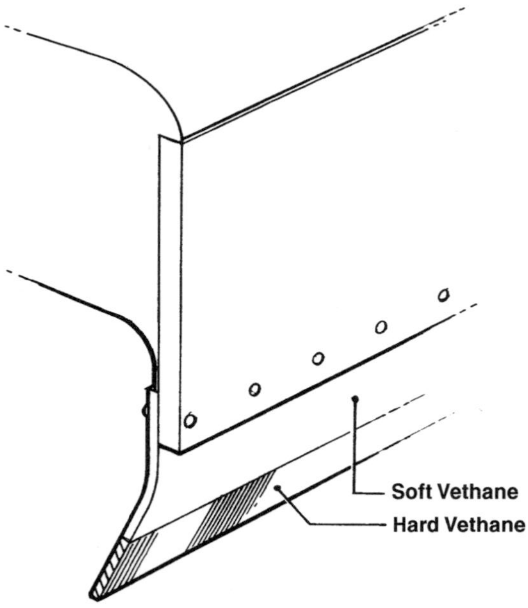

Fig. A.10 Vethane skirt.

the first races in 1983 with a fully active suspension T92. Two events led to this technology being shelved by Team Lotus until it reappeared in 1987. First, the day the active T93 first ran at Snetterton, Colin Chapman died. He never saw it run. The second event was that, only three months before the start of the season, FISA finally banned skirts and shaped undertrays. The era of flat bottoms had arrived.

A flat plate, at incidence to the airflow, will generate lift or downforce. Nose down, near the ground, it will be affected by ground effect: the downforce will be increased, L/D will be improved, and it will become very critical to the flow separating from the flat plate. Since 1983, Formula 1 aerodynamicists had been struggling to gain the most from this mechanism, with more or less success. In 1987, active suspension technology and the increasing need to control the attitude of the flat bottom came together on the Lotus and Williams cars. Williams continued to develop the technology, and the remainder of the teams were forced to follow suit as its superiority became apparent.

In 1993, the FIA banned all active suspensions as part of a package of measures designed to slow the cars, reduce costs, and reinstate the role of the driver. Steady development had increased downforce to levels greater than had ever been achieved in the days of venturis and skirts, with the power needed to push the high-drag cars through the air. In the aftermath of Ayrton Senna's accident at Imola in 1994, the FIA brought emergency measures to drastically reduce performance. Initially, the diffuser was shortened, the car was raised by attaching a wooden "plank" to the underside of the monocoque, and finally the sides of the car were raised an additional 50 mm (2 in.). The effect of these measures was to reduce downforce by approximately 40%. However, the relentless march of development has increased it steadily and will continue to do so, barring further major measures against it.

Ground effect has produced a revolution in race car performance. It has raised aerodynamics to the same level as power, tires, and weight in the hierarchy of parameters that influence performance. Cornering and braking levels have risen from slightly more than 1g to more than 5g and have affected the development of all aspects of the car—from its structure to its brakes. With the increase in cornering speeds and the shortening of braking distances, a whole new breed of drivers has developed, stronger and fitter, to sustain the loads imposed on them. More controversial has been the effect on safety and the necessary changes in circuit layout to contain an out-of-control car. The FIA now works closely with the teams' technical personnel to bring in regulation changes that keep in step with technical development, maintaining a constant performance at minimum cost to car and circuit owners.

Perhaps the biggest challenge that ground effect has wrought is to develop a racing environment in which cars can overtake each other. Ground effect is a natural phenomenon. Having found a way to harness its power to make a car go faster, it now is impossible to undiscover it.

Appendix B

Carbon-Fiber Reinforced Composites

Life exists in the universe only because the carbon atom possesses certain exceptional properties.
—James Jeans

Anyone who has held a piece of wood in his hands and has marveled at the way nature has grown fibers embedded in a matrix of lignum to give a supple, strong, and enduring material understands composites. Skilled craftsmen have used wood to build airplanes, ships, houses, tools, fishing rods, and even cars, having learned to select the wood according to its properties from a vast naturally grown range. These craftsmen align and join the parts for the best use of these properties, to give strength and stiffness where needed. Less than 50 years ago, man made a type of material that approached the natural properties of wood—composites.

The word "composites" means "made up of two or more materials" and covers a vast range of materials. In motorsport, it has come to mean materials composed of man-made fibers embedded in synthetic resin. It includes more complex composites that encompass core materials faced with load-bearing skins, making up a sandwich construction.

The advantages that composites have over homogeneous materials such as metals and plastics are twofold. First, certain materials (e.g., glass, Kevlar®, carbon, and boron) have extremely good stiffness and strength-to-weight ratios when formed into long fibers. Second, if these fibers are aligned with the direction in which the main stress is applied, they can be used most efficiently. This is exactly what a tree accomplishes, growing its fibers along the roots, flowing into and up the trunk, and then dividing again to flow along the branches. Fibers alone are most useful structurally if they are straight; however, structures are not composed of only straight lines, and stresses "flow" around corners. By including fibers in a matrix of a suitable material, the fibers are stabilized into the desired shape and can transfer loads among individual fibers. In this way, fibers can be laid-up into complex shapes with curved stress paths and can withstand compression loads without buckling.

The first composites were composed of glass fibers (manufactured by drawing fine fibers from a pool of molten glass) embedded in polyester resin. When the resin is in liquid form, the fibers (in the form of unidirectional bundles or sheets, woven cloth, or randomly chopped mat) can be draped over complex curved surfaces of a mold and wetted out with the resin. On curing of the resin and release from the mold, the laminate becomes a strong and flexible component. Glass has limitations in both strength and stiffness and is not light enough for many applications. The search began for higher-performance fibers.

The three fibers that emerged from R&D laboratories and have found extensive applications are carbon fiber, Du Pont Kevlar®, and boron. Of these, carbon fibers were the structural choice for race cars, although Kevlar® has specialized applications because of its extreme toughness, good tensile properties, and ability to resist tearing. Carbon fibers were first made at the Royal Aircraft Establishment at Farnborough, by a process of cooking rayon fibers in a carefully controlled, oxygen-free atmosphere—a process quite similar to making charcoal from wood. Because of the ability of carbon to form long-chain molecules, carbon is extremely suitable for making high-modulus strong fibers. These properties led to the extensive use of carbon fiber for race car applications.

First applications, and still by far the largest use of carbon fibers today, were in the leisure industry, for items such as fishing rods, tennis rackets, golf clubs, and skis. At the same time, the aerospace industry began investigating the use of carbon fibers for nonstructural components. Carbon fibers are now used widely in aerospace, the second largest user, for all main structural components in which temperature is not a limitation. The third major industry to take up carbon fibers was motorsport.

Specialised Mouldings was the first company to use carbon fibers to stiffen and reinforce the glass fiber body panels it manufactured for race car manufacturers such as Lola, McLaren, and Chevron. The big Can-Am cars of the early 1970s, with their large flat body panels, were ready applications. Initially, carbon fiber was available only as individual tows—3 mm (0.12 in.) bundles of fibers—and these were laid-up as a 50 mm (2.0 in.) square grid on the back of the glass fiber panels. As soon as the textile companies such as Courtaulds began to weave carbon fiber, the material was used to make whole panels, replacing glass fiber in the lay-up of composite body panels.

Before carbon fiber was used in high-load-bearing parts of the structure of race cars, two other technologies were needed: (1) honeycomb-core sandwich construction, and (2) pre-impregnated fibers. In the mid-1970s, aluminum-skinned, aluminum honeycomb-cored sandwich panels were starting to be used for the construction of monocoques in Formula 1. Thin, single aluminum skins were able to carry the loads involved; however, they tended to oilcan or even buckle when loaded in compression. Splitting the skins to form a sandwich produced a stiff, stable panel that had the added advantage of being better at energy absorption in an accident. When the techniques were developed for joining panels and feeding in concentrated loads, sandwich construction became widely accepted and prepared the way for replacing the aluminum skins with lighter carbon-fiber ones.

Hand lay-up, the process of wetting out the fibers with liquid resin, results in the maximum weight of fiber being approximately 30–35% of the total. To optimize for stiffness and strength, the fiber proportion of the laminate weight should be as high as possible, as long as there is enough resin to wet the fibers and bond them together. The process of pre-impregnating the fibers with resin was developed, in a machine-controlled process, and can achieve fiber weight of 65% or more. The resin is semi-solid when cold, and it flows and then cures only when heated. When the process for laying-up with "pre-preg" materials was developed, carbon fiber was ready to take over much of the primary structure of a race car.

The first carbon-fiber monocoques from McLaren and Lotus appeared within a week of each other in early 1981 (Fig. B.1). John Barnard's McLaren MP 4/1 used a monocoque molded from pre-impregnated carbon-fiber reinforced plastic (CFRP)/aluminum honeycomb by Hercules Corporation in the United States. Hercules is a leading aerospace composites manufacturer and used the most advanced techniques available to mold the monocoques around an internal aluminum mandrel, and then cured it under pressure in an autoclave.

Fig. B.1 Keith Duckworth (right) of Cosworth inspects the first composite McLaren monocoque. (Courtesy of Phipps/Sutton)

The in-house approach of Lotus was very different for the T88 monocoque. A large sheet of carbon/Nomex® honeycomb sandwich was hand laid-up on a flat table (actually a sheet of plate glass). When cured, the inner skin was grooved and the sheet was cut out and folded into the monocoque around a jig. The whole approach was similar to making a card model—"score here; cut here; fold along A–A; glue B to C." When the joints were reinforced and machined aluminum bulkheads were bolted in, it formed a stiff and lightweight structure (Figs. B.2 through B.4).

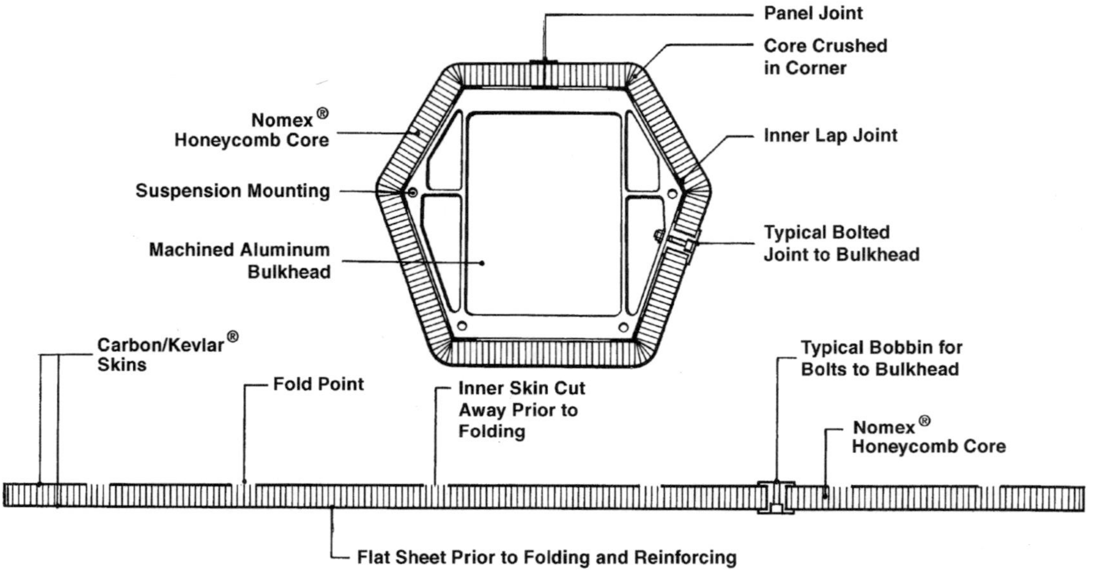

Fig. B.2 Construction technique used for the first Lotus composite monocoque.

The skins of the Lotus were not solely carbon because of concern over the crash performance of carbon fiber. Just prior to this development time, the Rolls-Royce aero-engine company had been forced to restructure financially, due to problems with the bird-strike resistance of the carbon-fiber fan blades of its new RB211 high-bypass engine. As a result, carbon fiber had gained a poor reputation for sustaining impacts.

A simple impact test rig was built at Lotus to evaluate various lay-ups and compare them with conventional aluminum construction. A 100 kg (220 lb) concrete block was hoisted to the top of a 10 m (32.8 ft) pole and allowed to fall onto the end of a 0.5 m (1.64 ft) square tube of the sample construction. The results were compared, and a carbon-fiber/Kevlar® mixed woven cloth with a Nomex®-honeycomb core was found to perform as well as aluminum/aluminum-honeycomb in the impact test. This is why the monocoques of the T88 (also used on the Lotus T87) have a black (carbon) and yellow (Kevlar®) woven pattern. This construction was used by Lotus until 1984. Of the seventeen monocoques made over four years, none was destroyed in an accident.

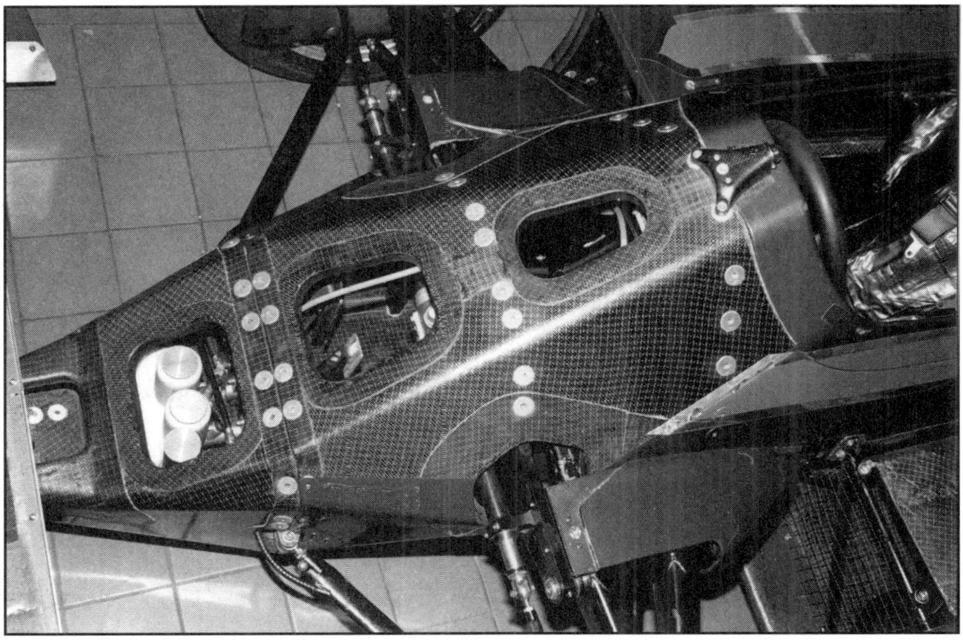

Fig. B.3 Lotus T87 monocoque.

However, the techniques used on the McLaren showed the future. Today, all race car composites are made from pre-pregs, autoclave-cured in molds, or on internal mandrels. The advantages of being able to mold the outer surface of the monocoque were soon exploited by being able to delete bodywork. The specification and performance of the fibers and resin systems, and the extent to which CFRP continues to be applied to more components on cars are the greatest indicators of the progress made in composites during the past 20 years.

The process from precursor (the base fiber from which carbon fibers are made) to molded component passes through a number of stages, each usually handled by a different company (Fig. B.5).

Carbon Fiber Manufacture

Most types of carbon fibers used in Formula 1, and indeed other classes of motor racing, are made by Toray in Japan. The precursor used is polyacrylonitrile (PAN), and a wide range of fibers with differing properties are produced (Fig. B.6), trading strength against stiffness. The most commonly used fibers are:

Fig. B.4 The inside of the Lotus T87 monocoque, showing mini-bulkheads fitted.

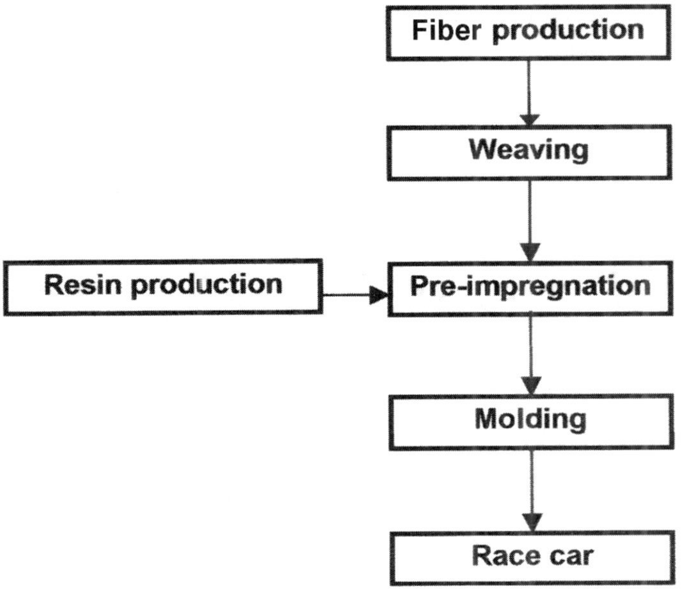

Fig. B.5 CFRP composite component manufacturing stages.

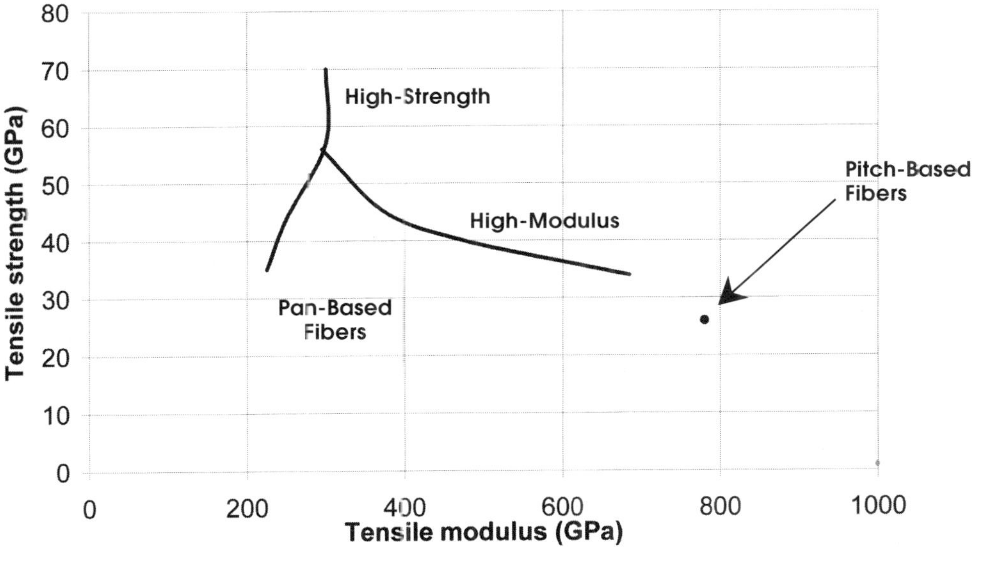

Fig. B.6 Properties of carbon fibers. (Author's illustration; data courtesy of Toray Ltd.)

- T800—High-strength applications
- M46J—Mixed properties applications
- M55J—High-stiffness applications

Even stiffer, pitch-based fibers are becoming available, and no doubt the compromise between stiffness and strength will improve steadily.

Some designers prefer to use even stiffer fibers in certain applications. However, the impact strength falls, and the cost rises sharply. At one time, some Formula 1 teams (notably Benetton and Ferrari) made monocoques from high-modulus fibers, in the search for torsional stiffness and low weight. Problems with passing the FIA crash tests and the scrapping of monocoques after minor impacts, even with curbs, has resulted in the tougher fibers being used more widely.

Textile Weavers

Individual carbon fibers are made up into tows, similar to what is done with any textile fiber. To enable the carbon to be pre-impregnated and laid-up, it is then processed into cloth or mat. The straighter the fibers, the better they are able to carry tensile loads. However, weaving tends to put crimps into the fibers, detracting from optimum strength and stiffness. The number of tows in any given direction (i.e., warp or weft) determines the number and extent of crimping and hence the properties of the resultant laminate in that direction. This ability to dictate the properties in a given direction is one of the greatest assets of CFRP.

A number of different weaves have been developed, with different structural properties and capabilities. In increasing order of strength/stiffness, the weaves most commonly used in race car components are:

- Plain weave 1×1
- Twill 2×2
- Satin unidirectional 4×1 or 5×1
- Unidirectional

For ultimate strength and stiffness, unidirectional fiber is used. Very high performance is possible in the direction of the fibers, but there is little strength in other directions. A typical application is in the spar-caps of wings.

Resins and Pre-Impregnation

Resin systems used in motor racing are almost exclusively epoxies. Since the early days, the temperature characteristics have changed the most. Initially, 100°C (212°F) was roughly the maximum usable temperature, and the component had to be cured in the mold to that temperature to achieve this heat-deflection temperature. Now, toughened epoxies can be cured at low temperatures and can achieve up to 400°C (752°F) by being heated to that temperature in use. Typically, monocoques use a resin system that has good properties up to 150°C (302°F) to cater for the temperatures at the rear of the structure where it is joined to the engine. Cytec 950-1 is one of the more popular monocoque resins. The high-temperature resins are used around the exhaust system and for what is one of the more demanding applications—the gearbox (e.g., Ferrari, TWR, and Stewart).

Pre-impregnation of the cloth with the chosen resin is a highly controlled process, ensuring a precise resin-to-fiber ratio. The resin is stable for several months at low temperatures (pre-pregs are stored in a deep freeze) but cures slowly at room temperature. Elevated temperatures are used for actual curing.

In racing applications, the other common material used as a matrix for carbon fibers is carbon itself (i.e., carbon-carbon), as used in brake discs and clutch plates. This is made by cooking either a laminate of carbon fiber and pitch, or a hydrocarbon such as epoxy, over a long period in a carbon-rich atmosphere at 1500°C (2732°F). Alternative methods have been developed involving vapor deposition of carbon on the fibers. Current processes are slow, expensive, and critical, and much research is being applied to find alternative, quicker, and cheaper processes. Carbon-carbon maintains its strength in excess of 2000°C (3632°F).

Molding

Tooling for CFRP components must be strong enough to withstand the temperature and pressure of the curing cycle, have similar thermal expansion to CFRP, and be easy to shape and finish to a high gloss. Materials include:

- Aluminum—Easy to machine and strong, but offers greater that CFRP thermal expansion
- Graphite—Easy to machine but structurally poor, with similar thermal expansion
- CFRP—Used for molded tools, strong with the same thermal expansion

Laminate patterns are either hand cut to templates drawn on CAD, or they are cut on CNC pattern-cutting tables, directly from CAD. The whole process is almost identical to tailoring, except for the stickiness of the cloth.

When the component is laid-up in the tool, with each layer of cloth or unidirectional fibers aligned according to the specification laid down by the designer, aluminum inserts are incorporated, and core materials are placed where required. It is then consolidated and cured. A monocoque may comprise dozens of individually tailored pieces of various cloths and unidirectional fibers. Cores are typically aluminum honeycomb (good shear strength and impact absorption) or Nomex® honeycomb (good drapeability but used mainly for nonstructural components). Depending on the resin system, a glue film may be placed between the honeycomb and the skins to ensure adequate bonding.

Prior to curing, the whole lay-up is covered with a release film (perforated with a number and size of holes according to how much resin bleed is required), followed by a breather/bleed mat to spread the vacuum and absorb the resin bleed, and finally the vacuum bag which is sealed to the mold. If no resin bleed is required, a gas-permeable PTFE release film is used.

Vacuum is applied to consolidate the laminate prior to curing in the autoclave. An autoclave is in effect a large oven that is able to withstand pressures in excess of 8 bar. Pressures of 7 bar for solid laminates and approximately 3 bar for sandwich panels (to prevent collapse of the core) are applied to further consolidate the laminate and promote resin flow. The temperature is raised progressively at approximately 2°C (36°F) per minute, to approximately 80°C (176°F), at which point the resin becomes liquid and flows, ensuring good fiber wet-out and interlayer adhesion. It is held at 80°C (176°F) for 20–30 minutes, and then ramped up to typically around 135°C (275°F) and held for 1.5 hours for full curing. On cooling, the laminate can be released from the mold, trimmed, machined, and finished.

Applications in Formula 1

The first application for both glass-fiber and carbon-fiber laminates was to replace aluminum fabrications, particularly bodywork. The vibration levels on race cars are such that all fabrications, whether aluminum or steel, tend to fatigue and crack after a time. The almost infinite fatigue life of composites makes them an ideal replacement when combined with greater panel stiffness and lower weight. Now it is more a question of which parts are not made of CFRP. Some limitations have been imposed by the FIA, encouraged by the teams on the basis of cost. These include brake calipers, wheels, and engine components—pistons, cylinder heads and blocks, cranks, and camshafts. Apart from these, those parts subjected to concentrated loads, high rubbing speeds, and/or high temperatures (e.g., gears, shafts, uprights, exhausts, and bearings) are still metallic.

Recent applications include suspension members, including their flexure pivot attachments to the monocoque, and the gearbox, probably the most challenging application yet. The gear case must accurately locate the gear shafts and other assemblies, while subjected to high loads and temperatures up to 150°C (302°F). Ferrari required several months to resolve the problems with its combined titanium and CFRP gearbox. Both TWR and Stewart suffered many problems with their new, all-carbon-fiber gearboxes, and Stewart (now Jaguar) has abandoned its carbon-fiber gearbox.

The biggest influence on the future application of CFRP in Formula 1 is the continuously improving FEA of composite structures. Conventional FEA of metallic structures uses uniform properties for the material. With composites, it is possible to vary the directional properties of the material, vary the distribution of properties throughout the structure, and even vary the properties within the thickness of a skin. This makes for a much more difficult computational analysis and many more options for the design of the component. Computer codes are becoming highly developed, and designers are learning to use and apply these powerful tools to explore more innovative uses of composites and to use their unique properties to realize more ambitious designs.

Predicting the impact performance of a composites structure is even more complex and an area that the creators of FEA software are now tackling. Metals absorb energy by taking up a permanent deformation, whereas composites do so by breaking. As Rolls-Royce discovered, the way in which CFRP fails during an impact is extremely complex, depending both on the structural shape as well as the fiber and resin details.

The process from design to composite part is long and involves much manual work. CAD/CAM has already speeded the manufacture of patterns and molds, using five-axis NC machines, rapid prototyping of patterns, and CNC fabric cutting. Machine lay-up of aircraft parts is now common, but the intricacies and low-volume production of race car parts demand manual techniques (Fig. B.7).

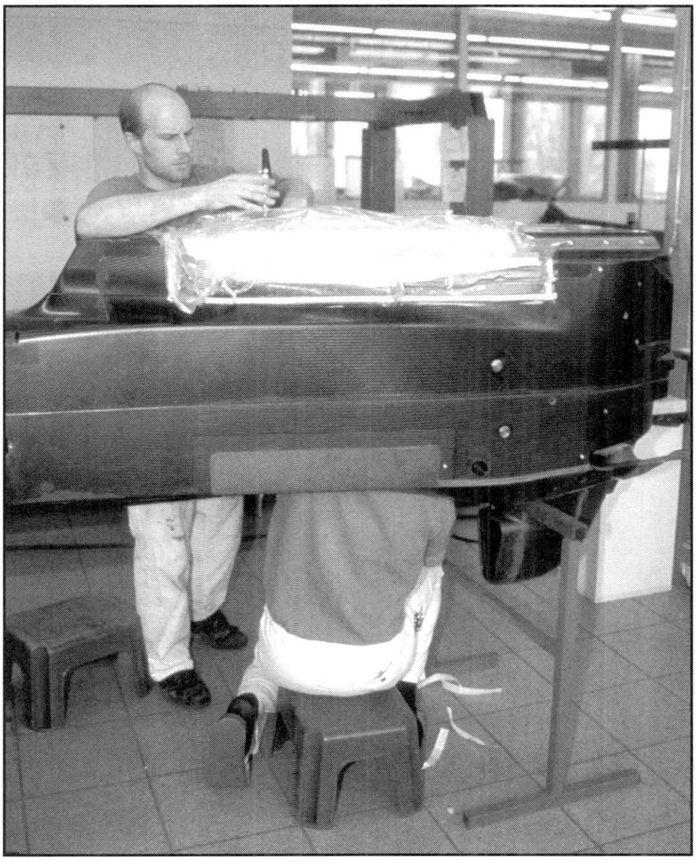

Fig. B.7 A CFRP monocoque being manufactured. (Courtesy of Sutton Motorsport Images)

Perhaps one of the most interesting carbon fiber applications will be in the use of carbon-carbon for exhaust systems. Rumors suggest that a fiber manufacturer is developing such a system for one of the Formula 1 teams. The highly complex shape is manufactured by casting the exhaust system in a high-temperature salt and then wrapping the pattern in carbon fibers, impregnated with pitch. Cooked at 1500°C (2732°F), the salt is melted out to leave a lightweight tubular exhaust system. Reputed to be showing similar durability to Inconel exhaust systems on the test bed, it cannot be long before they appear on the track.

There seems to be no limit to how these wonderful black fibers, made from one of the most abundant materials on earth, can be used. Composites have enabled parts to be shaped according to ideal structural design, aerodynamics, or aesthetics, rather than the constraints of the processes involved in manufacturing them as with sheet metal fabrications, castings, or machined parts. Structures are becoming more "organic," similar to a tree made from nature's composite. CNC manufacture of patterns and tools, combined with the stability and stiffness of cured CFRP, has provided a precision that enables fits between parts such as bodywork to equal or exceed those achieved on production cars. It sometimes appears that Formula 1 cars are hewn from granite. It is impossible to imagine race cars today without carbon fibers.

Appendix C

Active Suspension

The only difference between men and boys
is the cost of their toys!
—Anonymous and Colin Chapman

The characteristics of a ground-effect race car are such that the development of active suspension was inevitable. The effect on the performance of the car also made it almost inevitable that active suspension would be banned.

Appendix A describes the way in which the magnitude and distribution of pressure on the underside of a car in ground effect varies with the attitude of the car relative to the ground. As the ride height of the car diminishes, the pressure initially becomes lower, generating increased downforce. At some point, however, the flow separates and downforce is reduced. As the car pitches, the geometry of the venturi-shaped flow path under the car changes, and the pressure and hence force distributions change. Nose-down pitch moves the center of pressure forward, putting more of the downforce on the front axle, and vice versa. The magnitudes of the pressure and forces also increase as the square of the airspeed, and the large resulting forces generated at high speeds lower the car on its springs, thereby further increasing the downforce due to the reduced ride height. To control the magnitude and distribution of forces as the car pitches and bounces due to the inertia forces, during acceleration and braking, road undulations, and speed-squared effects, a suspension medium is needed that does not have the fixed load-extension characteristics of a spring. To suspend the car with a system that can raise it against an increasing load requires the addition of energy to the system—hence, active suspension.

The specification for an active suspension system will depend on the functions it is expected to perform. The vertical loads on a race car suspension have five main components, and each operates in a particular range of frequencies:

Component of load	Frequency
Weight change (fuel)	Very low
Aerodynamic downforce	<1 Hz
Inertial loads	2–3 Hz
Sprung mass oscillations	3–7 Hz
Unsprung mass oscillations	12–15 Hz

The choice of which of these components are to be controlled determines the specification of the system: full-active or semi-active (Figs. C.1 through C.3).

A full-active system employs servo-actuators to serve all the functions of the springs, dampers, and anti-roll bars. A spring may be mounted in parallel with the actuator to generate a load offset, and to reduce the force capability and hence the size and power consumption of the actuator. In this case, the spring characteristics have no significant effect on the system characteristics.

If the servo-actuator employs electro-hydraulics, as described in Chapter 11 on control systems, the bandwidth can be in the range 20–30 Hz and thus is capable of meeting the requirements of all the previously mentioned components of load. By measuring the load on the actuator and its displacement, and controlling displacement proportional to load, a spring can be synthesized in the software of the control computer. The spring can have any desired rate from infinitely stiff (if the actuator has sufficient authority for the maximum load to be carried) to zero rate, and even a negative rate. The latter is a dangerous beast, because any increase in load causes the actuator to extend, while the inertia force generated by the accelerating actuator increases the load and causes it to extend faster—maximum velocity is soon reached! Differentiating displacement gives velocity, and thus the force-velocity characteristics of a hydraulic damper can be synthesized by also controlling velocity.

A full-active suspension can simulate a spring-damper, with any rate, damping, free play, and stiction characteristics, all of which can be varied continuously. Used as the suspension medium of a vehicle, it is also possible (by measuring the vertical accelerations of the sprung and unsprung masses) to alter the apparent masses in the system and to provide inertial damping for either mass. In this way, the sprung and unsprung mass can be assigned a frequency and damping, the latter combining velocity and inertial damping. It is possible to treat the suspension as four individual wheels or to treat it modally, setting the frequency and damping for each of the four modes: heave, pitch, roll, and warp (twist). Further control laws deal with the other components of load in the following ways.

The effect of the change in the weight of the car as the fuel load is used is trimmed using a very slow self-leveling loop on the average static ride height.

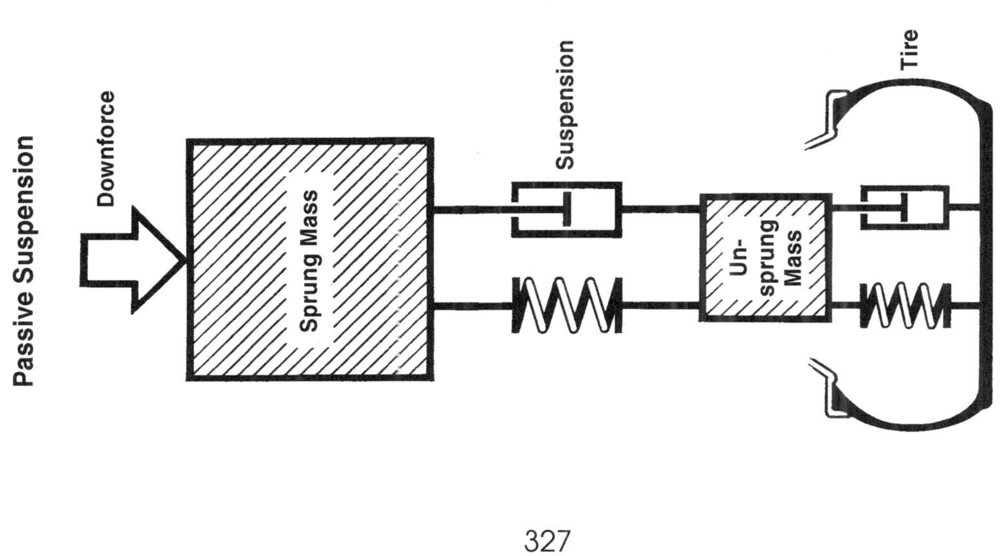

Fig. C.1 Passive suspension schematic.

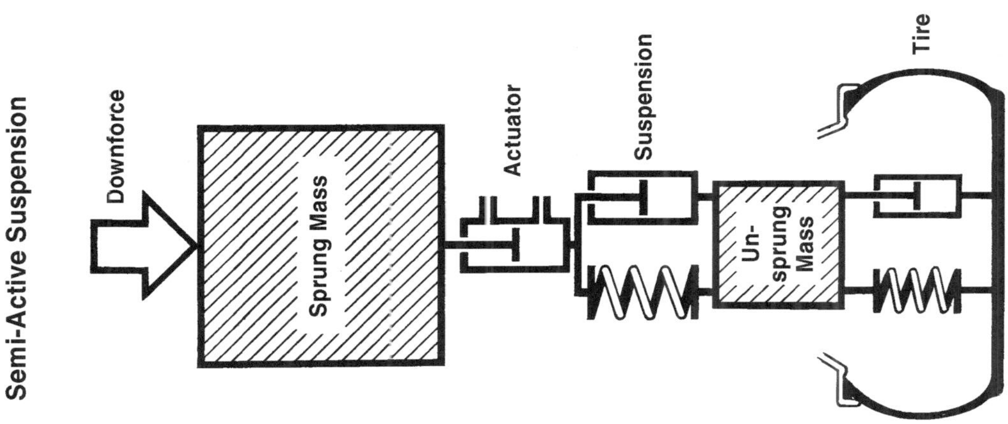

Fig. C.2 Semi-active suspension schematic.

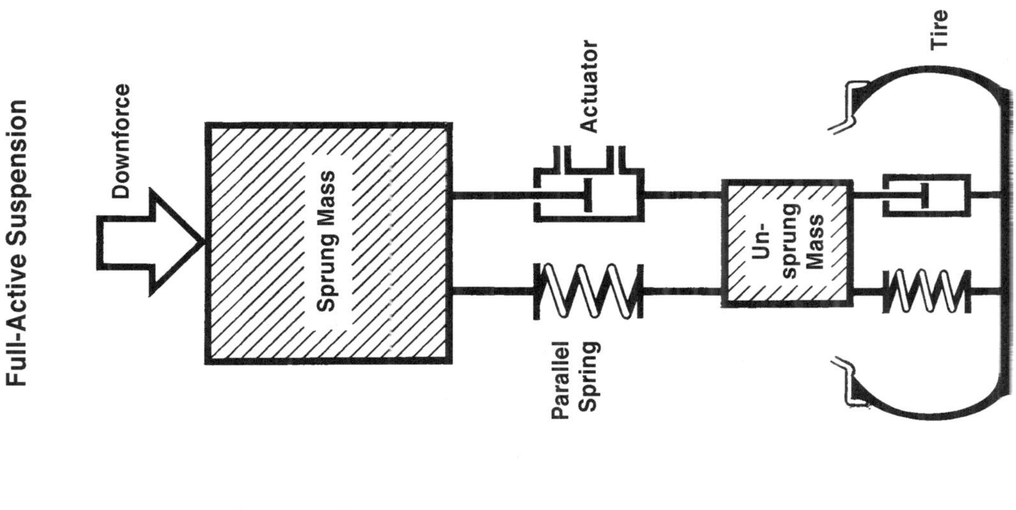

Fig. C.3 Full-active suspension schematic.

Aerodynamic loads are estimated from aero-mapping in the wind tunnel or on the track, and the total pressure, measured with a pitot tube, used to calculate the individual wheel load components due to downforce. Each total wheel load is offset by this value, such that the suspension does not respond to changes in downforce.

The components of wheel load due to the inertial loads caused by acceleration, braking, and cornering are derived from longitudinal and lateral accelerometers. The calculated wheel loads are used as offsets of the total load, such that the suspension does not respond to them and the car does not pitch or roll. One function of anti-roll bars is to set the lateral load transfer distribution, and this is accomplished by factoring the load offsets appropriately, to simulate greater roll stiffness at one axle than the roll moment distribution would dictate.

Of course, no response to aerodynamic loads, no pitch, and no roll may be undesirable responses, in which case each of these can be varied according to any parameters that are measured or derived. Car speed can be used to set the attitude of the car in any corner and to configure it for minimum drag on the straights. Individual corners can be identified and the car set up appropriately. Driver controls can be used to vary the gain of any control loop or to switch between algorithms.

Such a comprehensive approach, providing the chassis engineer with almost infinite and decoupled ways of setting up suspension characteristics, is also somewhat complex. Many of the control loops have high gains and are prone to instability, caused by feedback of resonances in the structure to which the actuators and sensors are mounted. The suspension of the car depends entirely on the correct functioning of the system, and it is able to distribute loads among the four wheels, in a way that can have a strong influence on directional control and stability. If the system malfunctions, the car is left with either no suspension or one that takes control of the car.

A simpler approach is semi-active suspension, in which the springs, dampers, and usually the anti-roll bars are retained, and servo-actuators are mounted in series with the spring/damper units.

The semi-active suspension units can, in theory, perform all the duties of a full-active system. However, in practice, the bandwidth of the actuator and spring in series is unlikely to be greater than 3 Hz. Thus, the suspension is unable to control the sprung and unsprung masses and is at its performance limits in terms of controlling inertial loads on a race car. (Semi-active systems are more capable of fulfilling this function on a road car because the requirement is nearer to 1 Hz than 3 Hz.) The springs and dampers provide control of road inputs, and the semi-active system handles weight changes and aerodynamic loads, and may be used to partially control inertia loads.

This simpler approach does not require load measurement and the high gains that cause instability in the system. Instead, the actuators are position servos, and the displacement required is calculated knowing the spring rate and its change in length for any load variation. Thus, the spring deflections can be partially, fully, or over-compensated to control the attitude of the car relative to the road, as required by the aeromap and setup.

The need to control the aerodynamic forces became apparent in 1979, but it was not until 1983 that active systems appeared on Formula 1 cars—a full-active system on the Lotus T92 and a semi-active system (based on Citroën road car technology) on the Ligier JS19. Even then, another four years passed before active suspension was routinely raced and winning Grand Prix. By 1992, active suspension was an essential part of a competitive Formula 1 car. At the end of 1993, it was banned, along with a host of control systems that helped the driver to control the car—namely, driver aids. A decade is a long time in motor racing, and it is unusual for a new technology to take so long to mature and to show its potential. Turbocharging had a similarly protracted gestation period, and both technologies suffered for coming along too early, before enabling technologies were ready.

Both active suspension and turbocharged engines rely heavily on the digital computer for control and analysis. To understand the functioning of both of these complex dynamic systems requires modeling and validation on comprehensively instrumented rigs, neither of which are possible without computers. Both require on-car computers to control them. Renault struggled from 1977 to tame the turbo-engine. However, significantly, the first five World Championships of the turbo era (1983–1987) were won by either BMW or TAG-Porsche engines, both of which were controlled by Bosch digital engine management systems.

The problems Lotus experienced with the T80 in 1979, already described in Appendix A, led to twin approaches to solving the aerodynamic pitching and heaving instability. Plan A was the twin-chassis T88, which was banned before it could race. Plan B was active suspension. David Williams, head of the Flight Instrumentation Department at the Cranfield College of Aeronautics, had introduced me to electro-hydraulic servo systems with load feedback when he demonstrated an artificial-feel control column for an aircraft. The ability to synthesize any spring, damper, or mass in the system was convincing, and all I had to do was to convince Colin Chapman to start an R&D program, without having any idea of what it would take to produce a viable system. At that time, the computer we had used at Cranfield College for data analysis filled a room, and Team Lotus had only one desktop computer in the two-person R&D department. Chapman's response to my proposal was to assign a Lotus Esprit Turbo to the project and tell me to have it running as an active suspension car in six months. His understanding of the need to solve the aerodynamic problem and his belief that the future lay with computers (although I do not believe he really liked them) epitomized Colin Chapman as one of the greatest motor racing visionaries of all time.

The Esprit was modified at Cranfield College with off-the-shelf, double-acting hydraulic cylinders and servo-valves replacing the springs and dampers. No springs were fitted in parallel, and thus the system had 100% authority. When switched off, the car subsided onto its bump rubbers. The computer consisted of the signal conditioning and data acquisition system used on the Formula 1 cars, as well as analog control loops. To change the loops required a soldering iron, but gains and offsets were adjustable, via sliding potentiometers mounted in a panel in front of the passenger. A gear pump driven by the engine supplied hydraulic power.

Initial problems centered around instability of the system. As soon as load feedback gains ("spring" stiffness) were increased to give rates equivalent to those of a normal car, the system excited the structure of the car at its natural frequency in torsion, and the whole car disappeared into a blur with a roaring sound. At the same time, the windshield wipers switched themselves on and then fell off. Fortunately, the hydraulic system was unable to keep up with the demands of this level of activity, and the pressure died before the fatigue life of the structure was completely exhausted. David Williams solved these problems by installing carefully designed electronic notch filters into the feedback loops; they were the key to making the system work (Fig. C.4).

In an early demonstration to Chapman, in the workshops at the Team Lotus' base in Ketteringham Hall, I attempted to show him how easy it was to adjust the attitude of the car, by moving the appropriate slider potentiometer while seated in the car. Unfortunately, I raised the car too fast, and the resulting vertical acceleration was such that it forced my hand down again on the potentiometer. The car attempted to lower itself extremely rapidly, and onlookers claimed that daylight could be seen under all four wheels. Chapman left the scene, shaking his head, and he was heard to mutter the quotation at the beginning of this appendix.

Track testing progressed to the point where Lotus drivers Nigel Mansell and Elio De Angelis were able to drive the car on test circuits. The first time that De Angelis drove the car at Snetterton, he was able to describe what he wanted from the handling of the car in such a way that I was immediately able to make adjustments from the passenger seat, while on the circuit. It took only five laps to tune the car to his liking, and I learned more about setting up a car for a driver at that session than I had ever learned in the past. By the end of the day, I was able to adjust the car between entry to the corner and exit, and to vary its characteristics to optimize the car at each stage to suit the driver. The concept of adaptive algorithms was conceived then, although the need for a digital computer also became obvious. The ability and speed with which different settings could be tried also became apparent—procedures that normally required pit stops and work by mechanics on a conventional car, by which time the driver had often forgotten how the car felt, or else the track had changed.

Chapman was driven by De Angelis that day and also drove the car. He immediately gave the go-ahead on a Formula 1 version of the system, only 12 months from his original go-ahead on the project.

Another 12 months later saw the first test of the Cosworth DFV-powered T92, again at Snetterton. The car had purpose-designed actuators, manufactured by Dowty, and a hybrid analog-digital computer, allowing loop gains and offsets to be set up via a plug-in laptop computer. Dave Scott, a young test driver, was to give the car its initial shakedown. While driving the active Esprit with me as the passenger and to familiarize himself with the active suspension characteristics, he lost control at the 160 kph (100 mph) corner before the pits, hitting the bank

Fig. C.4 (a) Passive suspension Esprit on Lotus' Hethel test track. (b) Active suspension Esprit prototype, at the same corner at Hethel.

hard but without injuring either of us. We had just returned the wreck to the garage when Peter Warr, team manager of Team Lotus, appeared to announce that Colin Chapman had died the previous night. He told us to carry on; however, although the car ran, it was a very bad day.

The car raced in the first two Grand Prix of 1983, performing reasonably against the turbos, although Nigel Mansell was extremely wary of it. I vividly remember checking the car on the grid in Rio de Janeiro and discovering that the memory of the car's computer was blank! As the minutes to the start ticked by, I reprogrammed the computer, entry by entry. The sight of an engineer with a laptop plugged into a car on the grid was extremely rare at that time. Now, each car has two or three laptops attached whenever it is stationary.

When Team Lotus converted to the Renault turbo-engine after a couple of races, active suspension was discarded, lacking Chapman's support. The two-man team developing active technology moved to Lotus Engineering to further develop the system for road cars. During the design phase of the T92, the Technical Regulations had changed, mandating flat bottoms for the first time. The T92 had been designed with high downforce, full ground-effect side pods, with aggressive pitch characteristics, which the active suspension would tame! Flat-bottom aerodynamics was in its infancy, and the need for active suspension greatly diminished. However, by 1986, they had been developed to generate considerable downforce and in the process had gained considerable pitch sensitivity. Team Lotus requested Lotus Engineering to supply an active suspension system for racing in 1987.

The 1987 system benefited from the development work of the intervening years, during which the technology had had a chance to mature under less demanding pressure. Considerable rig development work had been done, and a fully digital computer had been designed specifically for active suspension, by the team under David Williams at Cranfield College. The actuators were fitted with parallel springs, reducing the power needed to drive the hydraulic system and providing a measure of safety if the system depressurized. A new system was fitted to the Honda turbo-powered T99 (Fig. C.5) and included an integral data acquisition system supported by data analysis software, essential for diagnosing and developing the system and for tuning car handling characteristics.

Ayrton Senna brought his incredibly focused approach to testing and was instrumental in achieving two Grand Prix wins and finishing third in the World Championship. Williams also raced an active suspension car in certain races that year, winning one Grand Prix. The handwriting was on the wall.

Williams' system was based on an electronic variation of the Automotive Products semi-active system. It employed pneumatic springs and actuators in series and used accelerometers to compensate and distribute inertia forces (Fig. C.6).

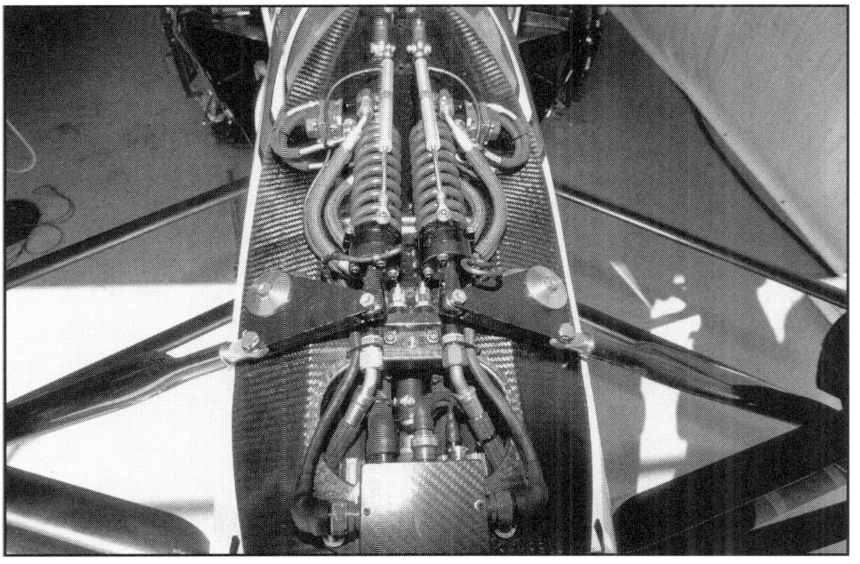

Fig. C.5 Active suspension fitted to the Lotus T99-Honda. (Courtesy of Sutton Motorsport Images)

Team Lotus was unable to fund further development of the Lotus system, and Williams continued to develop its system into a dominant race-winning technology. The attitude control aspect of active suspension, with its effects on optimizing the aerodynamics under all conditions, is the major benefit of the system in racing. This can be achieved with a semi-active system. Control of inertia loads prevents transient aerodynamic effects under braking and acceleration, but roll control has only a small advantage in steering response. Control of the masses by actively simulating springs and dampers appears to have been a mixed blessing. The effect on the tire performance was such that in events where tires were on the thermal limit, full-active was less hard on the tires. This characteristic was exploited by Ayrton Senna during the two races he won. However, when it was difficult for drivers to raise the tires to their optimum temperature, it became a disadvantage. This showed up particularly during qualifying, when it is necessary to bring the tires in quickly, and a stiffly suspended, aggressive car often benefits. A much greater understanding of the interaction between active suspension and the tires was needed to fully exploit the potential of a full-active system.

The benefits of Williams' continuous steady development of its system showed in the way it dominated Formula 1 in the final two years before the technology was banned. Active suspension was lost to racing at the same time as all the driver aids were banned, although active suspension fell under the regulation on movable aerodynamic devices. Many believe this was unfortunate because it is not a driver aid, and the technology has real value to the motor industry as a whole—a racing contribution that is all too rare now.

Fig. C.6 Semi-active suspension fitted to the Williams FW15C-Renault.

When Lotus Engineering took over the ongoing development of its full-active system in 1983, it set out to demonstrate the technology to the motor industry with a view to building vehicles suitable for research into suspension and handling characteristics. The system provides a method of exploring a wide range of characteristics in a short time, many of which are beyond that of which a passive system is capable. Two of the first demonstrations, using the original active Esprit development car, were to Mercedes and to General Motors Research. It transpired that Mercedes had been investigating active suspension for eight years and had built a single-corner test rig. Mercedes wished to know whether it should progress to the next stage of building a car, and its test rides in the Esprit convinced the company that it should!

General Motors had been modeling active suspension for almost ten years and was reputedly quite upset when it heard that Lotus was actually racing a full-active car in 1983. General Motors sent some researchers to Rio for the Brazilian Grand Prix to see whether it was real. A close technical relationship developed between Lotus and General Motors, in which Lotus built a number of active technology vehicles. This pioneering cooperation eventually contributed to General Motors purchasing Lotus in 1986.

Active suspension is one of the proofs that Formula 1 can originate and pioneer technologies of significant benefit to the motor industry, taking the technical risks necessary to bring the technology to fruition quickly. Lotus Engineering went on to build numerous active suspension and active technology vehicles, and to spread the technology widely throughout the industry (Fig. C.7).

Fig. C.7 An array of active technology research vehicles, including a tracked vehicle, lined up at Lotus.

Appendix D

2001 Formula One Technical Regulations

Fédération Internationale de l'Automobile [Ref. D.1]

Summary

Article 1: Definitions

Article 2: General Principles

2.1 Role of the FIA
2.2 Amendments to the regulations
2.3 Dangerous construction
2.4 Compliance with the regulations
2.5 Measurements
2.6 Driving
2.7 Duty of Competitor

Article 3: Bodywork and Dimensions

3.1 Wheel centre line
3.2 Height measurements
3.3 Overall width
3.4 Width ahead of the rear wheel centre line
3.5 Width behind the rear wheel centre line
3.6 Overall height
3.7 Front bodywork height
3.8 Height in front of the rear wheels
3.9 Bodywork between the rear wheels
3.10 Height behind the rear wheel centre line
3.11 Bodywork around the front wheels
3.12 Bodywork facing the ground
3.13 Skid block
3.14 Overhangs
3.15 Aerodynamic influence
3.16 Upper bodywork

Article 4: Weight

4.1 Minimum weight
4.2 Ballast
4.3 Adding during the race

Article 5: Engine

5.1 Engine specification
5.2 Other means of propulsion
5.3 Temperature and pressure of the charge
5.4 Exhaust system
5.5 Engine materials
5.6 Starting the engine
5.7 Throttle control
5.8 Engine control
5.9 Stall prevention systems
5.10 Engine rev limiters

5.11 Car speed limiter
5.12 Induction system

Article 6: Fuel System

6.1 Fuel tanks
6.2 Fittings and piping
6.3 Crushable structure
6.4 Fuel tank fillers
6.5 Refuelling
6.6 Fuel sampling

Article 7: Oil and Coolant Systems

7.1 Location of oil tanks
7.2 Longitudinal location of oil system
7.3 Catch tank
7.4 Transversal location of oil system
7.5 Oil replenishment
7.6 Coolant header tank
7.7 Cooling systems
7.8 Oil and coolant lines

Article 8: Electrical Systems

8.1 Cockpit controls
8.2 Software validation
8.3 Fault or error detection
8.4 Accident data recorders
8.5 Marshal information display

Article 9: Transmission System

9.1 Transmission types
9.2 Propulsion
9.3 Clutch control
9.4 Gear changing
9.5 Gear ratios
9.6 Reverse gear
9.7 Electronically controlled differentials

Article 10: Suspension and Steering Systems

10.1 Sprung suspension
10.2 Suspension geometry
10.3 Suspension members
10.4 Steering

Article 11: Brake System

11.1 Brake circuits and pressure distribution
11.2 Brake calipers
11.3 Brake discs
11.4 Air ducts
11.5 Brake pressure modulation
11.6 Liquid cooling

Article 12: Wheels and Tyres

12.1 Location
12.2 Number of wheels
12.3 Wheel material
12.4 Wheel dimensions

Article 13: Cockpit

13.1 Cockpit opening
13.2 Steering wheel
13.3 Internal cross section
13.4 Position of the driver's feet

Article 14: Safety Equipment

14.1 Fire extinguishers
14.2 Master switch
14.3 Rear view mirrors
14.4 Safety belts
14.5 Rear light
14.6 Cockpit padding
14.7 Wheel retention
14.8 Seat fixing and removal

Article 15: Safety Structures

15.1 Materials
15.2 Roll structures
15.3 Structure behind the driver
15.4 Survival cell specifications
15.5 Survival cell safety requirements

Article 16: Impact Testing

16.1 Conditions applicable to all impact tests
16.2 Frontal test
16.3 Side test

Article 1: Definitions

1.1 Formula One car:
An automobile designed solely for speed races on circuits or closed courses.

1.2 Automobile:
A land vehicle running on at least four non-aligned complete wheels, of which at least two are used for steering and at least two for propulsion.

1.3 Land vehicle:
A locomotive device propelled by its own means, moving by constantly taking real support on the earth's surface, of which the propulsion and steering are under the control of a driver aboard the vehicle.

1.4 Bodywork:
All entirely sprung parts of the car in contact with the external air stream, except cameras and the parts definitely associated with the mechanical functioning of the engine, transmission and running gear. Airboxes, radiators and engine exhausts are considered to be part of the bodywork.

1.5 Wheel:
Flange and rim.

1.6 Complete wheel:
Wheel and inflated tyre.

1.7 Automobile make:
In the case of Formula racing cars, an automobile make is a complete car. When the car manufacturer fits an engine which it does not manufacture, the car shall be considered a hybrid and the name of the engine manufacturer shall be associated with that of the car manufacturer. The name of the car manufacturer must always precede that of the engine manufacturer. Should a hybrid car win a Championship Title, Cup or Trophy, this will be awarded to the manufacturer of the car.

1.8 Event:
An event shall consist of official practice and the race.

1.9 Weight:
Is the weight of the car with the driver, wearing his complete racing apparel, at all times during the event.

1.10 Racing weight:
Is the weight of the car in running order with the driver aboard and all fuel tanks full.

1.11 Cubic capacity:
The volume swept in the cylinders of the engine by the movement of the pistons. This volume shall be expressed in cubic centimetres. In calculating engine cubic capacity, the number Pi shall be 3.1416.

1.12 Supercharging:
Increasing the weight of the charge of the fuel/air mixture in the combustion chamber (over the weight induced by normal atmospheric pressure, ram effect and dynamic effects in the intake and/or exhaust system) by any means whatsoever. The injection of fuel under pressure is not considered to be supercharging.

1.13 Cockpit:
The volume which accommodates the driver.

1.14 Sprung suspension:
The means whereby all complete wheels are suspended from the body/chassis unit by a spring medium.

1.15 Survival cell:
A continuous closed structure containing the fuel tank and the cockpit.

1.16 Camera:
Television cameras, the dimensions of which are defined in Drawing 6.

1.17 Camera housing:
A device which is identical in shape and weight to a camera and which is supplied by the relevant Competitor for fitting to his car in lieu of a camera.

1.18 Cockpit padding:
Non-structural parts placed within the cockpit for the sole purpose of improving driver comfort and safety. All such material must be quickly removable without the use of tools.

1.19 Brake caliper:
All parts of the braking system outside the survival cell, other than brake discs, brake pads, caliper pistons, brake hoses and fittings, which are stressed when subjected to the braking pressure. Bolts or studs which are used for attachment are not considered to be part of the braking system.

1.20 Automatic gearbox:
One in which gears may be changed and used without each one being requested by the driver.

Article 2: General Principles

2.1 Role of the FIA:
The following technical regulations for Formula One cars are issued by the FIA.

2.2 Amendments to the regulations:
Amendments to these regulations will be made in accordance with the Concorde agreement.

2.3 Dangerous construction:
The Stewards of the Meeting may exclude a vehicle whose construction is deemed to be dangerous.

2.4 Compliance with the regulations:
Automobiles must comply with these regulations in their entirety at all times during an Event.

Should a Competitor feel that any aspect of these regulations is unclear, clarification may be sought from the FIA Formula One Technical Department. If clarification relates to any new design or system, correspondence must include:

- A full description of the design or system;
- Drawings or schematics where appropriate;
- The Competitor's opinion concerning the immediate implications on other parts of the car of any proposed new design;
- The Competitor's opinion concerning any possible long-term consequences or new developments which may come from using any such new designs or systems;
- The precise way or ways in which the Competitor feels the new design or system will enhance the performance of the car.

2.5 Measurements:
All measurements must be made while the car is stationary on a flat horizontal surface.

2.6 Driving:
The driver must drive the car alone and unaided.

2.7 Duty of Competitor:
It is the duty of each Competitor to satisfy the FIA technical delegate and the Stewards of the Meeting that his automobile complies with these regulations in their entirety at all times during an Event.

Article 3: Bodywork and Dimensions

3.1 Wheel centre line:
The centre line of any wheel shall be deemed to be halfway between two straight edges, perpendicular to the surface on which the car is standing, placed against opposite sides of the complete wheel at the centre of the tyre tread.

3.2 Height measurements:
All height measurements will be taken normal to and from the reference plane.

3.3 Overall width:
The overall width of the car, including complete wheels, must not exceed 1800 mm with the steered wheels in the straight ahead position.

3.4 Width ahead of the rear wheel centre line:

3.4.1 Bodywork width ahead of the rear wheel centre line must not exceed 1400 mm.

3.4.2 No lateral extremity of any bodywork forward of the front wheels may deflect more than 5 mm vertically when a 50 kg mass is placed on it. During such a test the centre of area of the mass will be placed 700 mm forward of the front wheel centre line with its outer edge 700 mm from the car centre line.

The precise dimensions of the mass which will be used are available from the FIA Technical Department.

3.4.3 In order to prevent tyre damage to other cars, the top and forward edges of the lateral extremities of any bodywork forward of the front wheels must be at least 10 mm thick with a radius of at least 5 mm.

3.5 Width behind the rear wheel centre line:
Bodywork width behind the rear wheel centre line must not exceed 1000 mm.

3.6 Overall height:
No part of the bodywork may be more than 950 mm above the reference plane.

3.7 Front bodywork height:
All bodywork situated forward of a point lying 330 mm behind the front wheel centre line, and more than 250 mm from the centre line of the car, must be no less than 100 mm and no more than 300 mm above the reference plane.

3.8 Height in front of the rear wheels:

3.8.1 Other than the rear view mirrors, each with a maximum area in plan view of 9000 mm^2, no bodywork situated more than 330 mm behind the front wheel centre line and more than 330 mm forward of the rear wheel centre line, which is more than 600 mm above the reference plane, may be more than 300 mm from the centre line of the car.

3.8.2 No bodywork between the rear wheel centre line and a line 800 mm forward of the rear wheel centre line, which is more than 500 mm from the centre line of the car, may be more than 500 mm above the reference plane.

3.8.3 No bodywork between the rear wheel centre line and a line 400 mm forward of the rear wheel centre line, which is more than 500 mm from the centre line of the car, may be more than 300 mm above the reference plane.

3.9 Bodywork between the rear wheels:

3.9.1 No bodywork situated between the rear wheel centre line and a point lying 330 mm forward of it may be more than 600 mm above the reference plane.

3.9.2 No bodywork situated between the rear wheel centre line and a point lying 150 mm behind it may be more than 450 mm above the reference plane.

3.10 Height behind the rear wheel centre line:

3.10.1 Any part of the car more than 150 mm behind the centre line of the rear wheels must not be more than 800 mm above the reference plane.

3.10.2 No bodywork behind the centre line of the rear wheels, and more than 150 mm each side of the longitudinal centre line of the car, may be less than 300 mm above the reference plane.

3.10.3 Any bodywork more than 150 mm behind the rear wheel centre line which is more than 300 mm above the reference plane, amd between 75 mm and 480 mm from the car centre line, must lie in one of two areas when viewed from the side of the car. These areas are situated from 300 mm to 375 mm and 600 mm to 800 mm above the reference plane. When these areas are viewed from the side of the car, no longitudinal cross section may have more than three closed sections in the upper area or more than one in the lower.

3.11 Bodywork around the front wheels:
With the exception of brake cooling ducts, in plan view, there must be no bodywork in the area formed by two longitudinal lines parallel to and 400 mm and 900 mm from the car centre line and two transversal lines, one 350 mm forward of and one 800 mm behind the front wheel centre line.

3.12 Bodywork facing the ground:

3.12.1 All sprung parts of the car situated more than 330 mm behind the front wheel centre line and more than 330 mm forward of the rear wheel centre line, and which are visible from underneath, must form surfaces which lie on one of two parallel planes, the reference plane or the step plane. This does not apply to any parts of rear view mirrors which are visible, provided each of these areas does not exceed 9000 mm^2 when projected to a horizontal plane above the car. The step plane must be 50 mm above the reference plane.

3.12.2 The surface formed by all parts lying on the reference plane must:

- Extend from a point lying 330 mm behind the front wheel centre line to the centre line of the rear wheels;
- Have minimum and maximum widths of 300 mm and 500 mm, respectively;
- Be symmetrical about the centre line of the car;
- Have a 50 mm radius ($\pm$2 mm) on each front corner when viewed from directly beneath the car, this being applied after the surface has been defined.

3.12.3 The surface lying on the reference plane must be joined at its extremities to the surfaces lying on the step plane by a vertical transition. If there is no surface visible on the step plane vertically above any point around the extremity of the reference plane, this transition is not necessary.

3.12.4 The peripheries of the surfaces lying on the reference and step planes may be curved upward with maximum radii of 25 and 50 mm, respectively. Where the vertical transition meets the surfaces on the step plane a radius, no greater than 25 mm, is permitted.

A radius in this context will be considered as an arc applied perpendicular to the periphery and tangential to both surfaces.

The surface lying on the reference plane, the surfaces lying on the step plane and the vertical transitions between them, must first be fully defined before any radius can be applied or the skid block fitted. Any radius applied is still considered part of the relevant surface.

3.12.5 All parts lying on the reference and step planes, in addition to the transition between the two planes, must produce uniform, solid, hard, continuous, rigid (no degree of freedom in relation to the body/chassis unit), impervious surfaces under all circumstances.

Fully enclosed holes are permitted in these surfaces provided no part of the car is visible through them when viewed from directly below.

3.12.6 To help overcome any possible manufacturing problems, and not to permit any design which may contravene any part of these regulations, dimensional tolerances are permitted on bodywork situated between a point lying 330 mm behind the front wheel centre line and the rear wheel centre line. A vertical tolerance of $\pm$5 mm is permissible across the surfaces lying on the reference and step planes, and a horizontal tolerance of 5 mm is permitted when assessing whether a surface is visible from beneath the car.

3.12.7 All sprung parts of the car situated behind a point lying 330 mm forward of the rear wheel centre line, which are visible from underneath and are more than 250 mm from the centre line of the car, must be at least 50 mm above the reference plane.

3.13 Skid block:

3.13.1 Beneath the surface formed by all parts lying on the reference plane, a rectangular skid block, with a 50 mm radius ($\pm$2 mm) on each front corner, must be fitted. This skid block may comprise more than one piece but must:

a) Extend longitudinally from a point lying 330 mm behind the front wheel centre line to the centre line of the rear wheels.

b) Be made from a homogeneous material with a specific gravity between 1.3 and 1.45.

c) Have a width of 300 mm with a tolerance of $\pm$2 mm.

d) Have a thickness of 10 mm with a tolerance of $\pm$1 mm.

e) Have a uniform thickness when new.

f) Have no holes or cut outs other than those necessary to fit the fasteners permitted by 3.13.2 or those holes specifically mentioned in g) below.

g) Have six precisely placed holes in order that its thickness can be measured at any time. These holes must be 50 mm in diameter and must be placed in the positions detailed in Drawing 1. In order to establish the conformity of the skid block after use, its thickness will only be measured in these holes.

h) Be fixed symmetrically about the centre line of the car in such a way that no air may pass between it and the surface formed by the parts lying on the reference plane.

3.13.2 Fasteners used to attach the skid block to the car must:

a) Have a total area no greater than 40,000 mm^2 when viewed from directly beneath the car;

b) Be no greater than 2000 mm^2 in area individually when viewed from directly beneath the car;

c) Be fitted in order that their entire lower surfaces are visible from directly beneath the car.

When the skid block is new, ten of the fasteners may be flush with its lower surface, but the remainder may be no more than 8 mm below the reference plane.

3.13.3 The lower edge of the periphery of the skid block may be chamfered at an angle of 30° to a depth of 8 mm. The trailing edge, however, may be chamfered over a distance of 200 mm to a depth of 8 mm.

3.14 Overhangs:

No part of the car shall be more than 500 mm behind the centre line of the rear wheels or more than 1200 mm in front of the centre line of the front wheels.

No part of the bodywork more than 200 mm from the centre line of the car may be more than 900 mm in front of the front wheel centre line.

All overhang measurements will be taken parallel to the reference plane.

3.15 Aerodynamic influence:

Any specific part of the car influencing its aerodynamic performance (with the exception of the cover described in Article 6.5.2 in the pit lane only):

- Must comply with the rules relating to bodywork.
- Must be rigidly secured to the entirely sprung part of the car (rigidly secured means not having any degree of freedom).
- Must remain immobile in relation to the sprung part of the car.

In order to ensure that this requirement is respected, the FIA reserves the right to introduce load/deflection tests on any part of the bodywork which appears to be (or is suspected of) moving whilst the car is in motion.

Any device or construction that is designed to bridge the gap between the sprung part of the car and the ground is prohibited under all circumstances.

No part having an aerodynamic influence and no part of the bodywork, with the exception of the skid block in 3.13 above, may under any circumstances be located below the reference plane.

3.16 Upper bodywork:

3.16.1 With the exception of the opening described in Article 3.16.3, when viewed from the side, the car must have bodywork in the triangle formed by three lines, one vertical passing 1330 mm forward of the rear wheel centre line, one horizontal 550 mm above the reference plane and one diagonal which intersects the vertical at a point 940 mm above the reference plane and the horizontal 330 mm forward of the rear wheel centre line.

The bodywork over the whole of this area must be arranged symmetrically about the car centre line and must be at least 200 mm wide when measured at any point along a second diagonal line parallel to and 200 mm vertically below the first.

Furthermore, over the whole area between the two diagonal lines, the bodywork must be wider than a vertical isosceles triangle lying on a lateral plane which has a base 200 mm wide lying on the second diagonal line.

3.16.2 When viewed from the side, the car must have no bodywork in the triangle formed by three lines, one vertical 330 mm forward of the rear wheel centre line, one horizontal 950 mm above the reference plane, and one diagonal which intersects the vertical at a point 600 mm above the reference plane and the horizontal at a point 1030 mm forward of the rear wheel centre line.

3.16.3 In order that a car may be lifted quickly in the event of it stopping on the circuit, the principal rollover structure must incorporate a clearly visible unobstructed opening designed to permit a strap, whose section measures 60 mm × 30 mm, to pass through it.

Article 4: Weight

4.1 Minimum weight:
The weight of the car must not be less than 600 kg.

4.2 Ballast:
Ballast can be used provided it is secured in such a way that tools are required for its removal. It must be possible to fix seals if deemed necessary by the FIA technical delegate.

4.3 Adding during the race:
With the exception of fuel, nitrogen and compressed air, no substance may be added to the car during the race. If it becomes necessary to replace any part of the car during the race, the new part must not weigh any more than the original part.

Article 5: Engine

5.1 Engine specification:

5.1.1 Only four-stroke engines with reciprocating pistons are permitted.

5.1.2 Engine capacity must not exceed 3000 cc.

5.1.3 Supercharging is forbidden.

5.1.4 All engines must have 10 cylinders, and the normal section of each cylinder must be circular.

5.1.5 Engines may have no more than five valves per cylinder.

5.2 Other means of propulsion:

5.2.1 The use of any device, other than the 3 litre, four-stroke engine described in 5.1 above, to power the car, is not permitted.

5.2.2 The total amount of recoverable energy stored on the car must not exceed 300 kJ, any which may be recovered at a rate greater than 2 kW must not exceed 20 kJ.

5.3 Temperature and pressure of the charge:

5.3.1 Any device, system, procedure, construction or design the purpose and/or effect of which is any decrease whatsoever of the temperature of the intake air and/or of the charge (air and/or fuel) of the engine is forbidden.

5.3.2 Internal and/or external spraying of water or any substance whatsoever is forbidden (other than fuel for the normal purpose of combustion in the engine).

5.4 Exhaust system:

Variable geometric length exhaust systems are forbidden.

5.5 Engine materials:

5.5.1 The basic structure of the crankshaft and camshafts must be made from steel or cast iron.

5.5.2 Pistons, cylinder heads and cylinder blocks may not be composite structures which use carbon or aramid fibre reinforcing materials.

5.6 **Starting the engine:**
A supplementary device temporarily connected to the car may be used to start the engine both on the grid and in the pits.

5.7 **Throttle control:**

5.7.1 Other than the specific exceptions mentioned below in 5.7.2, there must be a fixed relationship between the position of the throttle pedal and the engine throttles. This relationship need not be linear, but the position of the engine throttles may not be influenced by anything other than movement of the throttle pedal when operated by the driver.

This relationship must remain fixed whilst the car is in motion subject only to Article 8.3.

5.7.2 The relationship between the throttle pedal and engine throttles may alter during one or more of the following operations:

- Idle control;
- Stall prevention;
- Gear changing;
- Car speed limiting.

5.7.3 Engine throttles may be modulated by no more than two individual operating mechanisms.

5.8 **Engine control:**
Ignition and fuel settings must maintain the same relationship with engine speed and throttle position whilst the car is in motion, with the following specific exceptions:

- Compensation for throttle acceleration;
- Driver adjustable fuel mixture control with a maximum of three settings;
- Compensation for changes in engine intake air temperature and pressure, engine pressures, or engine temperatures;
- Open or closed loop detonation and lambda control.

No engine parameter may be altered so as to diminish the degree of control the driver has over the propulsion system.

5.9 **Stall prevention systems:**

5.9.1 The sole purpose of such systems is to prevent the engine stalling when a driver loses control of the car.

5.9.2 Each time such a system is activated, the clutch must be fully dis-engaged and must remain so until the driver de-activates it by manually operating the clutch with the normal proportional request.

5.9.3 To avoid the possibility of a car involved in an accident being left with the engine running, all such systems must be configured to stop the engine no more than ten seconds after activation.

5.10 Engine rev limiters:
With the exception of the car speed limiter below and subject to Article 8.3, engine rev limits may vary for differing conditions provided all are significantly above the peak of the engine power curve.

5.11 Car speed limiter:

5.11.1 The purpose of the speed limiter is to improve safety by ensuring a driver is less likely to exceed the pit lane speed limit.

5.11.2 The car speed limiter may be operated only by the driver when he needs it and must be deactivated by him at the time it is no longer required.

5.11.3 Car speed limiters may only operate in first, second and third gears and may only be activated in the pit lane.

5.12 Induction system:
The length and volume of the engine induction system may be modulated by no more than two individual operating mechanisms.

Article 6: Fuel System

6.1 Fuel tanks:

6.1.1 The fuel tank must be a single rubber bladder conforming to or exceeding the specifications of FIA/FT5-1999; the fitting of foam within the tank, however, is not mandatory.

6.1.2 All the fuel stored on board the car must be situated between the front face of the engine and the driver's back when viewed in lateral projection. When establishing the front face of the engine, no parts of the fuel, oil, water or electrical systems will be considered.

Furthermore, no fuel can be stored more than 300 mm forward of the highest point at which the driver's back makes contact with his seat. However, a maximum of 2 litres of fuel may be kept outside the survival cell, but only that which is necessary for the normal running of the engine.

6.1.3 Fuel must not be stored more than 400 mm from the longitudinal axis of the car

6.1.4 All rubber bladders must be made by manufacturers recognised by the FIA. In order to obtain the agreement of the FIA, the manufacturer must prove the compliance of his product with the specifications approved by the FIA. These manufacturers must undertake to deliver to their customers exclusively tanks complying to the approved standards.

A list of approved manufacturers is available from the FIA.

6.1.5 All rubber bladders shall be printed with the name of the manufacturer, the specifications to which the tank has been manufactured and the date of manufacture.

6.1.6 No rubber bladders shall be used more than 5 years after the date of manufacture.

6.2 Fittings and piping:

6.2.1 All apertures in the fuel tank must be closed by hatches or fittings which are secured to metallic or composite bolt rings bonded to the inside of the bladder.

Bolt hole edges must be no less than 5 mm from the edge of the bolt ring, hatch or fitting.

6.2.2 All fuel lines between the fuel tank and the engine must have a self-sealing breakaway valve. This valve must separate at less than 50% of the load required to break the fuel line fitting or to pull it out of the fuel tank.

6.2.3 No lines containing fuel may pass through the cockpit.

6.2.4 All lines must be fitted in such a way that any leakage cannot result in the accumulation of fuel in the cockpit.

6.3 Crushable structure:

The fuel tank must be completely surrounded by a crushable structure, which is an integral part of the survival cell and must be able to withstand the loads required by the tests in Articles 18.2.1 and 18.3.

6.4 Fuel tank fillers:

Fuel tank fillers must not protrude beyond the bodywork. Any breather pipe connecting the fuel tank to the atmosphere must be designed to avoid liquid leakage when the car is running and its outlet must not be less than 250 mm from the cockpit opening.

All fuel tank fillers and breathers must be designed to ensure an efficient locking action which reduces the risk of an accidental opening following a crash impact or incomplete locking after refuelling.

6.5 Refuelling:

6.5.1 All refuelling during the race must be carried out using equipment which has been supplied by the FIA designated manufacturer. This manufacturer will be required to supply identical refuelling systems, the complete specification of which will be available from the FIA no later than one month prior to the first Championship Event.

Any modifications to the manufacturer's specification may only be made following written consent from the FIA.

6.5.2 A cover must be fitted over the car connector at all times when the car is running on the track. The cover and its attachments must be sufficiently strong to avoid accidental opening in the event of an accident.

6.5.3 Before refuelling commences, the car connector must be connected electrically to earth.

All metal parts of the refuelling system from the coupling to the supply tank must also be connected to earth.

6.5.4 Refuelling the car on the grid may only be carried out by using an unpressurized container which is no more than 2 metres above the ground.

6.5.5 No fuel on board the car may be more than ten degrees centigrade below ambient temperature.

6.5.6 The use of any device on board the car to decrease the temperature of the fuel is forbidden.

6.6 Fuel sampling:

6.6.1 Competitors must ensure that a one litre sample of fuel may be taken from the car at any time during the Event.

6.6.2 All cars must be fitted with a –2 'Symetrics' male fitting in order to facilitate fuel sampling. If an electric pump on board the car cannot be used to remove the fuel, an externally connected one may be used provided it is evident that a representative fuel sample is being taken. If an external pump is used, it must be possible to connect the FIA sampling hose to it and any hose between the car and pump must be –3 in diameter and not exceed 2 m in length.

6.6.3 The sampling procedure must not necessitate starting the engine or the removal of bodywork (other than the cover over the refuelling connector).

Article 7: Oil and Coolant Systems

7.1 Location of oil tanks:
All oil storage tanks must be situated between the front wheel axis and the rearmost gearbox casing longitudinally, and must be no further than the lateral extremities of the survival cell are from the longitudinal axis of the car.

7.2 Longitudinal location of oil system:
No other part of the car containing oil may be situated behind the complete rear wheels.

7.3 Catch tank:
In order to avoid the possibility of oil being deposited on the track, the engine sump breather must vent into the main engine air intake system.

7.4 Transversal location of oil system:
No part of the car containing oil may be more than 700 mm from the longitudinal centre line of the car.

7.5 Oil replenishment:
No oil replenishment is allowed during a race.

7.6 Coolant header tank:
The coolant header tank on the car must be fitted with an FIA approved pressure relief valve which is set to a maximum of 3.75 bar gauge pressure. If the car is not fitted with a header tank, an alternative position must be approved by the FIA.

7.7 Cooling systems:
The cooling systems of the engine must not intentionally make use of the latent heat of vaporisation of any fluid.

7.8 Oil and coolant lines:

7.8.1 No lines containing coolant or lubricating oil may pass through the cockpit.

7.8.2 All lines must be fitted in such a way that any leakage cannot result in the accumulation of fluid in the cockpit.

7.8.3 No hydraulic fluid lines may have removable connectors inside the cockpit.

Article 8: Electrical Systems

8.1 Cockpit controls:

8.1.1 With the exception of the car speed limiter described in Article 5.11, the cover referred to in Article 6.5.2 and during gear changes, no driver operated cockpit control may carry out more than one function at any one time.

8.1.2 There must be no significant delay between a driver requested action and the associated actuation.

8.2 Software validation:

8.2.1 Prior to the start of each season, the complete electrical system on the car must be examined and all onboard and communications software must be validated by the FIA Technical Department.

The FIA must be notified of any changes prior to the Event at which such changes are intended to be implemented.

8.2.2 All microprocessors and their enclosures will be classified as either:

- Sealed and not re-programmable via any external connector;
- Re-programmable via a direct connection but limited by an approved mechanism.
- Not re-programmable at an Event. This classification will be given if the microprocessor has no direct communication link to the external connectors of the unit that are capable of being used for re-programming during an Event.

8.2.3 All re-programmable microprocessors must have a mechanism that allows the FIA to accurately identify the software version loaded.

8.2.4 Reprogramming of electronic units during an event will be restricted by an approved mechanism that has been established before the electronic unit is first used at an event.

8.2.5 All setup and calibration data stored in microprocessor memory must be off-loadable by the FIA at any time. Appropriate communications equipment, software and analysis tools must be supplied by the team for FIA use.

8.2.6 The FIA will seal and identify all electronic units on the car that contain a programmable device.

8.2.7 All sealed units must be presented for inspection at the end of an Event.

8.2.8 No version of software will be approved for use at an Event if it is found to be capable of controlling any system on the car in a manner inconsistent with these technical regulations, even if the relevant control software may be disabled.

8.3 Fault or error detection:
If faults or errors are detected by the driver or by onboard software, backup sensors may be used and different settings may be manually or automatically selected. However, any backup sensor or new setting chosen in this way must not enhance the performance of the car, and the original setting may only be restored when the car is stationary in the pits.

8.4 Accident data recorders:
The recorder must be fitted:

- Symmetrically about the car centre line and with its top facing upward;
- With each of its 12 edges parallel to an axis of the car;
- Less than 50 mm above the reference plane;
- In a position which is normally accessible at the start and finish of an Event;
- In order that the entire unit lies between 40% and 60% of the wheelbase of the car;
- With its main connector facing forward;
- In order that its status light is visible when the driver is in the cockpit;
- In order that the download connector is easily accessible without the need to remove bodywork.

8.5 Marshal information display:
All cars must be fitted with cockpit lights to give drivers information concerning track signals or conditions. The precise specification of the lights and related components are available from the FIA Technical Department.

Article 9: Transmission System

9.1 Transmission types:

9.1.1 No transmission system may permit more than two wheels to be driven.

9.1.2 Automatic gearboxes are forbidden.

9.2 Propulsion:

9.2.1 No car may be equipped with a system or device which is capable of preventing the driven wheels from spinning under power or of compensating for excessive throttle demand by the driver.

9.2.2 Any device or system which notifies the driver of the onset of wheel spin is not permitted.

9.3 Clutch control:

9.3.1 A system which compensates for clutch wear is permissible provided it is clear that this is its sole function.

9.3.2 Except during gear changes and stall prevention, or as a result of compensation for wear, the amount by which the clutch is engaged must be controlled solely and directly by the driver at all times.

The way in which the clutch is re-engaged during gear change must be such that it is clear Article 9.2 cannot be contravened.

9.3.3 Other than wear compensation, or if a fault condition is detected (see Article 8.3), the relationship between the clutch operating device in the cockpit and the amount of clutch engagement may be non-linear but must remain fixed whilst the engine is running.

9.3.4 Partial clutch re-engagement is permitted during gear change sequences described under 9.4.3 below.

9.3.5 All cars must be fitted with a means of disengaging the clutch for a minimum of fifteen minutes in the event of the car coming to rest with the engine stopped. This system must be in working order throughout the Event even if the main hydraulic, pneumatic or electrical systems on the car have failed.

In order that the driver or a marshal may activate the system in less than five seconds, the switch or button which operates it must:

- Face upward and be positioned on the survival cell no more than 150 mm from the car centre line;
- Be less than 150 mm from the front of the cockpit opening;
- Be marked with a letter "N" in red inside a white circle of at least 50 mm diameter with a red edge.

9.4 Gear changing:

9.4.1 For the purposes of and only during gear changing, the clutch and throttle need not be under the control of the driver.

9.4.2 Each individual gear change must be initiated by the driver and, within the mechanical constraints of the gearbox, the requested gear must be engaged immediately unless over-rev protection is used.

9.4.3 Multiple gear changes may be made following one driver request provided they are not made before he needs the destination gear and that the car is not driven by any of the intermediate gears during the sequence. If for any reason the sequence cannot be completed, the car must be left in neutral or the original gear.

9.4.4 If a gear change fails for mechanical reasons (as opposed to the predicted engine revs in the target gear being too high), further attempts to engage the gear may be made automatically without the driver having to make a new request.

9.4.5 If an over-rev protection strategy is used, this may only prevent engagement of the target gear; it must not induce any significant delay. If a gear change is refused in this way, engagement may only follow a new and separate request made by the driver.

9.5 Gear ratios:
The minimum number of forward gear ratios is 4 and the maximum is 7.

9.6 Reverse gear:
All cars must have a reverse gear operable any time during the Event by the driver when the engine is running.

9.7 Electronically controlled differentials:

9.7.1 The design and control of the differential may not permit a greater ratio of torque distribution than the ratio of grip between the inner and outer driven wheels.

9.7.2 If a differential is controlled electronically, it may only use instantaneous values of one or more of the following parameters for control purposes:

- Measured and/or derived input torque;
- The difference between the rear wheel speeds;
- The difference between the output shaft torque.

In the case of measured and/or derived input torque, only measured engine torque, gear ratio, engine speed and throttle position may be used; it must also be clear that this figure is a genuine representation of the input torque.

9.7.3 Subject to Article 8.3, the driver may only make changes to the setup of an electronically controlled differential whilst the car is stationary.

Article 10: Suspension and Steering Systems

10.1 Sprung suspension:
Cars must be fitted with sprung suspension. The springing medium must not consist solely of bolts located through flexible bushes or mountings.

There must be movement of the wheels to give suspension travel in excess of any flexibility in the attachments.

The suspension system must be so arranged that its response is consistent at all times and results only from changes in vertical load applied to the wheels save only for movement permitted by inherent and fixed physical properties.

10.2 Suspension geometry:
Suspension geometry must remain fixed at all times.

10.3 Suspension members:

10.3.1 Each member of every suspension component must be made from material whose cross section has an aspect ratio no greater than 3.5:1. All suspension components may, however, have sections with an aspect ratio greater than 3.5:1 provided these are adjacent to their inner and outer attachments and form no more than 25% of the total distance between the attachments of the relevant member.

All measurements will be made perpendicular to a line drawn between the inner and outer attachments of the relevant member.

10.3.2 No major axis of a cross section of a suspension member may subtend an angle greater than 5° to the reference plane when measured parallel to the centre line of the car.

10.3.3 Non-structural parts of suspension members are considered bodywork.

10.3.4 In order to help prevent a wheel from becoming separated in the event of all suspension members connecting it to the car failing, two cables, each with separate attachments, must be fitted to connect each wheel/upright assembly to the main structure of the car. The cables and their attachments must be designed in order to help prevent a wheel making contact with the driver's head during an accident.

The length of each cable should be no longer than that required to allow normal suspension movement.

Each complete cable restraint system, including their attachments, must have a minimum tensile strength of 50 kN, and each cable must be flexible with a minimum diameter of 8 mm.

10.4 Steering:

10.4.1 Any steering system which permits the re-alignment of more than two wheels is not permitted.

10.4.2 Power-assisted steering systems are permitted but may not carry out any function other than to reduce the physical effort required to steer the car.

If an electronically controlled power steering system is used, the only permissible inputs are steering torque, steering angle and car speed. Additionally, other than under Article 8.3, the settings may not be changed whilst the car is in motion.

10.4.3 No part of the steering wheel or column, nor any part fitted to them, may be closer to the driver than a plane formed by the entire rear edge of the steering wheel rim.

10.4.4 The steering wheel, steering column and steering rack assembly must pass an impact test; details of the test procedure may be found in Article 16.5.

Article 11: Brake System

11.1 Brake circuits and pressure distribution:

11.1.1 All cars must be equipped with one brake system which has two separate hydraulic circuits operated by one pedal, one circuit operating on the two front wheels and the other on the two rear wheels. This system must be designed so that if a failure occurs in one circuit, the pedal will still operate the brakes in the other.

11.1.2 The brake system must be designed in order that brake caliper pressures in each circuit are the same at all times.

11.1.3 Any powered device which is capable of altering the configuration of the brake system whilst it is under pressure is forbidden.

11.1.4 Any change to, or modulation of, the brake system whilst the car is moving must be made by the driver's direct physical input, may not be pre-set and must be under his complete control at all times.

11.2 Brake calipers:

11.2.1 All brake calipers must be made from aluminium materials with a modulus of elasticity no greater than 80 Gpa.

11.2.2 No more than two attachments may be used to secure each brake caliper to the car.

11.2.3 No more than one caliper, with a maximum of six pistons, is permitted on each wheel.

11.2.4 The section of each caliper piston must be circular.

11.3 Brake discs:

11.3.1 No more than one brake disc is permitted on each wheel.

11.3.2 All discs must have a maximum thickness of 28 mm and a maximum outside diameter of 278 mm.

11.3.3 No more than two brake pads are permitted on each wheel.

11.4 Air ducts:

Air ducts for the purpose of cooling the front and rear brakes shall not protrude beyond:

- A plane parallel to the ground situated at a distance of 140 mm above the horizontal centre line of the wheel;

- A plane parallel to the ground situated at a distance of 140 mm below the horizontal centre line of the wheel;
- A vertical plane parallel to the inner face of the wheel rim and displaced from it by 120 mm toward the centre line of the car.

Furthermore, when viewed from the side the ducts must not protrude forward beyond the periphery of the tyre or backward beyond the wheel rim.

11.5 Brake pressure modulation:

11.5.1 No braking system may be designed to prevent wheels from locking when the driver applies pressure to the brake pedal.

11.5.2 No braking system may be designed to increase the pressure in the brake calipers above that achievable by the driver applying pressure to the pedal under static conditions.

11.6 Liquid cooling:
Liquid cooling of the brakes is forbidden.

Article 12: Wheels and Tyres

12.1 Location:
Wheels must be external to the bodywork in plan view, with the rear aerodynamic device removed.

12.2 Number of wheels:
The number of wheels is fixed at four.

12.3 Wheel material:
All wheels must be made from a homogeneous metallic material.

12.4 Wheel dimensions:

12.4.1 Complete wheel width must lie between 305 and 355 mm when fitted to the front of the car and between 365 and 380 mm when fitted to the rear.

12.4.2 Complete wheel diameter must not exceed 660 mm when fitted with dry-weather tyres or 670 mm when fitted with wet-weather tyres.

12.4.3 Complete wheel width and diameter will be measured horizontally at axle height when fitted with new tyres inflated to 1.4 bar.

12.4.4 Wheel bead diameter must lie between 328 and 332 mm.

Article 13: Cockpit

13.1 Cockpit opening:

13.1.1 In order to ensure that the opening giving access to the cockpit is of adequate size, the template shown in Drawing 2 will be inserted into the survival cell and bodywork.

During this test the steering wheel, steering column, seat and all padding required by Articles 14.6.1–6 (including fixings) may be removed and:

- The template must be held horizontal and lowered vertically from above the car until its lower edge is 525 mm above the reference plane;
- Referring to Drawing 2, the edge of the template which lies on the line d–e must be no less than 1800 mm behind the line A–A shown in Drawing 5.

Any measurements made from the cockpit entry template (when referred to in Articles 13.1.3, 15.2.2, 15.4.5, 15.4.6, 15.5.4, 16.3, and 18.4) must also be made whilst the template is held in this position.

13.1.2 The forward extremity of the cockpit opening, even if structural and part of the survival cell, must be at least 50 mm in front of the steering wheel.

13.1.3 The driver must be able to enter and get out of the cockpit without it being necessary to open a door or remove any part of the car other than the steering wheel. When seated normally, the driver must be facing forward and the rearmost part of his crash helmet may be no more than 125 mm forward of the rear edge of the cockpit entry template.

13.1.4 From his normal seating position, with all seat belts fastened and whilst wearing his usual driving equipment, the driver must be able to remove the steering wheel and get out of the car within 5 seconds and then replace the steering wheel in a total of 10 seconds.

For this test, the position of the steered wheels will be determined by the FIA technical delegate and after the steering wheel has been replaced steering control must be maintained.

13.2 Steering wheel:

The steering wheel must be fitted with a quick release mechanism operated by pulling a concentric flange installed on the steering column behind the wheel.

13.3 Internal cross section:

13.3.1 A free vertical cross section, which allows the outer template shown in Drawing 3 to be passed vertically through the cockpit to a point 100 mm behind the face of the rearmost pedal when in the inoperative position, must be maintained over its entire length.

The only things which may encroach on this area are the steering wheel and any padding that is required by Article 14.6.7.

13.3.2 A free vertical cross section, which allows the inner template shown in Drawing 3 to be passed vertically through the cockpit to a point 100 mm behind the face of the rearmost pedal when in the inoperative position, must be maintained over its entire length.

The only thing which may encroach on this area is the steering wheel.

13.3.3 The driver, seated normally with his seat belts fastened and with the steering wheel removed, must be able to raise both legs together so that his knees are past the plane of the steering wheel in the rearward direction. This action must not be prevented by any part of the car.

13.4 Position of the driver's feet:

13.4.1 The survival cell must extend from behind the fuel tank in a rearward direction to a point at least 300 mm in front of the driver's feet, with his feet resting on the pedals and the pedals in the inoperative position.

13.4.2 When he is seated normally, the soles of the driver's feet, resting on the pedals in the inoperative position, must not be situated forward of the front wheel centre line.

Article 14: Safety Equipment

14.1 Fire extinguishers:

14.1.1 All cars must be fitted with a fire extinguishing system which will discharge into the cockpit and into the engine compartment.

14.1.2 Any AFFF which has been specifically approved by the FIA is permitted.

14.1.3 The quantity of extinguishant may vary according to the type of AFFF used. A list is available from the FIA.

14.1.4 When operated, the fire extinguishing system must discharge 95% of its contents at a constant pressure in no less than 10 seconds and no more than 30 seconds.

If more than one container with extinguishant is fitted, they must be released simultaneously.

14.1.5 Each pressure vessel must be equipped with a means of checking its pressure which may vary according to the type of AFFF used. A list is available from the FIA.

14.1.6 The following information must be visible on each container with extinguishant:

a) Type of extinguishant

b) Weight or volume of the extinguishant

c) Date the container must be checked which must be no more than two years after the date of filling.

14.1.7 All parts of the extinguishing system must be situated within the survival cell, and all extinguishing equipment must withstand fire.

14.1.8 Any triggering system having its own source of energy is permitted, provided it is possible to operate all extinguishers should the main electrical circuits of the car fail.

The driver must be able to trigger the extinguishing system manually when seated normally with his safety belts fastened and the steering wheel in place.

Furthermore, a means of triggering from the outside must be combined with the circuit breaker switch described in Article 14.2.2. It must be marked with a letter "E" in red inside a white circle of at least 100 mm diameter with a red edge.

14.1.9 The system must work in any position, even when the car is inverted.

14.1.10 All extinguisher nozzles must be suitable for the extinguishant and be installed in such a way that they are not directly pointed at the driver.

14.2 Master switch:

14.2.1 The driver, when seated normally with the safety belts fastened and the steering wheel in place, must be able to cut off the electrical circuits to the ignition, all fuel pumps and the rear light by means of a spark-proof circuit breaker switch.

This switch must be located on the dashboard and must be clearly marked by a symbol showing a red spark in a white edged blue triangle.

14.2.2 There must also be an exterior switch, with a horizontal handle, which is capable of being operated from a distance by a hook. This switch must be situated at the base of the main rollover structure on the right-hand side.

14.3 Rear view mirrors:
All cars must have at least two mirrors mounted so that the driver has visibility to the rear and both sides of the car.

The reflective surface of each mirror must be at least 120 mm wide, this being maintained over a height of at least 50 mm. Additionally, each corner may have a radius no greater than 10 mm.

The FIA technical delegate must be satisfied by a practical demonstration that the driver, when seated normally, can clearly define following vehicles.

For this purpose, the driver shall be required to identify any letter or number, 150 mm high and 100 mm wide, placed anywhere on boards behind the car, the positions of which are detailed below:

Height:	From 400 mm to 1000 mm from the ground.
Width:	2000 mm either side of the centre line of the car.
Position:	10 m behind the rear axle line of the car.

14.4 Safety belts:
It is mandatory to wear two shoulder straps, one abdominal strap and two straps between the legs. These straps must be securely fixed to the car and must comply with FIA standard 8853/98.

14.5 Rear light:

All cars must have a red light in working order throughout the Event which:

- Has been manufactured as specified by the FIA;
- Faces rearward at 90 degrees to the car centre line and the reference plane;
- Is clearly visible from the rear;
- Is not mounted more than 100 mm from the car centre line;
- Is positioned in order to comply with Article 3.10.3;
- Is no less than 450 mm behind the rear wheel centre line measured parallel to the reference plane;
- Can be switched on by the driver when seated normally in the car.

The three measurements above will be taken to the centre of the rear face of the light unit.

14.6 Cockpit padding:

14.6.1 All cars must be equipped with three areas of padding for the driver's head which:

- Are so arranged that they can be removed from the car as one part;
- Are located by two horizontal pegs behind the driver's head and two fixings, which are clearly indicated and easily removable without tools, at the front corners;
- Are made from a material specified by the FIA;
- Are fitted with a cover manufactured from 60-240 gsm materials which use suitable thermosetting resin systems;
- Are positioned so as to be the first point of contact for the driver's helmet in the event of an impact projecting his head toward them during an accident.

14.6.2 The first area of padding for the driver's head must be positioned behind him and be between 75 mm and 90 mm thick over an area of at least 40,000 mm^2.

14.6.3 The two further areas of padding for the driver's head must be positioned directly alongside each side of his helmet. The upper surfaces of these areas of padding must be at least as high as the survival cell over their entire length.

Each area of padding must be between 75 mm and 90 mm thick over an area of at least 25,000 mm^2 and may have a radius of 10 mm along its upper inboard edge. When calculating their area, any part which is greater than 75 mm thick and which lies between the front face of the rear area of padding and the forwardmost part of the driver's helmet whilst he is seated normally, will be taken into account (area 'B' in Drawing 4). The thickness will be measured perpendicular to the car centre line.

14.6.4 Forward of the side areas of padding, further cockpit padding must be provided on each side of the cockpit rim. The purpose of the additional padding is to afford protection to the driver's head in the event of an oblique frontal impact and must therefore be made from the same material as the other three areas of padding.

These extensions must:

- Be symmetrically positioned about the car centre line and a continuation of the side areas of padding;
- Be positioned with their upper surfaces at least as high as the survival cell over their entire length;
- Have a radius on their upper inboard edge no greater than 10 mm;
- Be positioned in order that the distance between the two is no less than 360 mm;
- Be as high as practicable within the constraints of driver comfort.

14.6.5 All of the padding described above must be so installed that if movement of the driver's head, in any expected trajectory during an accident, were to compress the foam fully at any point, his helmet would not make contact with any structural part of the car.

Furthermore, for the benefit of rescue crews, all of the padding described above must be installed using the FIA approved system. The method of removal must also be clearly indicated.

14.6.6 No part of the padding described above may obscure sight of any part of the driver's helmet when he is seated normally and viewed from directly above the car.

14.6.7 In order to minimise the risk of leg injury during an accident, additional areas of padding must be fitted each side of, and above, the driver's legs.

These areas of padding must:

- Be made from a material specified by the FIA;
- Be no less than 25 mm thick over their entire area;
- Cover the area situated between points lying 50 mm behind the centre of the point at which the second roll structure test is carried out and 100 mm behind the face of the rearmost pedal when in the inoperative position, as shown in Drawing 4;
- Cover the area above the line A–A shown in Drawing 3.

14.7 Wheel retention:

All cars, whilst under their own power, must be fitted with devices which will retain any wheel in the event of it coming loose.

After the wheel nut is fastened, these devices must be manually fitted in a separate action to that of securing the wheel nut.

14.8 Seat fixing and removal:

14.8.1 In order that an injured driver may be removed from the car in his seat following an accident, all cars must be fitted with a seat which, if it is secured, must be done so with no more than two bolts. If bolts are used, they must:

- Be clearly indicated and easily accessible to rescue crews;
- Be fitted vertically;
- Be removable with the same tool for all Teams and which is issued to all rescue crews.

14.8.2 The seat must be equipped with receptacles which permit the fitting of belts to secure the driver and one which will permit the fitting of a neck support.

14.8.3 The seat must be removable without the need to cut or remove any of the seat belts.

14.8.4 Details of the tool referred to above, the belt receptacles and the neck support are available from the FIA Technical Department.

Article 15: Safety Structures

15.1 Materials:

15.1.1 The use of magnesium sheet less than 3 mm thick is forbidden.

15.1.2 No parts of the car may be made from metallic materials which have a specific modulus of elasticity greater than 40 GPa/(g/cm^3).

15.2 Roll structures:

15.2.1 All cars must have two roll structures which are designed to help prevent injury to the driver in the event of the car becoming inverted.

The principal structure must be at least 940 mm above the reference plane at a point 30 mm behind the cockpit entry template. The second structure must be in front of the steering wheel but no more than 250 mm forward of the top of the steering wheel rim in any position.

The two roll structures must be of sufficient height to ensure the driver's helmet and his steering wheel are at least 70 mm and 50 mm, respectively, below a line drawn between their highest points at all times.

15.2.2 The principal structure must pass a static load test, details of which may be found in Article 17.2. Furthermore, each Team must supply detailed calculations which clearly show that it is capable of withstanding the same load when the longitudinal component is applied in a forward direction.

15.2.3 The second structure must pass a static load test, details of which may be found in Article 17.3.

15.2.4 Both roll structures must have minimum structural cross sections of 10,000 mm^2, in vertical projection, across a horizontal plane 50 mm below their highest points.

15.3 Structure behind the driver:

The parts of the survival cell immediately behind the driver which separate the cockpit from the car's fuel tank, and which lie less than 150 mm from the centre line of the car, may be situated no further forward than the line a–b–c–d–e shown in Drawing 2.

15.4 Survival cell specifications:

15.4.1 Every survival cell must incorporate three FIA supplied transponders for identification purposes. These transponders must be a permanent part of the survival cell, be positioned in accordance with Drawing 7 and must be accessible for verification at any time.

15.4.2 The survival cell must have an opening for the driver, the minimum dimensions of which are given in Article 13.1. Any other openings in the survival cell must be of the minimum size to allow access to mechanical components.

15.4.3 An impact absorbing structure must be fitted in front of the survival cell. This structure need not be an integral part of the survival cell but must be solidly attached to it.

Furthermore, it must have a minimum external cross section, in horizontal projection, of 9000 mm^2 at a point 50 mm behind its forward-most point.

15.4.4 Referring to Drawing 5:

The external width of the survival cell between the lines B–B and C–C must be no less than 450 mm and must be at least 60 mm per side wider than the cockpit opening when measured normal to the inside of the cockpit aperture. These minimum dimensions must be maintained over a height of at least 350 mm.

The width of the survival cell may taper forward of the line B–B but, if this is the case, it must do so at a linear rate to a minimum of 300 mm at the line A–A.

Between the lines A–A and B–B the width of the survival cell must be greater than the width defined by the two lines a–b. This minimum width must be arranged symmetrically about the car centre line, must be maintained over a height of at least 400 mm at the line B–B and may taper at a linear rate to 275 mm at the line A–A. When assessing the minimum external cross sections of the survival cell, radii of 50 mm at the line B–B, and reducing at a linear rate of 25 mm at the line A–A, will be permitted.

The minimum height of the survival cell between the lines A–A and B–B need not be arranged symmetrically about the horizontal centre line of the relevant section but must be maintained over its entire width.

The minimum height of the survival cell between the lines B–B and C–D is 550 mm.

15.4.5 When the test referred to in Article 13.1.1 is carried out and the template is in position with its lower edge 525 mm above the reference plane, the shape of the survival cell must be such that no part of it is visible when viewed from either side of the car.

The parts of the survival cell which are situated each side of the driver's helmet must be no more than 550 mm apart and, in order to maintain good lateral visibility the driver, when seated normally with his seat belts fastened and looking straight ahead, must have his eyes above the sides of the survival cell.

15.4.6 In order to give additional protection to the driver in the event of a side impact, a flat test panel of uniform construction, which is designed and constructed in order to represent a section of survival cell sides, must pass a strength test. Details of the test procedure may be found in Article 18.6.

Referring to Drawing 5, with the exception of local reinforcement and/or inserts, all parts of the survival cell which are as wide or wider than the minimum widths stipulated in Article 15.4.4, including any radii applied, must be manufactured to the same specification as the panel tested under Article 18.6. Furthermore, parts of this tested specification must cover an area which:

- Begins at least 250 mm high at line A–A;
- Tapers at a linear rate to at least 400 mm high at line B–B and which remains at this height to the rear of the survival cell;
- Is no less than 100 mm above the reference plane between the line B–B and the rear of the survival cell.

15.5 Survival cell safety requirements:

15.5.1 The survival cell and frontal absorbing structure must pass an impact test against a solid vertical barrier placed at right angles to the centre line of the car. Details of the test procedure may be found in Article 16.2.

15.5.2 Between the front and rear roll structures, on each side of the survival cell, impact absorbing structures must be fitted and must be solidly attached to it. The purpose of these structures is to protect the driver in the event of a lateral impact and, in order to ensure this is the case, a lateral strength test in the vicinity of the driver's seating position must be carried out successfully. Details of the test procedure may be found in Article 18.2.2.

The survival cell and one of these impact absorbing structures must pass an impact test. Details of the test procedure may be found in Article 16.3. If these structures are not designed and fitted symmetrically about the car centre line, a successful impact test must be carried out on them both.

15.5.3 An impact absorbing structure must be fitted behind the gearbox symmetrically about the car centre line with its rearmost point no less than 480 mm behind the rear wheel centre line. It must also have a minimum external cross section, in horizontal projection, of 9000 mm^2 at a point 50 mm forward of its rearmost point. When calculating this area, only those parts situated less than 100 mm from the car centre line may be considered and the cross section may not diminish forward of this point.

This structure must pass an impact test and must be constructed from materials which will not be substantially affected by the temperatures it is likely to be subjected to during use. Details of the test procedure may be found in Article 16.4.

15.5.4 The survival cell must also be subjected to five separate static load tests:

1) On a vertical plane passing through the centre of the fuel tank;
2) On a vertical plane passing through the rearmost point at which the outer end of the front wheel tether would make contact with the survival cell when swung about the inner attachment;
3) On a vertical plane 375 mm forward of the rear edge of the cockpit entry tem plate;
4) From beneath the fuel tank;
5) On each side of the cockpit opening.

Details of the test procedures may be found in Article 18.2.

15.5.5 To test the attachments of the frontal impact absorbing structure to the survival cell, a static side load test must be carried out. Details of the test procedure may be found in Article 18.5.

Article 16: Impact Testing

16.1 Conditions applicable to all impact tests:

16.1.1 All tests must be carried out in accordance with FIA Test Procedure 01/99, in the presence of an FIA technical delegate and by using measuring equipment which has been calibrated to the satisfaction of the FIA technical delegate.

16.1.2 Any significant modification introduced into any of the structures tested shall require that part to pass a further test.

16.2 Frontal test:

All parts which could materially affect the outcome of the test must be fitted to the test structure which must be solidly fixed to the trolley through its engine mounting points but not in such a way as to increase its impact resistance.

The fuel tank must be fitted and must be full of water.

A dummy weighing at least 75 kg must be fitted with safety belts described in Article 14.4 fastened. However, with the safety belts unfastened, the dummy must be able to move forward freely in the cockpit.

The extinguishers, as described in Article 14.1, must also be fitted.

For the purposes of this test, the total weight of the trolley and test structure shall be 780 kg and the velocity of impact 14.0 metres/sec.

The resistance of the test structure must be such that during the impact:

- The average deceleration over the first 150 mm of deformation does not exceed 5g;
- The average deceleration of the trolley does not exceed 40g;
- The peak deceleration in the chest of the dummy does not exceed 60g for more than a cumulative 3 ms, this being the resultant of data from three axes.

Furthermore, there must be no damage to the survival cell or to the mountings of the safety belts or fire extinguishers.

This test must be carried out on the survival cell subjected to the higher loads in the tests described in Articles 18.2–4, and on a frontal impact absorbing structure identical to the one which was subjected to the test described in Article 18.5.

16.3 Side test:

All parts which could materially affect the outcome of the test must be fitted to the test structure which must be solidly fixed to the ground and a solid object, having a mass of 780 kg and travelling at a velocity of 10 m/s, will be projected into it.

The object used for this test must

- Incorporate an impactor assembly specified by the FIA which is fitted in accordance with their instructions;
- Be positioned in order that its centre of area strikes the structure 300 mm above the reference plane and at a point 500 mm forward of the rear edge of the cockpit opening template.

During the test the striking object may not pivot in any axis, and the survival cell may be supported in any way provided this does not increase the impact resistance of the parts being tested. The impact axis must be perpendicular to the car centre line and parallel to the ground.

The resistance of the test structure must be such that during the impact:

- The average deceleration of the object, measured in the direction of impact, does not exceed 20g;
- The force applied to any one of the four impactor segments does not exceed 80 kN for more than a cumulative 3 ms;
- The energy absorbed by each of the four impactor segments must be between 15% and 35% of the total energy absorption.

Furthermore, all structural damage must be contained within the impact absorbing structure.

This test must be carried out on the survival cell subjected to the higher loads in the tests described in Articles 18.2–4

16.4 Rear test:

All parts which will be fitted behind the rear face of the engine and which could materially affect the outcome of the test must be fitted to the test structure. If suspension members are to be mounted on the structure, they must be fitted for the test. The structure and the gearbox must be solidly fixed to the ground and a solid object, having a mass of 780 kg and travelling at a velocity of 12 m/s, will be projected into it.

The object used for this test must be flat, measure 450 mm wide by 550 mm high and may have a 10 mm radius on all edges. Its lower edge must be at the same level as the car reference plane and must be so arranged to strike the structure vertically and at 90° to the car centre line.

During the test, the striking object may not pivot in any axis, and the crash structure may be supported in any way provided this does not increase the impact resistance of the parts being tested.

The resistance of the test structure must be such that during the impact:

- The average deceleration of the object does not exceed 35g;
- The maximum deceleration does not exceed 60g for more than a cumulative 3 ms, this being measured only in the directon of impact.

Furthermore, all structural damage must be contained within the area behind the rear wheel centre line.

16.5 Steering column test:

The parts referred to in Article 10.4.4 must be fitted to a representative test structure. Any other parts which could materially affect the outcome of the test must also be fitted. The test structure must be solidly fixed to the ground, and a solid object, having a mass of 8 kg and travelling at a velocity of 7 m/s, will be projected into it.

The object used for this test must be hemispherical with a diameter of 165 mm.

For the test, the centre of the hemisphere must strike the structure at the centre of the steering wheel along the same axis as the main part of the steering column.

During the test the striking object may not pivot in any axis, and the test structure may be supported in any way provided this does not increase the impact resistance of the parts being tested.

The resistance of the test structure must be such that during the impact, the peak deceleration of the object does not exceed 80g for more than a cumulative 3 ms, this being measured only in the direction of impact.

After the test, all substantial deformation must be within the steering column and the steering wheel quick release mechanism must still function normally.

Article 17: Roll Structure Testing

17.1 Conditions applicable to both roll structure tests:

17.1.1 Rubber 3 mm thick may be used between the load pads and the roll structure.

17.1.2 Under the load, deformation must be less than 50 mm, measured along the loading axis, and any structural failure limited to 100 mm below the top of the rollover structure when measured vertically.

17.1.3 Any significant modification introduced into any of the structures tested shall require that part to pass a further test.

17.2 Principal roll structure:
A load equivalent to 50 kN laterally, 60 kN longitudinally in a rearward direction and 90 kN vertically, must be applied to the top of the structure through a rigid flat pad which is 200 mm in diameter and perpendicular to the loading axis.

During the test, the roll structure must be attached to the survival cell which is supported on its underside on a flat plate, fixed to it through its engine mounting points and wedged laterally by any of the static load test pads described in Article 18.2.

17.3 Second roll structure:
A vertical load of 75 kN must be applied to the top of the structure through a rigid flat pad which is 100 mm in diameter and perpendicular to the loading axis.

During the test, the rollover structure must be attached to the survival cell which is fixed to a flat horizontal plate.

Article 18: Static Load Testing

18.1 Conditions applicable to the tests in 18.2–18.5:

18.1.1 All the following tests must be carried out on the survival cell subjected to the impact tests described in Article 16.

18.1.2 Every subsequent survival cell must also be subjected to all the following tests with peak loads reduced by 20%. During these subsequent tests (on deflections greater than 3.0 mm), the deflection across the inner surfaces must not exceed 120% of the deflection obtained at 80% of the peak load during the first test.

18.1.3 Deflections and deformations will be measured at the centre of area of circular load pads and at the top of rectangular pads.

18.1.4 All peak loads must be applied in less than three minutes, through a ball jointed junction at the centre of area of the pad, and maintained for 30 seconds.

18.1.5 In the tests described in 18.2, 18.3 and 18.4, permanent deformation must be less than 1.0 mm (0.5 mm in 18.3) after the load has been released for 1 minute.

18.1.6 All tests must be carried out by using measuring equipment which has been calibrated to the satisfaction of the FIA technical delegate.

18.1.7 A radius of 3 mm is permissible on the edges of all load pads and rubber 3 mm thick may be placed between them and the test structure.

18.1.8 For the tests described in 18.2, 18.3 and 18.4, the survival cells must always be produced in an identical condition in order that their weights may be compared. If the weight differs by more than 5% from the one subjected to the impact tests described in Articles 16.2 and 16.3, further frontal and side impact tests and roll structure tests must be carried out.

18.1.9 Any significant modification introduced into any of the structures tested shall require that part to pass a further test.

18.2 Survival cell side tests:

18.2.1 For test 1), referred to in Article 15.5.4, pads 100 mm long and 300 mm high, which conform to the shape of the survival cell, must be placed against the outermost sides of the survival cell with the lower edge of the pad at the lowest part of the survival cell at that section.

A constant transverse horizontal load of 25.0 kN will be applied and, under the load, there must be no structural failure of the inner or outer surfaces of the survival.

On every survival cell tested after that one, the same tests must be carried out but with a load of only 20.0 kN. During the tests, on deflections greater than 3.0 mm only, the deflection across the inner surfaces must not exceed 120% of the deflection obtained at 20.0 kN load during the first test.

18.2.2 For test 2), referred to in Article 15.5.4, pads 200 mm in diameter which conform to the shape of the survival cell must be placed against the outermost sides of the survival cell.

The centre of the pads must pass through the plane mentioned above and the midpoint of the height of the structure at that section.

A constant transverse horizontal load of 30.0 kN will be applied to the pads, and, under the load, there must be no structural failure of the inner or outer surfaces of the survival cell, and the total deflection must not exceed 15 mm.

18.2.3 For test 3), referred to in Article 15.5.4, pads 200 mm in diameter which conform to the shape of the survival cell, must be placed against the outermost sides of the survival cell.

The centre of the pads must be located 350 mm above the reference plane and on the vertical plane mentioned in Article 15.5.4.

A constant transverse horizontal load of 30.0 kN will be applied to the pads and, under the load, there must be no structural failure of the inner or outer surfaces of the survival cell, and the total deflection must not exceed 15 mm.

18.3 **Fuel tank floor test:**

A pad of 200 mm diameter must be placed in the centre of area of the fuel tank floor and a vertical upward load of 12.5 kN applied.

Under the load, there must be no structural failure of the inner or outer surfaces of the survival cell.

18.4 **Cockpit rim test:**

Two pads, each of which is 100 mm in diameter, must be placed on both sides of the cockpit rim with their upper edges at the same height as the top of the cockpit side with their centres at a point 200 mm forward of the rear edge of the cockpit opening template longitudinally.

A constant transverse horizontal load of 10.0 kN will then be applied at 90° to the car centre line, and, under the load, there must be no structural failure of the inner or outer surfaces of the survival cell, and the total deflection must not exceed 20 mm.

18.5 Nose push off test:
During the test the survival cell must be resting on a flat plate and secured to it solidly but not in a way that could increase the strength of the attachments being tested.

A constant transversal horizontal load of 40.0 kN must then be applied to one side of the impact absorbing structure, using a pad identical to the ones used in the lateral tests in Article 18.2.1, at a point 550 mm from the front wheel axis.

The centre of area of the pad must pass through the plane mentioned above and the midpoint of the height of the structure at the relevant section. After 30 seconds of application, there must be no failure of the structure or of any attachment between the structure and the survival cell.

18.6 Side intrusion test:
The precise test procedure is yet to be defined but will be based on the tests carried out recently on test panels supplied by four Teams.

Article 19: Fuel

19.1 Purpose of Article 19:

19.1.1 The purpose of this Article is to ensure that the fuel used in Formula One is petrol as this term is generally understood.

19.1.2 The detailed requirements of this Article are intended to ensure the use of fuels which are predominantly composed of compounds normally found in commercial fuels and to prohibit the use of specific power-boosting chemical compounds.

19.1.3 Additionally, and in order to encourage the development of future commercial fuels, those formulated to achieve one or more of the following objectives will be permitted:

a) Fuels needed to meet advanced passenger car engine designs;

b) Fuels formulated to minimise overall emissions;

c) Fuels suitable to be offered to the commercial market with some special feature permitting greater efficiency, better driveability or economy to the user;

d) Fuels developed through advances in refinery techniques and suitable for trial by the general public.

19.1.4 Any petrol which appears to have been formulated in order to subvert the purpose of this regulation will be deemed to be outside it.

19.2 Definitions:

Paraffins – Straight chain and branched alkanes.

Olefins – Straight chain and branched mono-olefins.
– Monocyclic mono-olefins (with five or more carbon atoms in the ring) and saturated aliphatic side chains.

Naphthenes – Monocyclic paraffins (with five or more carbon atoms in the ring) and saturated aliphatic side chains.

Aromatics – Monocyclic and bicyclic aromatic rings with and without saturated aliphatic side chains and/or fused naphthenic rings.

Oxygenates – Specified organic compounds containing oxygen.

19.3 Properties:

The only fuel permitted is petrol having the following characteristics:

Property	Units	Minimum	Maximum	Test Method
RON		95.0	102.0	ASTM D 2699-86
MON		85.0		ASTM D 2700-86
Oxygen	% m/m		2.7	Elemental analysis
Nitrogen	% m/m		0.2	ASTM D 3228
Benzene	% v/v		1.0	EN 238
RVP	hPa	450	600	ASTM D 323
Lead	g/l		0.005	ASTM D 3237
Density at 15°C	kg/m^3	720.0	775.0	ASTM D 4052
Oxidation stability	minutes	360		ASTM D 525
Existent gum	mg/100 ml		5.0	EN 26246
Sulphur	mg/kg		50	EN-ISO/DIS 14596
Copper corrosion	rating		C1	ISO 2160
Electrical conductivity	pS/m	200		ASTM D 2624

Distillation characteristics:

At E70°C	% v/v	20.0	48.0	ISO 3405
At E100°C	% v/v	46.0	71.0	ISO 3405
At E150°C	% v/v	75.0		ISO 3405
Final boiling point	°C		210	ISO 3405
Residue	% v/v		2.0	ISO 3405

The fuel will be accepted or rejected according to ASTM D 3244 with a confidence limit of 95%.

19.4 Composition of the fuel:

19.4.1 The petrol must consist solely of substances defined in 19.2 and 19.4.4, and whose proportions of aromatics, olefins and di-olefins, within the total petrol sample, comply with those detailed below:

	Units	Minimum	Maximum	Test Method
Aromatics	% v/v	0*	35*	ASTM D 1319
Olefins	% v/v	0	18*	ASTM D 1319
Total di-olefins	% m/m	0	1	GCMS
* Values when corrected for fuel oxygenate content.				

In addition, the fuel must contain no substance which is capable of exothermic reaction in the absence of external oxygen.

19.4.2 The total of individual hydrocarbon components present at concentrations of less than 5% m/m must be at least 30% m/m of the fuel.

19.4.3 The total concentration of each hydrocarbon group in the total fuel sample (defined by carbon number and hydrocarbon type, must not exceed the limits given in the table below:

% m/m	C4	C5	C6	C7	C8	C9+	Unallocated
Paraffins	10	30	25	25	55	20	–
Naphthenes	–	5	10	10	10	10	–
Olefins	5	20	20	15	10	10	–
Aromatics	–	–	1.2	35	35	30	–
Maximum	15	40	45	50	60	45	10

For the purposes of this table, a gas chromatographic technique should be employed which can classify hydrocarbons in the total fuel sample such that all those identified are allocated to the appropriate cell of the table. Hydrocarbons present at concentrations below 0.5% by mass which cannot be allocated to a particular cell may be ignored. However, the sum of the unallocated hydrocarbons must not exceed 10.0% by mass of the total fuel sample.

19.4.4 The only oxygenates permitted are:

Methanol (MeOH)
Ethanol (EtOH)
Iso-propyl alcohol (IPA)
Iso-butyl alcohol (IBA)
Methyl Tertiary Butyl Ether (MTBE)
Ethyl Tertiary Butyl Ether (ETBE)
Tertiary Amyl Methyl Ether (TAME)
Di-Isopropyl Ether (DIPE)
n-Propyl Alcohol (NPA)
Tertiary Butyl Alcohol (TBA)
n-Butyl Alcohol (NBA)
Secondary Butyl Alcohol (SBA)

Compounds normally found as impurities in any of the above oxygenates are permitted at concentrations below 0.8% m/m of the total petrol sample.

19.5 Air:

Only ambient air may be mixed with the fuel as an oxidant.

19.6 Safety:

19.6.1 Manganese based additives are not permitted.

19.6.2 All competitors must be in possession of a Material Safety Data Sheet for each type of petrol used. This sheet must be made out in accordance with EC Directive 93/112/EEC and all information contained therein strictly adhered to.

19.7 Fuel approval:

19.7.1 Before any fuel may be used in an Event, two separate five litre samples, in suitable containers, must be submitted to the FIA for analysis and approval.

19.7.2 No fuel may be used in an Event without prior written approval of the FIA.

19.8 Sampling and testing:

19.8.1 All samples will be taken in accordance with FIA Formula One fuel sampling procedures.

19.8.2 Fuel samples taken during an Event will be checked for conformity by using densitometry and a gas chromatographic technique which will compare the sample taken with an approved fuel.

19.9 Amendments to Article 19:

19.9.1 The physical and compositional properties of the fuel described in 19.3 and 19.4 incorporate the currently known limits for 2000, as laid out in European Fuels Directive 98/70/EC (13 October 1998).

19.9.2 When the Final Directive, as defined by the FIA, is adopted for 2005 (or such other date as the Directive may specify), the new values will replace those being used in 19.3 and 19.4 no later than one year after the figures are known.

Article 20: Television Cameras

20.1 Presence of cameras and camera housings:
All cars must be fitted with either two cameras, two camera housings or one of each at all times throughout the Event.

20.2 Location of camera housings:
Camera housings, when used, must be fitted in the same location as cameras.

20.3 Location of camera equipment:
All cars must be equipped with five positions in which cameras or camera housings can be fitted. Referring to Drawing 6, all cars must carry a camera or camera housing in position 4. The position of the remaining camera or camera housing will be determined by the FIA after consultation with the relevant Competitor.

Once positions are determined in the above manner, any decision as to whether a camera or camera housing is fitted in those positions will rest solely with the relevant Competitor.

20.4 Timing transponders

All cars must be fitted with a timing transponder supplied by the officially appointed timekeepers. This transponder must be fitted in strict accordance with the instructions of the FIA.

Article 21: Final Text

The final text for these regulations shall be the English version should any dispute arise over their interpretation.

Article 22: Changes for 2002

22.1 Changes to Article 15.5.5:

15.5.5 To test the attachments of the frontal and rear impact absorbing structures, static side load tests must be carried out. Details of these test procedures may be found in Articles 18.5 and 18.7.

22.2 Addition of an Article 18.7:

18.7 Rear impact structure push off test:

During the test, the gearbox and the structure must be solidly fixed to the ground, but not in a way that could increase the strength of the attachments being tested.

A constant transversal horizontal load of 40 kN must then be applied to one side of the impact absorbing structure, using a pad identical to the ones used in the lateral tests in Article 18.2.1 at a point 300 mm from the rear wheel axis.

The centre of area of the pad must pass through the plane mentioned above and the midpoint of the height of the structure at the relevant section. After 30 seconds of application, there must be no failure of the structure or of any attachment between the structure and the gearbox.

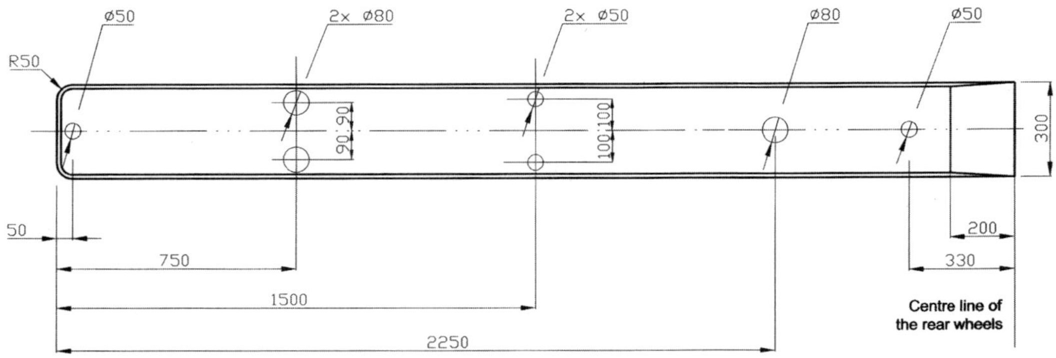

Drawing 1 Skid block dimensions.

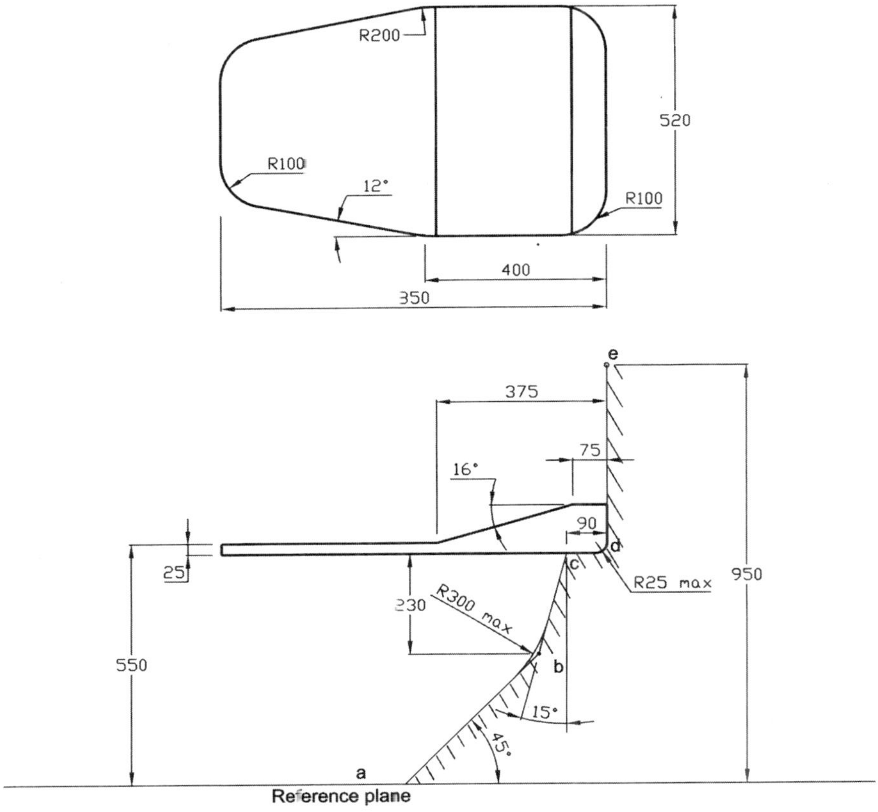

Drawing 2 Cockpit entry template.

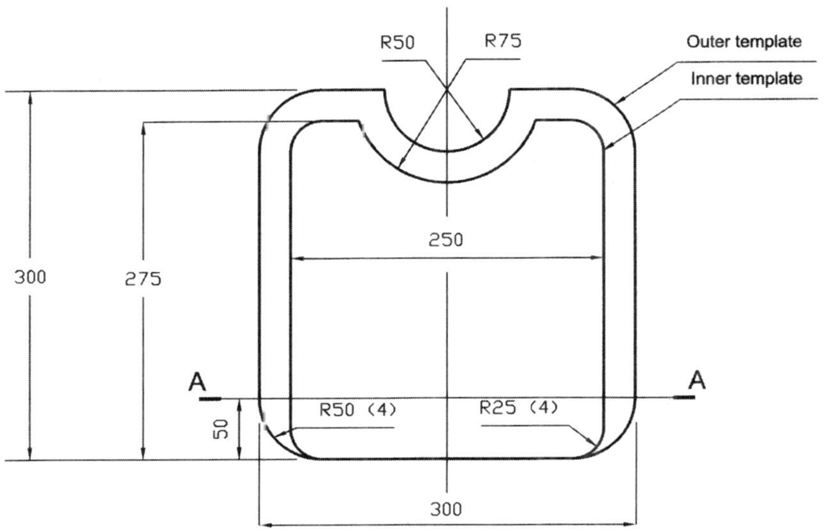

Drawing 3 Cockpit cross-section template.

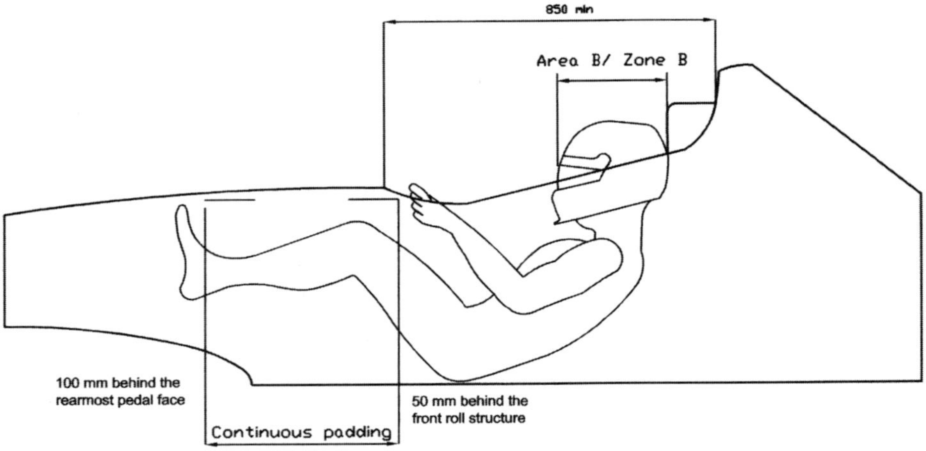

Drawing 4 Cockpit padding.

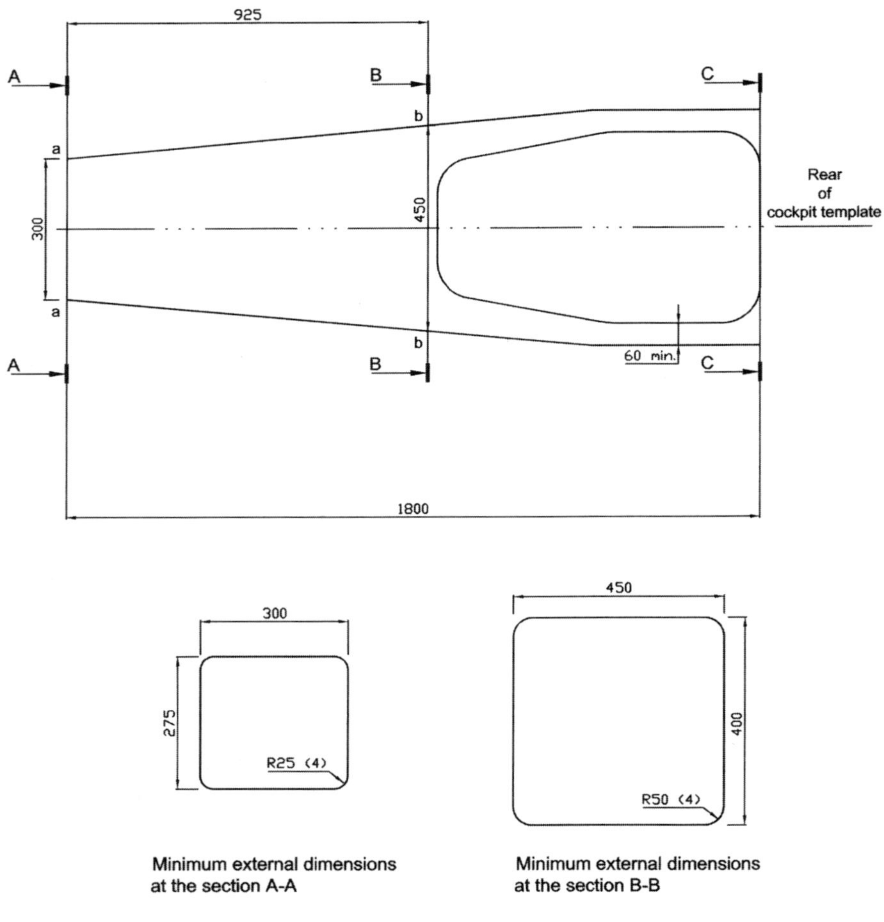

Drawing 5 Survival cell dimensions.

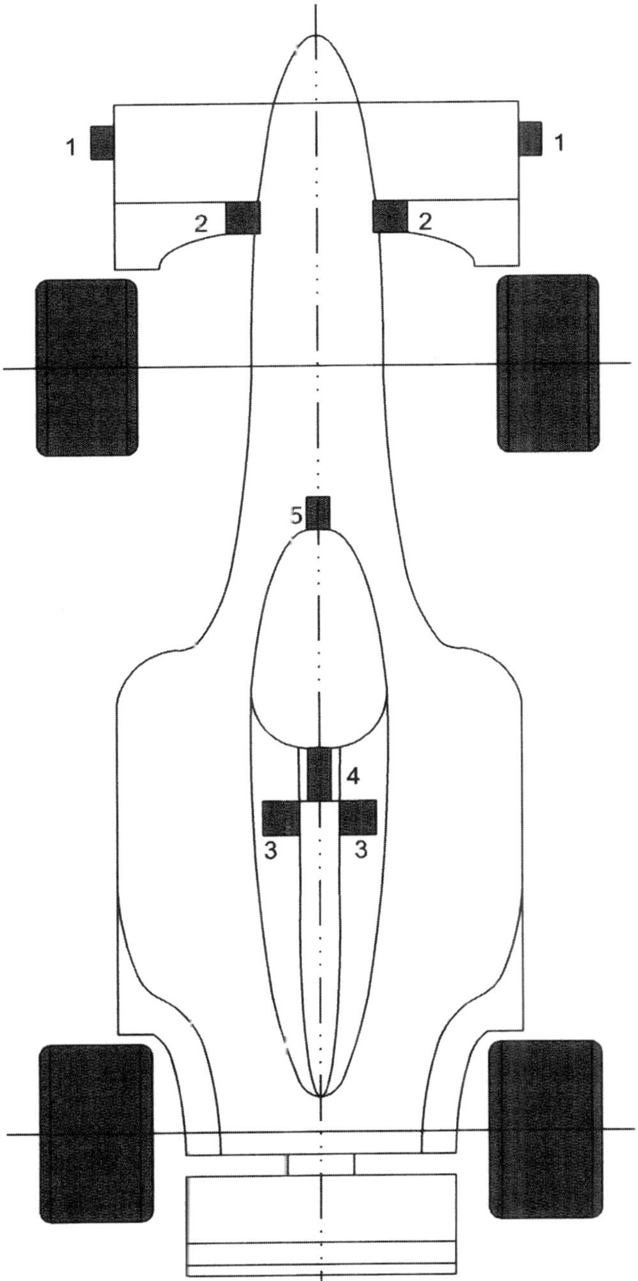

Drawing 6 Standard onboard camera positions.

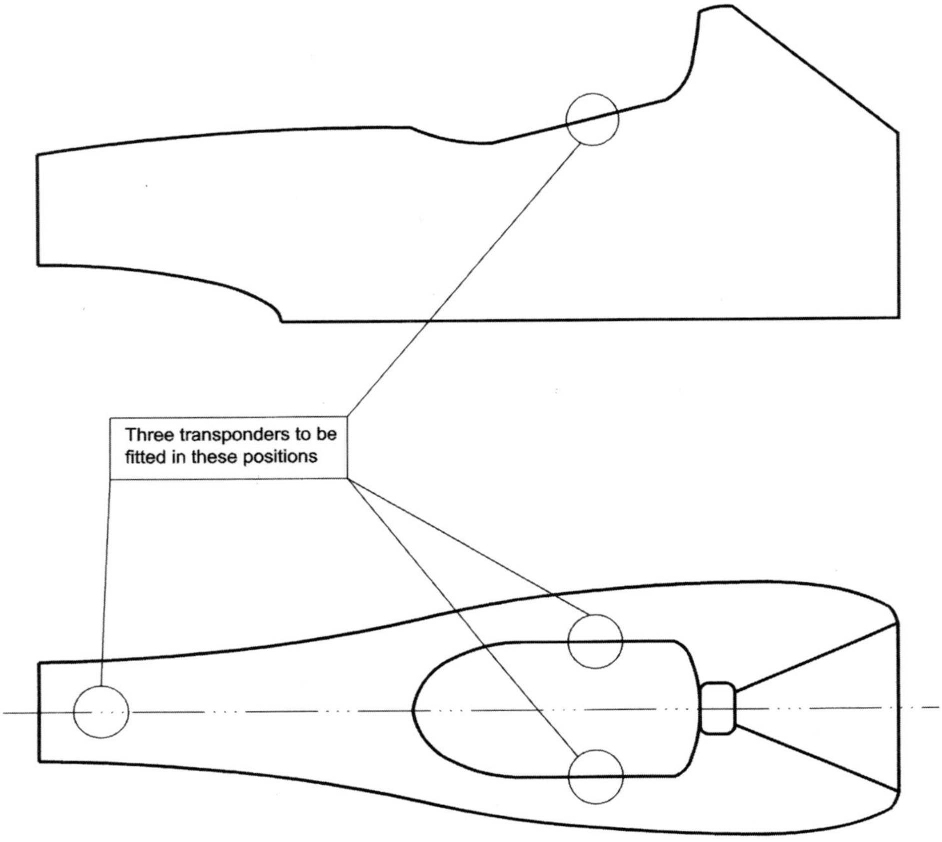

Drawing 7 Chassis transponder positions.

Appendix E

Relevant Sections of the 2001 Formula One Sporting Regulations

Fédération Internationale de l'Automobile

General Conditions

6) It is the Competitor's responsibility to ensure that all persons concerned by his entry observe all the requirements of the Agreement, the Code, the Technical Regulations and the Sporting Regulations. If a competitor is unable to be present in person at the Event, he must nominate his representative in writing. The person having charge of an entered car during any part of an Event is responsible jointly and severally with the Competitor for ensuring that the requirements are observed.

7) Competitors must ensure that their cars comply with the conditions of eligibility and safety throughout practice and the race.

8) The presentation of a car for scrutineering will be deemed an implicit statement of conformity.

9) All persons concerned in any way with an entered car or present in any other capacity whatsoever in the paddock, pits, pit lane, or track must wear an appropriate pass at all times.

Testing

65) No testing is permitted:

a) On any circuit outside Europe except general testing organised by the Commercial Rights Holder;

b) On any circuit during the seven days preceding and the day following race day at each Event except for a shakedown test not exceeding 50 km which must be supervised by the relevant ASN;

c) On any circuit between the last Event of the Championship and 31 December;

d) On any circuit between two Events in the month of August, provided such events are separated by two clear weekends, except for one shakedown test not exceeding which must be supervised by the relevant ASN;

e) On any circuit which has hosted, or will host, an Event during the Championship year other than those hosting the British, French, Italian and Spanish Grand Prix, provided that testing on the latter four circuits may not take place during the 28 days preceding the race day of the relevant Event in each case.

f) At sites which are not currently approved for Formula 1.

If any test is to be attended by more than one Team, it is the responsibility of the Teams concerned to inform the FIA. A representative will then be appointed by the FIA to attend the test and ensure that track safety standards are maintained and normal running procedures respected at all times.

Pit Lane

66) a) For the avoidance of doubt and for description purposes, the pit lane shall be divided into two lanes. The lane closest to the pit wall is designated the "fast lane," and the lane closest to the garages is designated the "inner lane." Other than when cars are at the pit exit under Articles 137 or 159, the inner lane is the only area where any work can be carried out on a car.

b) Competitors may not attempt to enhance the grip of the surface in the pit lane unless a problem has been clearly identified and a solution agreed by the FIA Safety Delegate.

c) Competitors must not paint lines on any part of the pit lane.

d) No equipment may be left in the fast lane. A car may enter or remain in the fast lane only with the driver sitting in the car behind the steering wheel in his normal position, even when the car is being pushed.

e) Team personnel are only allowed in the pit lane immediately before they are required to work on a car and must withdraw as soon as the work is complete.

f) It is the responsibility of the Competitor to release his car after a pit stop only when it is safe to do so.

Sporting Checks

67) Each competitor must have all relevant Super Licences available for inspection at any time during the Event.

68) At the first Event of each Championship, the FIA will check all licences.

69) No competitor, driver or other person concerned with a car can be required to sign any waiver or other document save as provided by the Agreement.

Scrutineering

70) Initial scrutineering of the car will take place three days (Monaco: four days) before the race between 10.00 and 16.00 in the garage assigned to each team.

71) Unless a waiver is granted by the Stewards, Competitors who do not keep to these time limits will not be allowed to take part in the Event.

72) No car may take part in the Event until it has been passed by the scrutineers.

73) The scrutineers may:

a) Check the eligibility of a car or of a Competitor at any time during an Event,

b) Require a car to be dismantled by the Competitor to make sure that the conditions of eligibility or conformity are fully satisfied,

c) Require a Competitor to pay the reasonable expenses which exercise of the powers mentioned in this Article may entail,

d) Require a Competitor to supply them with such parts or samples as they may deem necessary.

74) Any car which, after being passed by the scrutineers, is dismantled or modified in a way which might affect its safety or call into question its eligibility, or which is involved in an accident with similar consequences, must be re-presented for scrutineering approval.

75) The race director or the clerk of the course may require that any car involved in an accident be stopped and checked.

76) Checks and scrutineering shall be carried out by duly appointed officials who shall also be responsible for the operation of the *parc fermé* and who alone are authorised to give instructions to the competitors.

77) The Stewards will publish the findings of the scrutineers each time cars are checked during the Event. These results will not include any specific figure except when a car is found to be in breach of the Technical Regulations.

Supply of Tyres in the Championship and Tyre Limitation During the Event

78) Supply of tyres:

a) Any tyre company wishing to supply tyres to Formula One Teams must notify the FIA of its intention to do so no later than 1 January preceding the year during which such tyres will be supplied.

Any tyre company wishing to cease the supply of tyres to Formula One Teams must notify the FIA of its intention to do so no later than 1 January of the year preceding that in which such tyres were to be supplied.

b) No tyre may be used in the Championship unless the company supplying such tyre accepts and adheres to the following conditions:

- One tyre supplier present in the Championship: this company must equip 100% of the entered teams on ordinary commercial terms;
- Two tyre suppliers present: each of them must, if called upon to do so, be prepared to equip up to 60% of the entered teams on ordinary commercial terms;
- Three or more tyre suppliers present: each of them must, if called upon to do so, be prepared to equip up to 40% of the entered teams on ordinary commercial terms;
- Each tyre supplier must undertake to provide only two specifications of dry-weather tyre and three specifications of wet-weather tyre at each Event, each of which must be of one homogenous compound only;
- If, in the interests of maintaining current levels of circuit safety, the FIA deems it necessary to reduce tyre grip, it shall introduce such rules as the tyre suppliers may advise or, in the absence of advice which achieves the FIA's objectives, specify the maximum permissible contact areas for front and rear tyres.

79) Quantity and type of tyres:

a) During the Event, no driver may use more than forty dry-weather tyres and twenty-eight wet-weather tyres. From the forty dry-weather tyres, each driver will be allocated twelve (six front and six rear) for use on the first day of practice, these tyres may not be used at any other time during the Event. No more than eight (four front and four rear) of the twelve tyres allocated for the first day of practice may be of one specification.

Before qualifying practice begins, each driver must nominate which specification of tyre he will use for the remainder of the Event.

b) All dry-weather tyres must incorporate circumferential grooves square to the wheel axis and around the entire circumference of the contact surface of each tyre.

c) Each front dry-weather tyre, when new, must incorporate 4 grooves which are:

- Arranged symmetrically about the centre of the tyre tread;
- At least 14 mm wide at the contact surface and which taper uniformly to a minimum of 10 mm at the lower surface;
- At least 2.5 mm deep across the whole lower surface;
- 50 mm (±1.0 mm) between centres.

Furthermore, the tread width of the front tyres must not exceed 270 mm.

d) Each rear dry-weather tyre, when new, must incorporate 4 grooves which are:

- Arranged symmetrically about the centre of the tyre tread;
- At least 14 mm wide at the contact surface and which taper uniformly to a minimum of 10 mm at the lower surface;
- At least 2.5 mm deep across the whole lower surface;
- 50 mm (±1.0 mm) between centres.

The measurements referred to in c) and d) above will be taken when the tyre is fitted to a wheel and inflated to 1.4 bar.

e) A wet-weather tyre is one which has been designed for use on a wet or damp track.

All wet-weather tyres must, when new, have a contact area which does not exceed 280 cm^2 when fitted to the front of the car and 440 cm^2 when fitted to the rear.

Contact areas will be measured over any square section of the tyre which is normal to and symmetrical about the tyre centre line and which measures 200 mm × 200 mm when fitted to the front of the car and 250 mm × 250 mm when fitted to the rear. For the purposes of establishing conformity, only void areas which are greater than 2.5 mm in depth will be considered.

Prior to use at an Event, each tyre manufacturer must provide the technical delegate with a full-scale drawing of each type of wet-weather tyre intended for use.

With the exception of race day, wet-weather tyres may only be used after the track has been declared wet by the race director and, during the remainder of the relevant session, the choice of tyres is free.

f) Tyre specifications will be determined by the FIA no later than 1 September of the previous season. Once determined in this way, the specification of the tyres will not be changed during the Championship season without the agreement of the Formula One Commission.

80) Control of tyres:

a) All tyres which are to be used at an Event will be marked with a unique identification.

b) At any time during an Event, and at his absolute discretion, the FIA technical delegate may select the dry-weather tyres to be used by any Team from among the total stock of tyres which such Team's designated supplier has present at the Event.

c) From among the twenty-eight dry-weather tyres available to each driver following the first day of practice, the FIA technical delegate will choose at random sixteen tyres (eight front and eight rear) which are the only dry-weather tyres which such car may use in qualifying practice.

d) A competitor wishing to replace one unused tyre by another unused one must present both tyres to the FIA technical delegate.

e) The use of tyres without appropriate identification is strictly forbidden.

81) Wear of tyres:

The Championship will be contested on grooved tyres. The FIA reserves the right to introduce at any time a method of measuring remaining groove depth if performance appears to be enhanced by high wear or by the use of tyres which are worn so that the grooves are no longer visible.

Weighing

82) a) During qualifying practice, cars will be weighed as follows:

1) The FIA will install weighing equipment in an area as close to the first pit as possible, and this area will be used for the weighing procedure;

2) Cars will be selected at random to undergo the weighing procedure. The FIA technical delegate will inform the driver by means of a red light at the pit entrance that his car has been selected for weighing;

3) Having been signalled (by means of a red light) that his car has been selected for weighing, the driver will proceed directly to the weighing area and stop his engine;

4) The car will then be weighed and the result given to the driver in writing;

5) If the car is unable to reach the weighing area under its own power, it will be placed under the exclusive control of the marshals who will take the car to be weighed;

6) A car or driver may not leave the weighing area without the consent of the FIA technical delegate.

7) If a car stops on the circuit and the driver leaves the car, he must go to the weighing area immediately on his return to the pits in order for his weight to be established.

b) After the race, each car crossing the Line will be weighed. If a driver wishes to leave his car before it is weighed, he must ask the technical delegate to weigh him in that this weight may be added to that of the car.

c) Should the weight of the car be less than that specified in Article 4.1 of the Technical Regulations when weighed under a) or b) above, the car and the driver will be excluded from the Event save where the deficiency in weight results from the accidental loss of a component of the car due to *force majeure*.

d) No solid, liquid, gas or other substance or matter of whatsoever nature may be added to, placed on, or removed from a car after it has been selected for weighing or has finished the race or during the weighing procedure (except by a scrutineer when acting in his official capacity).

e) Only scrutineers and officials may enter the weighing area. No intervention of any kind is allowed there unless authorised by such officials.

83) Any breach of these provisions for the weighing of cars may result in the exclusion of the relevant car.

General Car Requirements

84) No signal of any kind may pass between a moving car and anyone connected with the car's entrant or driver save for the following:

a) Legible messages on a pit board.

b) Body movement by the driver.

c) Telemetry signals from the car to the pits.

d) Lap trigger signals from the pits to the car. Lap marker transmitters shall be battery powered and once operating must be freestanding (not attached to any other pit equipment by means of wires or optical fibres) and incapable of receiving external information.

Such lap triggers shall use a transmitter operating with a carrier frequency above 10 GHz (radio or optical) and a beam half angle of no more than 36° when measured at the 3 dB point, and shall not be used for the transmission of any data from pit to the car other than the lap mark. Lap mark data must be transmitted repeatedly and must be demonstrably consistent.

e) Verbal communication between a driver and his team by radio.

f) Electromagnetic radiation between 2.0 and 2.7 GHz is forbidden save with the written consent of the FIA.

85) Accident data recording:

a) Each car must be fitted with an FIA accident data recorder during each Event and all tests which are attended by more than one Team. Teams must use their best endeavours to ensure that the recorder is in working order at all times. The sole purpose of these units is to monitor and record data relevant to an accident or incident.

b) The recorder and any related sensors must be fitted and operated in accordance with the instructions of the FIA.

c) At any time following an accident or incident, competitors must make the data recorder available and accessible to the FIA. A representative of the Team concerned must be present when data relevant to an accident or incident is being uploaded from the recorder. A copy of the data will be made available to the Team.

d) Any conclusions as to the cause of an accident, or any data relevant to an accident, may only be published in the form of a report which has been agreed between the Team concerned and the FIA.

General Safety

89) Official instructions will be given to drivers by means of the signals laid out in the Code. Competitors must not use flags similar in any way whatsoever to these.

90) Drivers are strictly forbidden to drive their car in the opposite direction to the race unless this is absolutely necessary in order to move the car from a dangerous position. A car may only be pushed to remove it from a dangerous position as directed by the marshals.

91) Any driver intending to leave the track or to go to his pit or the paddock area must signal his intention to do so in good time, making sure that he can do this without danger.

92) During practice and the race, drivers may use only the track and must at all times observe the provisions of the Code relating to driving behaviour on circuits.

93) A driver who abandons a car must leave it in neutral or with the clutch disengaged and with the steering wheel in place.

94) Repairs to a car may be carried out only in the paddock, pits and on the grid.

95) The organiser must make at least two fire extinguishers of 5 kg capacity available at each such pit and ensure that they work properly.

96) Save as provided in Article 138, refuelling is allowed only in the pits.

97) The driver may remain in his car throughout refuelling but, unless an FIA approved race refuelling system is used, the engine must be stopped.

Race refuelling systems may not be used during, or immediately after, any practice session. Whilst being used in the race all team personnel working on the car must wear clothing which will protect all parts of their body from fire. The competitor must ensure that an assistant with an extinguisher (minimum capacity, 25 kg) ready to work is beside the car throughout all refuelling operations.

98) Oil replenishment is forbidden during the race. All orifices for oil filling must be designed in such a way that the scrutineers can seal them.

99) Save as specifically authorised by the Code or these Sporting Regulations, no one except the driver may touch a stopped car unless it is in the pits or on the starting grid.

100) At no time may a car be reversed in the pit lane under its own power.

101) During the periods commencing 15 minutes prior to and ending 5 minutes after every practice session and the period between the commencement of the formation lap which immediately precedes the race and the time when the last car enters the *parc fermé*, no one is allowed on the track, the pit entry, or the pit exit, with the exception of:

a) Marshals or other authorised personnel in the execution of their duty;

b) Drivers when driving or on foot, having first received permission to do so from a marshal;

c) Team personnel clearing equipment from the grid after all cars have left the grid on the formation lap;

d) Mechanics under Article 140 only.

102) During a race, the engine may only be started with the starter except:

a) In the pit lane where the use of an external starting device is allowed, or;

b) Under Article 145c) or d).

103) Drivers taking part in practice and the race must always wear the clothes and helmets specified in the Code.

104) A speed limit of 80 km/h in practice and 80 km/h during the Warm Up and the race (60 km/h in Monaco), or such other speed limits as the Permanent Bureau of the Formula One Commission may decide, will be enforced in the pit lane.

Except in the race, any driver who exceeds the limit will be fined US$250 for each km/h above the limit (this may be increased in the case of a second offence in the same Championship season). During the race, the Stewards may impose a time penalty on any driver who exceeds the limit.

105) If a driver has serious mechanical difficulties during practice or the race, he must leave the track as soon as it is safe to do so.

106) The car's rear light must be illuminated at all times when it is running on wet-weather tyres. It shall be at the discretion of the race director to decide if a driver should be stopped because his rear light is not working. Should a car be stopped in this way, it may re-join when the fault has been remedied.

107) Only six team members per participating car (all of whom shall have been issued with and wearing special identification) are allowed in the signalling area during practice and the race.

People under 16 years of age are not allowed in the pit area.

108) Animals, except those which may have been expressly authorised by the FIA for use by security services, are forbidden in the pit area and on the track and in any spectator area.

109) The race director, the clerk of the course or the FIA medical delegate can require a driver to have a medical examination at any time during an Event.

110) Failure to comply with the general safety requirements of the Code or these Sporting Regulations may result in the exclusion of the car and driver concerned from the Event.

References

0.1 Kingston, William, *Innovation*, John Calder, 1977.

0.2 Diamond, Jared, *Guns, Germs and Steel*, Vintage, London, 1998.

0.3 Van Valkenburgh, Paul, *Chevrolet—Racing? Fourteen Years of Raucous Silence! 1957–1970*, Society of Automotive Engineers, Warrendale, PA, 2000.

0.4 Milliken, William F., and Milliken, Douglas L., *Race Car Vehicle Dynamics*, Society of Automotive Engineers, Warrendale, PA, 1995.

0.5 Rudd, Tony, *It Was Fun!*, Patrick Stephens Ltd., Somerset, England, 1993.

1.1 Milliken, William F., and Douglas L.Milliken, *Race Car Vehicle Dynamics*, Society of Automotive Engineers, Warrendale, PA, 1995.

4.1 Harris, M.G., "Optic and Retinal Flow," *Visual Detection of Motion*, Academic Press Ltd., London, 1994.

6.1 Milliken, William F., and Milliken, Douglas L., *Race Car Vehicle Dynamics*, Society of Automotive Engineers, Warrendale, PA, 1995.

7.1 Milliken, William F., and Milliken, Douglas L., *Race Car Vehicle Dynamics*, Society of Automotive Engineers, Warrendale, PA, 1995.

8.1 Scibor-Rylski, A.J., *Road Vehicle Aerodynamics*, The City University, London, 1970.

8.2 Fackrell, J.E., and Harvey, J.K., "The Flow Field and Pressure Distribution of an Isolated Road Wheel," Paper 10, *Advances in Road Vehicle Aerodynamics*, BHRA Fluid Engineering, England, 1973.

8.3 Agathangelou, Ben, and Gascoyne, Mike, "Aerodynamic Design Considerations of a Formula 1 Racing Car," SAE Paper No. 980399, Society of Automotive Engineers, Warrendale, PA, 1998.

8.4 Crewe, C.M., Passmore, M.A., and Symonds, P., "Measurement of Formula One Car Drag Forces on the Test Track," SAE Paper No. 962517, Society of Automotive Engineers, Warrendale, PA, 1996.

13.1 Best, Tony, Neads, Steve J., Whitehead, John P., and Willows, Ian R., "Design and Operation of a New Vehicle Suspension Kinematics and Compliance Facility," SAE Paper No. 970096, Society of Automotive Engineers, Warrendale, PA, 1997.

13.2 Miliken, W.F., and D.L. Miliken, *Race Car Vehicle Dynamics*, Society of Automotive Engineers, Warendale, PA, 1995.

14.1 Pomeroy, Laurence, *The Grand Prix Car*, Temple Press Books Ltd., London, 1954.

15.1 Wright, Peter, and Andrew Mellor, "Barrier Testing," SAE Paper No. 983061, Society of Automotive Engineers, Warendale, PA, 1998.

15.2 Gramling, Hubert, Hodgman, Peter, and Hubbard, Robert, "Development of the HANS Head and Neck Support for Formula One," SAE Paper No. 983060, Society of Automotive Engineers, Warrendale, PA, 1998.

15.3 Rudd, Tony, *It Was Fun!*, Patrick Stephens Ltd., Somerset, England, 1993.

16.1 Azzoni, P, Moro, D., and Rizzoni, G., "Reconstruction of Formula 1 Engine Instantaneous Speed by Acoustic Emission Analysis," SAE Paper No. 983047, Society of Automotive Engineers, Warrendale, PA, 1998.

16.2 Crewe, C.M., Passmore, M.A., and Symonds, P., "Measurement of Formula One Car Drag Forces on the Test Track," SAE Paper No. 962517, Society of Automotive Engineers, Warrendale, PA, 1996.

17.1 *The National Survey of Motorsport Engineering and Services*—2000, Motorsport Industry Association, Warwickshire, England, 2001.

A.1 Milne-Thompson, L.M., *Theoretical Hydrodynamics*, McMillan, London, 1946.

A.2 Stollery, J.L., and Burns, W.K., "Forces on Bodies in the Presence of the Ground," *Road Vehicle Aerodynamics*, "The" City University, London, 1969.

D.1 *Year Book Automobile Sport 2000*, Fédération Internationale de l'Automobile, Geneva, 2000.

Index

About the Author

Peter Wright earned a Master of Arts degree studying mechanical sciences at the Cambridge University in England. Upon leaving the university, he joined BRM as a research engineer on chassis and aerodynamics.

Peter's subsequent jobs included work at Specialised Mouldings, where he designed and commissioned a one-quarter-scale wind tunnel for race car testing. At Specialised Mouldings, he also worked on the design and development of composite bodywork for a variety of race cars.

In 1974, Peter joined the Lotus group of companies, where he rose to become managing director of Lotus Engineering and subsequently technical director of Team Lotus. During his 21 years at Lotus, Peter worked on advanced composites, ground-effect aerodynamics, and active control systems, involving diverse vehicles which included road and racing cars, pleasure and racing boats, ultralight aircraft, tracked and off-road vehicles, and, for a short period, flying saucers.

In 1995, Peter joined the FIA as Technical Advisor, working on safety issues for road and racing cars. He is a member of the FIA Safety Commission and Research Group, and the EuroNCAP Assembly and Technical Working Group.

Peter writes technical articles for a number of motorsport publications worldwide and has had papers published on aerodynamics, active suspension, and safety. He lives in southwestern France and flies gliders for relaxation.